NATO Science for Peace and Security Series

This Series presents the results of scientific meetings supported under the NATO Programme: Science for Peace and Security (SPS).

The NATO SPS Programme supports meetings in the following Key Priority areas: (1) Defence Against Terrorism; (2) Countering other Threats to Security and (3) NATO, Partner and Mediterranean Dialogue Country Priorities. The types of meeting supported are generally "Advanced Study Institutes" and "Advanced Research Workshops". The NATO SPS Series collects together the results of these meetings. The meetings are co-organized by scientists from NATO countries and scientists from NATO's "Partner" or "Mediterranean Dialogue" countries. The observations and recommendations made at the meetings, as well as the contents of the volumes in the Series, reflect those of participants and contributors only; they should not necessarily be regarded as reflecting NATO views or policy.

Advanced Study Institutes (ASI) are high-level tutorial courses intended to convey the latest developments in a subject to an advanced-level audience

Advanced Research Workshops (ARW) are expert meetings where an intense but informal exchange of views at the frontiers of a subject aims at identifying directions for future action

Following a transformation of the programme in 2006 the Series has been re-named and re-organised. Recent volumes on topics not related to security, which result from meetings supported under the programme earlier, may be found in the NATO Science Series.

The Series is published by IOS Press, Amsterdam, and Springer, Dordrecht, in conjunction with the NATO Public Diplomacy Division.

Sub-Series

A.	Chemistry and Biology	Springer
B.	Physics and Biophysics	Springer
C.	Environmental Security	Springer
D.	Information and Communication Security	IOS Press
E.	Human and Societal Dynamics	IOS Press

http://www.nato.int/science
http://www.springer.com
http://www.iospress.nl

Series C: Environmental Security

Threats to Global Water Security

Edited by

J. Anthony A. Jones

IGU Commission for Water Sustainability
Institute of Geography and Earth Sciences
Aberystwyth University
Aberystwyth
United Kingdom

Trahel G. Vardanian

Department of Physical Geography
Yerevan State University
Yerevan
Armenia

and

Christina Hakopian

Department of Physical Geography
Yerevan State University
Yerevan
Armenia

 Springer

Published in cooperation with NATO Public Diplomacy Division

Proceedings of the NATO Advanced Research Workshop on
Natural Disasters and Water Security: Risk Assessment, Emergency Response
and Environmental Management
Yerevan, Armenia
18–22 October 2007

Library of Congress Control Number: 2009927446

ISBN 978-90-481-2343-8 (PB)
ISBN 978-90-481-2336-0 (HB)
ISBN 978-90-481-2344-5 (e-book)

Published by Springer,
P.O. Box 17, 3300 AA Dordrecht, The Netherlands.

www.springer.com

Printed on acid-free paper

TABLE OF CONTENTS

PREFACE

The UN designated the decade 2005–2015 as the International Decade for Action – Water for Life. The move was initiated at the third World Water Forum in Kyoto, 2003, and it could prove the most significant and effective outcome of the triennial series of World Water For a yet. Its major aims are: (1) to promote efforts to fulfil recent international commitments, especially in the Millennium Goals, (2) to advance towards a truly integrated, international approach to sustainable water management, and (3) to put special emphasis on the role of women in these efforts.

Even so, it faces tremendous and, as I write, increasing obstacles. The intense season of hurricanes and tropical storms in 2008 illustrated yet again not only the power of nature, but also the vulnerability of the poorer nations, like Haiti and Jamaica. New Orleans and Texas fared better, not because of the efforts of the International Decade for Natural Disasters (1990–2000) to increase preparedness, but more because the USA had learnt from its own experiences in Hurricane Katrina.

The biggest obstacle of all is the burgeoning world population. It took off last century, but it is predicted to reach unimaginable heights this century: at least 10 billion by 2050, maybe 20 billion by 2100. Governments are powerless to halt it, even the Chinese. Achieving water security globally against this backdrop will be a Herculean task.

Thus far, the 21st century has also seen three major additions to the problems facing attempts to improve water security: international conflicts, accelerating climate change and seismic shifts in global financial systems. The tremendous effort to install the new hydropower plant at Kajaki Dam in Afghanistan and the continuing cost of protecting it, seemingly in perpetuity, are just one illustration of the new threats from war and terrorism. The arrest of terrorists bent on poisoning Rome's water supply in 2002, and the daily conflicts over water in Israel and Palestine are lower profile, but remind us of the new global nature of terrorism and ideological clashes, and their potential to disrupt water security. In a remarkable gamble for a man with the scientific kudos of Lord Martin Rees, President of the Royal Society and British Astronomer Royal, he laid a bet of $1,000 in 2002 in the computer magazine *Wired* that by 2020 an instance of bio-error or bio-terror will have killed a million people. One has to hope that he loses his money, but he just has to be taken seriously.

The impacts of climate change are becoming more apparent, not only from climate change models, but now also from daily experience. The ten warmest years on record have occurred in the last two decades: normal statistical theory tells us this is not random. The 2007 reports from the International Panel on Climate Change predict more rapid and extreme changes than previous predictions. Melting glaciers and ice sheets reinforce this prospect. More rainfall and less snow in the world as a whole, more intense and frequent storms, floods and droughts and above all the gross redistribution of global water resources are set to radically change our assessment of risk. The time-honoured methods of risk assessment based on records from the recent past, as advocated in the World Meteorological Organization's *Guides to Hydrological Practices* and used by every professional designer of dams and public water supplies in the world, will no longer be applicable.

The climatic changes will be far from egalitarian, providing more water where it is least needed and ravaging "The South" with drought and the occasional extreme rainfall. More than a quarter-century since Willy Brandt's seminal report drew attention to the fragility of life in the developing world, coining the term "The South", and its sequel the UN Special Conference on the Least Developed Countries, began an explosion of Western aid for agriculture and water management, most of The South is even more fragile and faces a steadily worsening situation.

But this is not just due to climate change. A large part of the problem is poor governance and corruption: diversion of aid funds to political pockets, megaprojects aimed at prestige rather than improving living standards and creating unsustainable levels of national debt, disinheritance of the peasantry by agri-business, and straightforward design faults and failures.

And so we come to the third new challenge: financing the water. The nature of the problem has been changing rapidly in recent years. Problems began with the globalization of water companies, first, with the profit motivation of multinational enterprises and subsequently with the loss of national control. Tanzania had a bad experience. It then appeared that loss of control might be exacerbated by the proliferation of takeovers by sovereign wealth funds that might be politically motivated as well. Then just as calls for greater transparency in the operation of these funds seemed to be being heeded by some key players, the credit crunch and one of the most dangerous meltdowns in global financial institutions hit. Its full impact on the flow of international aid and on the financing of bank loans for water-related projects has yet to be felt. But it has the potential to set back all targets, possibly at least for the medium term.

The research papers presented in this volume arise from the NATO Advanced Research Workshop on Global Water Security held in Yerevan, Armenia, in October 2007. Over 40 scientists and water managers attended from Nato countries, Nato-partner countries and the Mediterranean Alliance.

Delegates shared the latest information on both natural and man-made threats to the security of water resources, focusing especially on issues of risk assessment, emergency response and environmental restoration. During the workshop, three Working Groups also discussed the key issues of terrorism, climate change and governance, and summaries of their deliberations are included here.

J.A.A. Jones
Aberystwyth University, UK

PART I:

OVERVIEW: THE MAJOR ISSUES

THREATS TO GLOBAL WATER SECURITY: POPULATION GROWTH, TERRORISM, CLIMATE CHANGE, OR COMMERCIALISATION?

J.A.A. Jones[*]

Chair, IGU Commission for Water Sustainability, Institute of Geography and Earth Sciences, Aberystwyth University, UK

Abstract. As the water crisis intensifies and climate change predictions suggest worse to come, no single factor is responsible. This paper looks at a number of sources of insecurity in water supplies. It concludes that population growth and its multiplier factors, such as irrigated agriculture, urbanisation and water pollution, are the prime causes. But major questions need to be addressed regarding financial and political control.

Keywords: Water crisis, population growth, climate change, terrorism, commercialisation

1. Introduction

Despite repeated efforts and resolutions by the international community, from G8 summits to the World Water Forum, aimed at improving water supply and sanitation around the world, we are losing ground. There are now more people without safe water and sound sanitation than there were a decade ago. The poorer countries are suffering most, and everywhere it is the poor, the young and the old that are most vulnerable. Growing millions of urban poor have no access to safe, centrally supplied water and are forced to use polluted surface or ground water, or else pay over the odds for water from commercial tankers. A third of the world population presently live in countries suffering moderate to severe water stress, but this is likely to rise to 2/3 of a much larger population by 2025.

2. Population growth

The burgeoning world population is the largest single cause of the water crisis (Jones, 1999). World population is due to double or even quadruple before stabilising by the end of the century. The best UN estimates suggest a 50% increase by mid-century, from 6 billion in 2000 to 9 billion in 2050. Worst of all from the water resources point of view, the geographical distribution of this population increase will not only be uneven, but largely concentrated

[*] To whom correspondence should be addressed. e-mail: jaj@aber.ac.uk

in the Developing World and especially in regions where water resources are already stretched, notably in sub-Saharan Africa, the Middle East and parts of the Indian subcontinent. More than a dozen countries in this region are expected to show population growths in excess of 50% between 2000 and 2030, and most others will experience increases of more than 30%. The Middle East is projected to have the fastest growing population outside Africa, doubling over the next 25 years. There is already a critical scarcity of resources in the Middle East and North Africa, where 11 of the 20 countries in the region currently use over 50% of their renewable resources. Libya and most of the Arabian Peninsula already use over 100% of their traditionally assessed resources with the support of desalination and fossil groundwater, some of which is 10,000 years old and has not received significant recharge since the Pluvial Period at the end of the last glacial period.

The most vulnerable areas are the semi-arid regions on the desert margins, where water management is complicated by three natural problems, often in combination: low average precipitation, high seasonality of rainfall and high inter-annual variability. In the true desert, the lack of water resources is plain and can be planned for, but in the semi-arid lands the unreliability of resources makes planning difficult, as floods alternate with droughts.

3. Waste of water

There are, however, many other significant driving factors besides population growth. Some of these have been with us for a while, like wasteful practices, especially in irrigated agriculture or with ageing and poorly maintained infrastructure. Over three-quarters of water supplies worldwide are used by agriculture, and traditional flood irrigation can waste 60–70% of this. Adoption of water-saving systems like drip or spray irrigation, which can reduce losses by more than 60%, is still very limited and is found mostly in Developed Countries, which *ab initio* are not the main users of irrigation waters by volume. Some of the largest per capita consumptions of water lie in the Middle East, where 95% is being used in agriculture, mostly very inefficiently and well in excess of requirements. It has been demonstrated that irrigation on peasant farms in Saudi Arabia is significantly less efficient than on enterprises run by large corporations: the problem is mainly due to ignorance of the optimum water requirements amongst the local farmers, supported by the cheapness of the water supply. Driven by this inefficient agriculture, per capita consumptions in the Middle East can be 10–20 times greater than in the United Kingdom. Raising water tariffs is an obvious way forward. Even communist China is now taking this route in the agricultural heartland of the Hwang-ho plains. But a kinder route is through education and providing farmers with appropriate information on crop requirements.

There are two additional concerns with irrigated agriculture. There is a growing tendency to use non-renewed "fossil" groundwater for irrigation on marginal farmland in arid regions. This is clearly unsustainable, even though there is considerable ignorance as to the exact amount of useable groundwater in many of these aquifers. It is also environmentally damaging, speeding salinisation of the soil due to the higher salt content of the groundwater, and lowering the water table often over wide areas, killing local vegetation. The second major concern is the rate at which water use in agriculture is increasing. Despite the demise of the grand Soviet plans to increase irrigation in the 1980s and predictions that the area of new land suitable for irrigated agriculture is nearing exhaustion, withdrawals for agricultural use are predicted to rise by a fifth between 2000 and 2025. Admittedly, this marks a slowdown from the 40% rise during the preceding quarter-century, but in terms of the absolute amount of cubic kilometres the predicted rise is equivalent to between 40% and 50% of current withdrawals for industrial and domestic uses *combined*, or some 500 km^3 (Struckmeier et al., 2005). By 2025, agriculture could be withdrawing three times as much water as industry and six times as much as domestic users.

Leakage from ageing pipes, even in Developed Economies, can be in excess of 25%, despite government drives to reduce them, like the UK National Leakage Control Initiative from the early 1990s. London built a new "ring main" for water to reduce losses by enabling pressures to be lowered in the old, corroding pipes (Jones, 1997). Even so, in 2006, London was still losing 915 l/day, accounting for a quarter of all leakage losses in the UK. The government watchdog Ofwat considered fining Thames Water for failing to meet targets and the Mayor of London refused the company planning permission to build a desalination plant until it has reduced water losses. Despite new technologies like *in situ* relining of pipes, most megacities face a major challenge in tackling old pipe systems.

4. Climate change

Many other factors, however, are new or are taking on greater importance in the 21st century. Climate change alone could add 100–170 million to severe water stress areas, using more than 40% of available resources, by 2080 (Arnell and King, 1997). Climate change appears to be increasing the frequency of extremes (see Working Group 1 Report, this volume). In 2007 alone, devastating and often record-breaking floods in India, China, North Korea, Britain, Texas and the Sudan, have been matched by continuing multi-year droughts in the Sahel, East Africa and Australia, and excessive summer heat in southeastern Europe. Many of these events have exceeded the capacity of both local and international authorities to cope with them. Recent evidence suggests that the intensity of tropical storms and hurricanes is increasing, as expected from warmer sea surface temperatures. Hurricanes are very rare in the South Atlantic, but in 2004 Brazil experienced the first hurricane to reach 30° South.

If global warming models and the latest IPCC (2007) report are correct, then the trend is likely to continue and get worse, and the problems caused by these extreme hydrological events, not least their effects on human health and nutrition, are set to intensify. Fears of resulting mass starvation, armed conflict and migration outlined by Tickell (1977, 1991) have been reinforced by the findings of Burke et al. (2006). Their analysis of global drought over the last 50 years shows that the area affected by extreme drought expanded from 1% of the land to 3%. However, this is as nothing compared with their analysis of output from the UK Met Office's HadCM3 model, which indicate an accelerated increase, especially during the second half of the century, so that by 2090 nearly one third of the land could be affected by extreme drought, 40% by severe drought and 50% by moderate drought. Post-2050, the frequency of drought events will increase slightly, but droughts will last much longer. Manabe et al. (2004) also predicted that soil moisture levels in semi-arid regions will fall markedly this century. Such trends will displace tens of millions in Africa alone.

The increased incidence of drought may be aggravated by deforestation. Kayane (1996) estimated that deforestation in the humid tropics reduces evapotranspiration by over 400 mm a^{-1}. A fifth of the original forest in the Amazon basin has been cut down in recent few decades and evidence seems to be emerging of a commensurate reduction in rainfall. Many parts of Amazonia suffered the worst drought in 50 years in 2005 and a state of emergency was announced as many river tributaries dried up. As drinking water became scarce, local inhabitants were reduced to using stagnant pools, increasing the risk of disease. Trucks and army helicopters supplied water to villages cut off because the rivers were no longer navigable.

Droughts in eastern Amazonia are normally linked to El Niño events in the Pacific Ocean, which bring counterbalancing heavy rains to the Pacific coastal lands. Global warming is likely to increase the intensity of El Niño events and with it drought in the Amazon. Drought also increases the risk of forest fires, which intensify the water shortage because the large numbers of condensation nuclei in the smoke inhibit rain formation by producing too many small cloud droplets rather than fewer larger raindrops. There is therefore a complex synergy here between the effects of climate change and commercial deforestation. The fact that some deforestation is to make room for biofuels to counter global warming is a most unfortunate irony.

Recent years have also seen a number of other severe natural events, like the earthquake in the Himalayan region of Pakistan in 2005, or the Indian Ocean tsunami of 2004, which have also exceeded the ability of emergency services to provide basic water and sanitation. These events underline the importance of a well-organised system of international aid, something which in spite of many undoubted successes in recent years, is still in need of better overall coordination and monitoring (see Working Group 2 Report, this volume). There have been too many reports of aid promised but never materialising, and of aid being looted or corruptly diverted.

5. Urbanisation, industrialisation and the changing geography of wealth

Recent climatic trends may be partly or wholly natural, but there is a new set of threats that are totally man-made. These range from terrorism to the apparently innocuous change of ownership of water resources. Somewhere in between are some critical socio-economic trends that have been intensifying during the early years of this century: the geography of wealth is changing.

More people now live in cities than in the countryside. By 2050 75% of world population are likely to be urban dwellers. Most of the new megacities are now in India, South America and Southeast Asia, especially in China. Shanghai has taken in 5 million rural immigrants in recent years, and China has produced the first "instant megacities" like Shenzhen, built on an unprecedented scale in just a few years.

Last century, people in cities tended to consume and waste considerably more water than the rural population: typically four to six times as much. But most of the new urban population is in the Least Economically Developed Countries, where a considerable portion lives in makeshift housing without mains water supplies. This reduces overall water consumption, but it also raises the risk of water-related diseases.

At least 50% of the world population currently suffers from water pollution, most of it directly or indirectly man-made. Historically, poor sanitation and hygiene have been major factors, and these are still affecting increasing numbers of people in the Developing World, especially in the shanty towns. In sub-Saharan Africa alone, 300 million lack access to safe water. Of the 3.4 million who die annually from water-related diseases, over 2.2 million are in Developing Countries. The poor, the young and the old are most at risk. Industrialisation and the development of new chemicals, notably petrochemical products, PCBs, herbicides and pesticides, have dramatically extended the repertoire of health risks. The Aral Sea disaster is the most graphic illustration of the problems caused by a chemically-sustained agriculture combined with profligate use of water resources (see Special Aral Sea section, this volume).

Perhaps the most worrying case of pollution that may be *indirectly* induced by human activity is that of arsenic in groundwater. The mobilisation of arsenic can be accelerated by drawdown of groundwater levels as a result of water abstraction, which allows oxidation of arsenic-bearing minerals. Arsenic may also be released as a result of chemical reactions caused by percolating sewage water. It is a serious problem in the Indian subcontinent, particularly in Rajasthan where surface water is in short supply. Recent international discussions have referred to the "global arsenic calamity". The causes are varied and are currently the focus of intense international research. Human settlements are clearly implicated in some cases, but the degree of human influence is still an open question in Botswana's Okavango Delta (Huntsman-Mapila et al., 2006). Climate change can also be cause of falling water tables, and the drylands are destined to be the first to feel the impact of reduced rainfall.

One of the greatest changes in recent years has been the rise of the Bric economies: Brazil, Russia, India and China, especially the last two. Water pollution has been developing apace with urbanisation and industrialisation in India and China. Just as the Developed Countries are beginning to get control of pollution – at least from urban-industrial sources if not from agriculture – Southeast Asia from the Indian subcontinent to China is losing control in the headlong rush to economic development and wealth creation. At their stage of economic development, the more wealthy they are, the more water they are likely to consume and waste. Every litre of polluted water returned to the environment destroys 10 l of freshwater resource.

Alongside the direct consumption of water amongst the rising and urbanising population, and the reduction in available water resources caused by pollution, come dietary shifts that also have implications for water security. This is not just the increased demand for foodstuffs, but particularly the shift towards increased consumption of meat. Whereas it takes 1 m^3 of water to produce 1,000 kcal of vegetarian food, meat requires 5 m^3. If that meat is produced in Brazil on land cleared of forest for grazing, then it has a multiplier effect in lost rainfall as well.

6. Terrorism and war

Yet another "new" factor is terrorism. This has really come to the fore since the events of 9/11, but in fact it is as old as warfare itself. Withholding or poisoning water supplies was a military tactic often used in the ancient world. As noted in the Working Group 1 Report (this volume), water is still being used as a weapon by both national armies and non-state terrorist organisations, from American rainmaking exercises to bog down the Vietcong in the Vietnamese war, and the recent US Air Force intelligence briefing "Owning the Weather in 2025", to the foiling of an attempt by Islamist terrorists to poison the water supply in Rome in 2002.

After 9/11 in 2001, the US Environmental Protection Agency (EPA) announced the establishment of a Water Protection Task Force. The President's Infrastructure Assurance Office concluded "The water supplied to US communities is potentially vulnerable to terrorist attacks ... The possibility of attack is of considerable concern ... These agents ... inserted at critical points in the system ... could cause a large number of casualties". And the EPA's National Homeland Security Research Center initiated the Environmental Technology Verification (ETV) Program to examine rapid screening tests that could detect water terrorist acts. In June 2002, the federal government passed the Public Health Security and Bioterrorism and Response Act (H.R. 3448), which requires all public water systems serving more than 3,300 people to submit a security assessment and an appropriate emergency response plan. Water companies removed sensitive information from their websites and tightened up on public access to infrastructure and the vetting of employees.

Britain has also spent tens of thousands of pounds on improving the security of water systems from attack. To date, there have been no success-ful attacks in the UK or elsewhere, and both companies and government have generally presented the argument to the public that such attacks are unlikely to be physically very successful, largely because of the sizeable amounts of poison needed. The real weapon is fear and the best defence is denial of publicity. This was the strategy adopted by the British govern-ment in 1999 when the Irish Republican Revenge Group threatened to introduce poison via public water hydrants (Working Group 1 Report, this volume).

The recent spate of terrorist threats is unlikely to subside until the root causes of discontent are removed or negated. This seems to have been largely achieved with the Irish question. It is a very long way from being achieved with the much larger Islamist issue. But from the water security point of view the real question revolves around the quantitative impact of such acts, and the answer, at least in terms of current technology, is that the effects are likely to be relatively limited and local.

More disruption is being caused inadvertently by wars, both international and civil. Water supply systems have been damaged on a grand scale in Iraq. Significant disruption was produced during the Balkans conflict in the 1990s from both collateral damage and deliberate bombing of dams. Nor is the "temporary" disruption of supplies the only issue: pollution of soils and groundwater by the products of weaponry, like depleted uranium, and the various chemicals released directly from damaged industrial plants or resulting from combustion can persist for decades.

This also raises the question of so-called "water wars", wars fought over access to water resources, and the broader issue of transboundary water resources. Worldwide, one third of river basins, amounting to some 261 systems, pass through more than one country. Whilst international collaborative agreements have been generally successfully established in North America and Europe, notably on the Rhine and, more recently, the Danube in Europe and the Columbia in North America, there are few such agreements in Africa and Asia. And Africa and Asia have the added problem that transboundary rivers are more common there, accounting for roughly two thirds of all river basins. Armed conflict has already occurred with Israeli warplanes attacking the proposed site of a Syrian dam on the Yarmuk River and subsequent acquisition of control of the other headwaters of the Jordan River by the occupation of the Golan Heights and part of southern Lebanon (Jones, 1997).

The UN-brokered Mekong agreement was heralded as a flagship success in 1994, but without the signatures of the headwater states, China and Burma, it could soon become worthless as China implements plans to divert Mekong water to supplement the Three Gorges Dam and aids Burma in building hydropower dams. Perhaps the greatest success story is the cooperative agreement of the Southern Africa Development Community signed in 1995, which is based on the best of the Helsinki Rules, the Principle of Equitable Utilization, and appears to be supporting economic development amongst the 12 nation signatories with exchanges of water and power.

Nearly all major aquifers are transboundary (cf. Struckmeier et al., 2004). As surface sources are becoming more stretched and water demand increases in the drylands, groundwater resources are attaining greater significance and increasingly a focus for dispute. Egypt has been watching Libya's exploitation of the Nubian Sandstone aquifer for fear of crossboundary effects (Jones, 1997). Palestinians are protesting about Israeli domination of the West Bank aquifers and especially the new security fence around the West Bank, that has cut many off from their traditional wells (Pearce, 2006). The International Law Association's (2004) "Berlin Rules" offer new guidelines on shared aquifers, which were not well covered previously (Dellapenna, 2004). However, there is still a need for more binding legislation and a strong means of enforcement based on the UN International Court of Justice system for transboundary waters, both surface and subsurface.

7. Commercialisation, privatisation and globalisation

Finally, security means more than just having the resources. It means having control of them as well (see Working Group 3 Report, this volume). The commercialisation, privatisation and globalisation of water resources began with the exploitation of water for profit, with water companies driven by the need to produce dividends for investors ahead of providing a service to customers. The World Trade Organisation long ago declared water a "commercial good" and supports foreign takeover of water services under international trade rules with specific enforcement procedures. Yet there are also fears that commercial interests and the needs of the consumer community can be at variance, even incompatible: that companies will not fund water supplies that are unprofitable, especially to the urban poor.

Since the turn of the millennium, global commercialisation has developed into international takeovers by large multinational corporations with diverse commercial interests. The latter development is perhaps the most worrying, especially for developing countries as they can lose control of their national resources. Tanzania famously terminated the contract of the British company Biwater in 2005 to regain control, just 2 years into its 10-year lifespan, after marked increases in tariffs and lack of progress in extending the water supply system. In Bolivia, after riots and deaths, the government finally ousted the US-Dutch-owned company Bechtel in 2000 for failing to fulfil its contract to provide affordable mains water supply to the poor areas of Cochabamba, which they claimed to have done. Other states have not been so pro-active. This is not surprising as the economic power of some of these companies can exceed that of a small state. Legal actions have proliferated and can be very expensive. Both Biwater and Bechtel subsequently sued the national governments. Conversely, the giant French companies, Vivendi and Suez, have been charged with corruption in Argentina and Indonesia. Indeed, corruption is another creeping disease that is affecting water security throughout the world, from the maintenance of infrastructure to the awarding of contracts. There have been numerous allegations of corrupt diversion of foreign aid in many African countries.

Dr. Peter Gleick, President of the Pacific Institute, has stated: "Our assessment shows that rigorous, independent reviews of water privatization efforts are necessary to protect the public. Water is far too important to human health and the health of our natural world to be placed entirely in the private sector". In its "Water Action Plan", the 2003 G8 summit in Evian gave its support for a compromise solution, public–private partnerships (PPPs), to utilise private sources of funds and commercial management expertise, whilst theoretically retaining governmental control. The big question is "can such partnerships work?". Given that the effectiveness of PPPs has already been questioned in the UK, for example in unsuccessful arrangements in London's transport system, the prospects for success in developing countries must be open to serious doubt. The grave danger is that the private sector takes any profits, but the public purse shoulders any losses and the job is not even satisfactorily completed.

The whole question of funding for water resources projects needs fundamental re-evaluation to ensure water security. This ranges from the long-held policies of the World Bank and IMF to encourage privatisation and PPPs through to the recent highly-leveraged forays of private equity into water companies.

Following the collapse of the easy credit market in the summer of 2007, a major new source of globalisation came forward with a switch to "sovereign wealth funds" (SWFs) to finance company takeovers. SWFs are state-owned investment companies, which have the power and stability of state finances behind them enabling them to outbid most competition. It is estimated that the global value of SWFs currently stands at $2,500 billion and could rise to $12,000 billion by 2016. Many SWFs are run by oil-producing countries, but the China Investment Corporation (CIC) is also a major player. SWFs have been criticised for lack of transparency in their operations. Serious concerns have been expressed over the potential for hidden political agendas. Political commentators have linked China's reluctance to support UN criticism of the Sudanese government's role in the Darfur crisis to the level of Chinese investment in Sudan, particularly in the oil industry. Takeovers by SWFs and state-controlled companies could add an element of foreign control that is even more difficult to shake off than the profit-oriented private companies.

International calls for greater transparency in the accounts of SWFs may be beginning to be heard, but the banking crisis and global depression that developed in 2008 out of the credit crunch may yet give more power to the seemingly sounder examples of SWFs compared with western banks.

8. Conclusion

Water resources are under sustained attack from a wide variety of sources. As ever, the developed world will be able to cope. These states are blessed with the financial resources and technological expertise, they have lower rates of population growth, some even falling, their urban and industrial development is almost complete, their regulatory machinery is well-developed, and climate change is likely to be kinder to them, even enhancing rainfall resources north of latitude 50° N. Water-related terrorism may be a mounting threat, but it appears to be a relatively minor one, especially compared with other targets. It is the Third World that is the worry. It has the very antitheses of *all* the factors that cushion the First World. African states in particular are all too easily prey to the many economic, social and political upheavals that so readily disrupt the provision of life's essentials of food and water. Even such an apparent paragon of stability as Kenya has recently suffered serious civil unrest. Most of all, it is population growth in both the Least Economically Developed Countries and the emerging economies that threatens their per capita resources. Some will cope, but many are destined to present the international community with desperate cries for help in coming decades.

References

Arnell, N. and King, R., 1997: The impact of climate change on water resources. In: *Climate change and its impacts: A global perspective*, Department of the Environment, Transport and the Regions & Met. Office, UK, pp. 10–11.

Burke, E., Brown, S.J. and Christidis, N., 2006: Modeling the recent evolution of global drought and projections for the twenty-first century with the Hadley Centre climate model. *Journal of Hydrometeorology*, 7: 1113–1125.

Dellapenna, J.W., 2004: Adapting the law of water management to global climate change and other hydropolitical stresses. In: Rodda, J.C. and Ubertini, L. (Eds.), *The basis of civilization – water science?* Rome: International Association of Hydrological Sciences Publication No. 286, pp. 291–299.

Huntsman-Mapila, P., Mapila, T., Letshwenyo, M., Wolski, P., Hemond, C. and Harry Oppenheimer Okavango Research Centre, 2006: Characterization of arsenic occurrence in the water and sediments of the Okavango Delta, NW Botswana. *Applied Geochemistry*, **21(2)**: 1376–1391.

International Law Association, 2004: The Berlin Rules on water resources. In: *Report of the Seventy-First Conference*, London: International Law Association, pp. 337–411.

Intergovernmental Panel on Climate Change (IPCC), 2007: *Climate change 2007: Synthesis report*. 4th Assessment Report, Intergovernmental Panel on Climate change, Cambridge, Cambridge University Press.

Jones, J.A.A., 1997: *Global hydrology: Processes, resources and environmental management*. Addison Wesley Longman, Harlow, 399pp.

Jones, J.A.A., 1999: Climate change and sustainable water resources: Placing the threat of global warming in perspective. *Hydrol. Sci. J.* **44(4)**: 541–557.

Kayane, I., 1996: An introduction to global water dynamics. In: Jones, J.A.A., Liu, C., Woo, M.-K. and Kung, H.-T. (Eds.), *Regional hydrological response to climate change*. Dordrecht: Kluwer, pp. 25–38.

Manabe, S., Wetherald, R.T., Milly, P.C.D., Delworth, T.L. and Stouffer, R.J., 2004: Century-scale change in water availability: CO_2-quadrupling experiment. *Climatic Change*, **64**: 59–76.

Pearce, F., 2006: *When the rivers run dry – water, the defining crisis of the twenty-first century*. London: Eden Books, 368pp.

Struckmeier, W., Rubin, Y. and Jones, J.A.A., 2005: *Groundwater – reservoir for a thirsty planet*. Leiden: Earth Sciences for Society Foundation, 13pp.

Struckmeier, W., Gilbrich, W.H., Richts, A. and Zaepke, M., 2004: *Groundwater resources of the world*. Special edition for 32nd International Geological Congress, Florence, August 2004, Hannover, BGR, and Paris, UNESCO.

Tickell, Sir C., 1977: *Climate change and world affairs*. Cambridge, MA: Harvard Studies in International Affairs No. 37, 76pp.

Tickell, Sir C., 1991: The human species: A suicidal success? *Geographical Journal*, **159(2)**: 219–226.

LESS IS MORE: APPROACHING WATER SECURITY AND SUSTAINABILITY FROM THE DEMAND SIDE

D.B. Brooks[1*] and J. Linton[2]

[1] Friends of the Earth Canada, Ottawa, Ontario, Canada
[2] Dept. of Geography, Queen's University, Kingston, Ontario Canada

Abstract. Limiting our demands for water is one of the best ways to reduce risk and increase security. The need for greater attention to water demand has come to be widely accepted over the past decade. However, it has not generally been considered in the context of natural disasters. This presentation looks at three time frames within which a role for reductions in demand for water can be explored: emergency (weeks to months) when the focus is on staying alive; medium term (next 10 to 20 years) when the focus is on efficiency of water services; and long term (beyond 20 years) when the focus is on sustainability. Reserve sectors and restrictions on use are appropriate for emergency situations. Conventional and extended forms of demand management are appropriate for the medium term. More transformative measures, which focus on changes in behaviour patterns and economic structure, are appropriate for the long term. These approaches make the efforts to reduce water demand less a choice of technology than a form of governance, and, therefore, all the elements of political economy come into play.

Keywords: Demand management, water security, emergency water, water soft paths

1. Introduction

This paper argues that the cheapest and longest lasting options to reduce risk and promote water security in modern society lie with reducing demands for water rather than increasing sources of supply. To the extent that a family, factory, or farm is dependent on large volumes of delivered water, so each is subject to risks in rough proportion to its dependence.

The concept underlying the argument is that water security is achieved less by technology and physical infrastructure than by social resilience in the face of hydrological challenges and economic losses that are mounting year by year (Kron, 2006). From this perspective, our approach fits comfortably with long-standing perspectives that sustainable water management depends critically on parallel efforts toward economic and gender equity and democratic decision-making, and more recent efforts to determine when water *demand* management does, or does not, support the same goals.

---

* To whom correspondence should be addressed. e-mail: dbrooks@foecanada.org

J.A.A. Jones et al. (eds.), *Threats to Global Water Security*,
© Springer Science + Business Media B.V. 2009

The disaster management literature does not generally address reductions in water use as a risk reduction strategy. Nor does demand reduction seem to be included in "non-structural" approaches to flood protection and sustainability (Kundzewicz, 2002). Some recent work, mainly in South Asia, focuses on shifting economies away from water dependent activities through diversification of livelihood strategies to include non-farm sources of income (Schreier et al., 2006), or even a complete shift away from farming (Moench and Dixit, 2004). These approaches can be useful, but they typically pose economic and social challenges that make them difficult to accept. Strategies to reduce water use can provide more easily implemented ways to improve water security.

We approach our task in three time frames:

1. Emergency (weeks to months), for which strategies to reduce demands for water are least relevant
2. Medium term (next 2 to 20 years), which is the reach of efficiency-focused water demand management (WDM)
3. Long term (beyond 20 years), which is the reach of an alternative approach to sustainable water planning known as the "water soft path" (WSP)

We will cite research results mainly from North America. Though this research will not be directly applicable elsewhere, it is indicative of what can be accomplished in all nations of moderate to high income. The conclusions can also be useful in lower income countries, but only if applied cautiously. Notably, in many semi-arid nations, significant reform of the irrigation sector will be necessary over the next couple of decades (Beaumont, 2002). However, irrigation and agriculture are so closely tied with culture, with lifestyles and with livelihoods that such measures cannot be considered without parallel efforts to promote social and economic equity.

Concepts of water demand management and of water soft paths are outlined in Section 2. Discussions of emergency, medium and longer time frames follow, respectively, in Sections 3, 4 and 5.

2. Promoting water security with demand management and soft paths

Tate (1993) defines water demand management (WDM) as "any socially beneficial action that reduces or reschedules average or peak water withdrawals or consumption from either surface or groundwater, consistent with the protection or enhancement of water quality." Brooks (2006) offers a 5-part operational definition of demand management that distinguishes between water quantity and water quality and emphasizes methods for providing water during dry spells. Both of these definitions provide scope for technological solutions but recognize that technologies must be understood as socially and politically embedded practices. In essence, therefore, demand management is a form of water governance in which the question of *how* water decisions are made is inseparable from the question of *what* decisions are made. Both Tait and Brooks use water demand management in a broad sense to include all measures that serve to reduce withdrawals of water. In this article, the term applies more narrowly to focus mainly on measures that are cost effective.

2.1. SEEKING EFFICIENCY: DEMAND MANAGEMENT

WDM seeks to achieve water efficiency at all stages from extraction to use. It is an application of environmental economics, the results of which can be measured by a physical ratio, as with litres of water applied per kilogram of crop produced, by an economic ratio, as with the cost of water applied per the value of crop produced, or by some combination of the two.

In contrast to supply management, which is generally centralized, demand management involves literally every household or firm or activity that uses water in a given location. Success with WDM therefore depends on the involvement of the people who manage water at these points of use. Thus, Trumbo and O'Keefe (2005) show how intentions and information interact to yield results in demand management campaigns, and Easter and Liu (2007) show that farmer involvement is critical when introducing water-saving irrigation methods.

In most developing countries, WDM is pursued less to save water than to reduce deficits incurred by government subsidies (Hellegers et al., 2006; Brooks et al., 2007). However, from our perspective, the promotion of water security requires that demand-side measures should also be supportive of socio-political goals including improvements in class and gender equity, wider public participation in decision-making and environmental sustainability. In particular, "pro-poor" approaches must be built in explicitly, just as do measures to assure that women and other disadvantaged groups get a fair share of the saved water (DAWN, 1985; van Koppen, 1999; Gender and Water Alliance, 2003).

2.2. SEEKING SUSTAINABILITY: WATER SOFT PATHS

The first steps toward a less risky and more secure water future are found with demand management, which use existing technologies and economic incentives to achieve water efficiency. However, because they begin from an anthropocentric rather than an ecosystem perspective, efficiency-oriented measures alone are not generally sufficient to achieve sustainable water management. In many places, we already withdraw so much water as to impair the ability of nature to provide ecological services (Postel and

Thompson, 2005). Water soft paths accept the importance of water efficiency, but go further by searching for changes in water use habits and water
management institutions that will promote long-term ecological and social
sustainability (Gleick, 2003; Wolfe and Brooks, 2003; Brooks, 2005). As
such, they are an application of ecological as opposed to environmental
economics.

The soft path approach changes the conception of "water." Instead of
being viewed as an end product, water becomes the means to accomplish
specific tasks, such as sanitation or agricultural production. Demand management asks the question "*how*": How can we get more from each drop of
water? Soft paths ask the question "*why*": Why should we use water to do
this at all? *Why*, for example, do we use water (and, commonly enough,
potable water) to carry away our waste? Demand management would urge
low-flow toilets, whereas soft paths might promote waterless or composting
systems in homes and on-site waste treatment and reuse for commercial
buildings. Irrigation is the largest use of water, accounting for around 70%
of water withdrawals world-wide and even greater proportions in low-
income developing countries. Demand management would urge more efficient technologies, such as drip systems with automatic shut-offs. A soft
path approach would ask whether irrigated agriculture might be replaced by
other modes of cropping. Many areas now under irrigation for international
export are capable of agricultural production with rainwater or at most
supplemental irrigation (adding small volumes of water to that provided by
rain or recessional flooding at those times critical to plant growth). In such
places, water security can be increased by promoting drought-resistant
crops for local consumption and crops that can be grown with low-quality
(e.g., saline) water.

Even more than with demand management, it is imperative that soft
paths be designed with social equity, democratic decision-making and
environmental protection in mind. To cite but one example, it might make
economic sense to reallocate water from agricultural to industrial production. However, such a move could hardly be advised if it promotes rural
migration to already over-crowded cities.

3. Response to emergency: staying alive

Reduced demand for water is typically a result of, not a solution for, the
problems that arise in the immediate aftermath of natural disasters or major
accidents. Appropriate measures cannot, for the most part, be brought into
operation in the time scale of weeks to months. To be part of the solution,
demand management must be in place before the event occurs.

There are two exceptions to the foregoing generalization, both of which
apply mainly to seasonal or exceptional droughts. First, many communities

in affluent countries have measures that limit optional uses of water, notably landscaping, when water resources are low. Lawns and home gardens can account for more than half of summertime water use, and the necessary regulations are self-enforced by the community. Second, in many countries, there is an assumed or legislated hierarchy for uses of water that typically puts drinking water in first place and commercial irrigation in last place. Whenever supplies are threatened, those uses deemed most important receive priority. Similar ranking appears in religious legislation going back to Biblical times (Hirsch, 1959). The role of agricultural water use as a reserve sector that can be called upon when needed to supply higher priority uses is well illustrated by the system in Israel (Allan, 1995; Ben-Zvi et al., 1998).

As the emergency and stabilization phases of disaster management shift to the recovery and rehabilitation phases, demand management and soft paths come back into consideration. If previous systems failed, it makes sense to consider how they can be redesigned to be less vulnerable. For example, construction costs can be minimized by placing pipelines in a wadi or on the bank of river, but such placement increases the risk of disruption. It is more expensive but more secure to place pipelines higher up on the slopes. Disasters are also an opportunity to introduce new forms of water management that promote reduced demand. Emphasis on local water supply and treatment rather than central plants will reduce the length and diameter of pipelines, which are both the most expensive and the most vulnerable component in urban systems (see for example Stimson and Peñon, 2004).

4. Promoting water security through demand management

With economic growth and the effects of climate change, reducing demand will, in most cases, be the cheapest and most readily available source of "new" water in most parts of the world. The amount of water that can be saved through demand management over the next 2 to 20 years is very great indeed.

4.1. PRICING AND COSTS AS THE CRITERION OF EFFICIENCY

Putting a price on water is standard economic wisdom. Pricing "works" in the sense that most household users and all commercial users of water, including farmers, adjust water use in response to price signals. Even in affluent and relatively water-rich countries such as Canada, cities that metre water show distinctly lower per capita rates of use than those that do not (Brandes and Ferguson, 2004). However, if pricing is to promote true water security, it must be implemented in concert with social and economic equity. For example, high prices for water are a fact of life for poor people, who commonly pay ten times as much per litre for water of questionable quality as do richer people for water of good quality. The inequity arises because the wealthy are able to bring public water services to their neighborhoods while those less fortunate have to rely on private vendors. Improvements in water security can only occur when increases in the price of water are coupled with ways to extend the benefits of public water to all.

Many cities in both industrialized and in developing nations have adopted increasing block prices, which serves several purposes. An idealized system might include three tiers or blocks. The lowest block for perhaps 10 to 20 m^3 a month is sold at a low price (sometimes called a "lifeline" or a "social" tariff) that provides enough water for the basic needs of low-income households. The price for the middle block is set to cover full costs of treatment and delivery (including capital costs), and provides enough water for all indoor needs. The price for the upper block is high enough to cover not only its own cost but also the loss entailed by the under-priced initial block, as well as to provide a strong incentive for conservation (Easter and Liu, 2007). Subsidies to higher income people can be avoided by making the second block non-marginal. Those who consume more than the initial block pay the higher rate for the full volume, not just the additional water.

The situation is very different in rural areas. Farmers are not generally expected to pay capital costs for irrigation systems, which implies a huge subsidy. As put succinctly in *Human Development Report 2006* (p. 191):

> Governments since the time of the ancient Egyptians have financed the capital costs of irrigation infrastructure out of general tax revenue.

Even in those countries where irrigation infrastructure is being extended to new areas and where commercial farmers are expected to contribute to investment costs, subsidies remain high. Nevertheless, Easter and Liu (2007) report considerable improvement around the world in cost recovery for irrigation systems that adopt some measure of market pricing. However, they caution that the extent of improvement varies widely depending on the role of farmers in designing the cost recovery system, the degree to which payments were used to maintain or improve the irrigation system, and the transparency of the collection process. A particular effective approach is to establish or strengthen community-based water users associations (Dinar and Mody, 2004).

4.2. WATER DEMAND MANAGEMENT IN PRACTICE

Recent research at the University of Victoria in Canada demonstrates the potential of demand management in urban areas of Canada, which may be generalized to many parts of the Global North (Brandes and Ferguson,

2004; Brandes et al., 2005). At almost no marginal cost, low-flow shower-heads reduce water for this use by half or more. Rooftop water collection is more costly but can reduce water for toilet flushing (the largest indoor water use in most buildings) by nearly as much. Some changes can occur surprisingly quickly. In the last decade, the number of Canadian house-holds with low-flow showerheads increased by 50%, and the number with low flow toilets tripled – despite low water prices (Statistics Canada, 2007).

The gold standard in determining cost effectiveness of water demand management has been set by the Pacific Institute in California with a pair of studies. The first study looked at urban (residential, commercial, institu-tional, and much industrial) water use in California and found that current use rates could be cut by 30% using off-the-shelf technologies (Gleick et al., 2003). Those savings are available at lower cost and in less time than any new supply project, and they would eliminate the need for California to provide additional capacity for at least several decades. The follow-up study, which added agricultural and rural uses of water, and which adopted the same population, housing, and economic projections as did the official California Water Plan, concluded (Gleick et al., 2005, p. 1):

> Under a High Efficiency scenario, total human use of water in California could decline by as much as 20 percent while still satisfy-ing a growing population, maintaining a healthy agricultural sector, and supporting a vibrant economy. Some of the water saved could be rededicated to agricultural production elsewhere in the state; support new urban and industrial activities and jobs; and restore California's stressed rivers, groundwater aquifers, and wetlands.

These conclusions are not based on expectations of new technology or higher prices than those already planned by officials in the state of California. Rather, they stem from wider and more rapid adoption of high-efficiency water technologies.

Of course, the High Efficiency scenario will not come about on its own. As the authors (Gleick et al., 2005, p. 7) comment:

> While we do not believe a highly efficient/water/future is necessarily easy to achieve, we think it will be easier, faster, and cheaper than any other option facing us.

We agree, but we caution that these studies apply particularly well in affluent societies where equity of access to water is taken for granted. The emphasis on efficiency needs to be tempered by social equity in cir-cumstances where there is uneven access to water resources, something that is common in the urban areas of many countries, and even more so in rural areas.

5. Promoting water security through water soft paths

There is as yet only limited experience with water soft path analysis; even less experience with its application in policy; and none at all as a way promote water security over the long term. Though there is a modest body of published material, the only study that incorporates all elements of soft path analysis was undertaken by Friends of the Earth Canada (2007).

The Canadian analysis involved water soft analysis at three scales: urban, watershed, and provincial. For each of these scales, the study developed three scenarios for water use over the next 30 or 40 years. A Business-as-Usual scenario (really a projection) was compared with the reductions in water use that could flow from policies under a WDM scenario and under a WSP scenario. The three scenarios all use the same "official" expectations of economic and population growth. For brevity, the summary of the urban case study focuses on residential and commercial uses of water, the water-shed study on agricultural uses, and the provincial study on industrial uses.

5.1. URBAN SCALE OF ANALYSIS

Results for the urban component of the water soft path study were deter-mined for a generalized urban centre in a semi-arid area with a base population of 200,000 in 2005 and 300,000 in 2050 (Brandes and Maas, 2007). As in other urban areas, the bulk of water use is for residential and commercial uses, secondarily for institutional and light industrial uses. The Demand Management scenario is based on fairly rapid uptake of readily available technologies and easily adopted practices, including low-flow and dual-flush toilets, efficient showers and faucets, and water-saving clothes washers. It results in significant savings but not enough to offset population growth. The Water Soft Path scenario adds adoption of composting toilets, waterless urinals, xeriscaping, widespread reuse and recycling, and rain-water harvesting. With these additions, water savings that are more than enough to overcome the expected population growth; indeed, water use in 2050 would be below that in 2000! Implications for improving water security in urban areas are therefore considerable.

5.2. WATERSHED SCALE OF ANALYSIS

The watershed component of the WSP study focused on the Annapolis Valley in the Province of Nova Scotia, which has a maritime climate (Isaacman and Daborn, 2007). Though a comfortable gap seems to exist between the amount of water available and the amount withdrawn, only 10% of the rain arrives between June and August when demands are highest. Withdrawals have exceeded the renewable flow in 12 of the last 40 years during the summer and three times on an annual basis.

Agriculture is the largest water using sector with over one third of total withdrawals. Problems are compounded by a strong tourism sector. Golf courses account for only 2% of annual water use, but, together with agriculture, two thirds of summer use. In the Business-as-Usual projection, surface water would be inadequate to meet demand in at least 1 year in 12 on an annual basis, and nearly every second summer.

Under a Demand Management scenario, annual withdrawal could be reduced by about 40% per year, and by nearly half in the summer. Though this scenario is characterized by widespread adoption of efficient household and agricultural technologies, the Annapolis Valley would still suffer from periods with insufficient surface and ground water. The Soft Path scenario, which adds runoff storage for crop and golf course irrigation, and wastewater recycling and reuse in food processing, results in annual water use that is only half of *current* summer use. With this scenario, the likelihood of insufficient supply is all but eliminated, and water securely is greatly enhanced.

5.3. PROVINCIAL SCALE OF ANALYSIS

Ontario, which is Canada's largest and most industrialized province, has a continental climate with wide extremes between summer and winter. Its industrial sector accounts for nearly half of provincial water use (excluding water withdrawn for cooling in the electrical power sector). Many industries are beginning to recycle water (which is already a larger component of gross use than is fresh intake), and, when this approach is extended under the Demand Management scenario, water consumption is cut by 25% in 2031 compared with Business-as-Usual, but is still one third greater than at present (Kay et al., 2007). Even under the Water Soft Path scenario, rates of use are 16% above present levels. Cutting these rates of use would require investment in new plants to incorporate best available technologies from the two biggest water using sub-sectors, transportation equipment and paper and allied products. Shifts in the structure of economic activity in Ontario would likely achieve reductions in water use than greater efficiency with the current industrial mix. Achieving water security through water soft paths requires close look at the most water-intensive industries.

5.4. FROM ANALYSIS TO PLANNING TOOL

The Canadian Water Soft Path study is the first test anywhere of the application of water soft path concepts to specific political jurisdictions in specific ecological and geographic settings. Though limitations in the study mean that are results are far from definitive, they are strong enough to indicate that water soft paths can achieve water savings that go well beyond those available with demand management. The strength of water soft paths is their ability to identify the potential for structural shifts in living patterns and the economy (and to incorporate climate change). However, such shifts pose a tremendous challenge, for they imply changes in socio-economic, political and perhaps even cultural relationships. Consideration of such changes illustrates the need to incorporate the politics of water and democratic decision-making in any strategy to improve water security.

6. Conclusion

The experiences and the research findings presented above provide ample evidence for reducing water demand as an effective strategy to increase water security in a wide variety of circumstances. Over and above the shorter term efficiency approach of demand management, water soft paths offer greater savings, and hence greater water security, within the context of a longer term search for a triple bottom line – economic growth, social equity, ecological sustainability.

Though many institutional obstacles hinder the promotion of water security from the demand side, many can be overcome through enlightened water policy reform that takes account of economic, social and gender equity. Approaches will vary depending on the conditions in each country and region. In all cases, the best and most general approach is to maximize our respect for water resources and to minimize our demands on them.

References

Allan, J.A., 1995: The Role of Drought in Determining the Reserve Water Sector in Israel. *Drought Network News*, **7(3)**: 21–23.

Beaumont, P., 2002: Water Policies for the Middle East in the 21st Century: The New Economic Realities. *International Journal of Water Resources Development*, **18(2)**: 315–334.

Ben-Zvi, A., Dlayahu, E., Gottesmann, M. and Passal, A., 1998: Evolution of the 1990–1991 Water Crisis in Israel. *Water International*, **23(2)**: 67–74.

Brandes, O.M. and Ferguson, K., 2004: *The Future in Every Drop: The Benefits, Barriers, and Practice of Urban Water Demand Management in Canada*. Victoria, BC, Canada: POLIS Project for Ecological Governance, University of Victoria.

Brandes, O.M., Ferguson, K., M'Gonigle, M. and Sandborn, C., 2005: *At a Watershed: Ecological Governance and Sustainable Water Management in Canada*. Victoria, BC, Canada: POLIS Project for Ecological Governance, University of Victoria.

Brandes, O.M. and Maas, T., 2007: Community Paths: Investigating BC's Urban Water Use. *Alternatives*, **33(4)**: 14.

Brooks, D.B., 2005: Beyond Greater Efficiency: The Concept of Water Soft Paths. *Canadian Water Resources Journal*, **30(1)**: 83–90.

Brooks, D.B., 2006: An Operational Definition of Water Demand Management. *International Journal of Water Resources Development*, **22(4)**: 521–528.

Brooks, D.B., Thompson, L. and Lamia El Fattal, 2007: Water Demand Management in the Middle East and North Africa: Observations from the IDRC Forums and Lessons for the Future. *Water International*, **32(2)**: 193–204.

DAWN (Development Alternatives with Women for a New Era), 1985: *Development Crises, and Alternative Visions: Third World Women's Perspectives*. New Delhi: DAWN Secretariat, Institute of Social Sciences Trust.

Dinar, A. and Mody, J., 2004: Irrigation Water Management Policies: Allocation and Pricing Principles and Implementation Experience. *Natural Resources Forum*, **28(2)**: 112–122.

Easter, W.K. and Liu, Y., 2007: Who Pays for Irrigation: Cost Recovery and Water Pricing? *Water Policy*, **9(3)**: 285–303.

Friends of the Earth Canada, 2007: *Lexicon of Water Soft Path Knowledge, Vol. 1* (CD-ROM). Ottawa; see also the July 2007 issue of *Alternatives*, **33(2)**.

Gender and Water Alliance, 2003: *Gender Perspectives on Policies in the Water Sector*. The Gender and Water Development Report 2003. Loughborough University, UK: WEDC.

Gleick, P., 2003: Global Freshwater Resources: Soft Path Solutions for the 21st Century. *Science* (14 November 2003).

Gleick, P.H., Haasz, D., Henges-Jeck, C., Srinivasan, V., Wolff, G., Cushing, K.K. and Mann, A., 2003: *Waste Not, Want Not: The Potential for Urban Water Conservation in California*. Oakland, CA: Pacific Institute for Studies in Development, Environment, and Security.

Gleick, P.H., Cooley, H. and Groves, D., 2005: *California Water 2030: An Efficient Future*. Oakland, CA: Pacific Institute for Studies in Development, Environment, and Security.

Hellegers, P.J.G.J. and Perry, C.J., 2006: Can Irrigation Water Use Be Guided by Market Prices? Theory and Practice. *International Journal of Water Resources Research*, **22(1)**: 79–86.

Hirsch, A., 1959: Water Legislation in the Middle East. *The American Journal of Comparative Law*, **8(2)**:168–186.

Isaacman, L. and Daborn, G., 2007: Watershed Paths: Application in the Annapolis Valley, NS. *Alternatives*, **33(4)**: 15.

Kay, P., Hendricks, E. and Rahman, N., 2007: Provincial Paths: Planning for Ontario's Future. *Alternatives*, **33(4)**: 16–17.

Kron, W., 2006: On the Brink of Disaster. *Stockholm Water Front*, No. 3: 16–18.

Kundzewicz, Z.W., 2002: Non-Structural Flood Protection and Sustainability. *Water International*, **27(1)**: 3–13.

Moench, M. and Dixit, A. (Eds.), 2004: *Adaptive Capacity and Livelihood Resilience: Adaptive Strategies for Responding to Floods and Droughts in South Asia*. Boulder, CO/Nepal: Institute for Social and Environmental Transition.

Postel, S. and Thompson, B.H., 2005: Watershed Protection: Capturing the Benefits of Nature's Water Supply Services. *Natural Resources Forum*, **29(2)**: 98–108.

Schreier, H., Brown, S. and MacDonald, J.R., 2006: *Too Little and Too Much: Water and Development in a Himalayan Watershed*. Vancouver, British Columbia, Institute for Resource and Environmental Studies, University of British Columbia.

Statistics Canada, 2007: Households and the Environment Survey. *The Daily* (Catalogue 11-001-XIE: 11 July 2007), 6–8.

Stimson, J. and Peñon, M.L., 2004: A Critique of Existing Water Supply System Reconstruction Policies in the Aftermath of Hurricane Mitch, Northwest Coast of Honduras. *Water International*, **29(1)**: 91–104.

Tate, D.M., 1993: *An Overview of Water Demand Management and Conservation.* In: *Vision 21: Water for People.* Geneva: Water Supply and Sanitation Collaborative Council.

Trumbo, C.W. and O'Keefe, G.K., 2005: Intention to Conserve Water: Environmental Values, Reasoned Action, and Information Effects Across Time. *Society & Natural Resources*, **18(6)**: 573–585.

van Koppen, B., 1999: Targeting Irrigation to Poor Women and Men. *International Journal of Water Resources Development*, **15(1/2)**: 121–140.

United Nations Development Programme, 2006: *Beyond Scarcity: Power, Poverty and the Global Water Crisis: Human Development Report (HDR) 2006.* New York: United Nations.

Wolfe, S., and Brooks, D.B., 2003: Water scarcity: an alternative view and its implications for policy and capacity building. *Natural Resources Forum*, **27(2)**: 99–107.

WATER SECURITY OF NATIONS: HOW INTERNATIONAL TRADE AFFECTS NATIONAL WATER SCARCITY AND DEPENDENCY

A.Y. Hoekstra
University of Twente, Enschede, The Netherlands

Abstract. Import of water in virtual form, i.e. in the form of agricultural and industrial commodities, can be an effective means for water-scarce countries to preserve their domestic water resources. On the other hand, export of water-intensive commodities will increase the use and thus the scarcity of water in the exporting countries. Another likely effect of international trade is that it increases dependency of nations that are net importers of water-intensive commodities on net exporters. This paper reviews the following questions: how relevant is international trade in magnifying water scarcity in some nations and in alleviating water scarcity in other nations, and to which extent does international trade contribute to national water dependency?

It is shown that the current global trade pattern significantly influences water use in most countries of the world, either by reducing domestic water use or by enhancing it. In the period 1997–2001, 16% of the water use in the world was not for producing products for domestic consumption but for making products for export. Many of the water problems in the export countries are partially related to their export position. On the other hand, domestic water savings can be enormous for the import countries, i.e. the countries that have partly externalized their water footprint. Jordan annually imports a virtual water quantity that is five times its own yearly renewable water resources. Although saving its domestic water resources, it makes Jordan heavily dependent on other nations, for instance the United States. Other water-scarce countries with high virtual water import dependency (25–50%) are for instance Greece, Italy, Portugal, Spain, Algeria, Libya, Yemen and Mexico.

It is suggested that future national and regional water policy studies include an assessment of the effects of trade on water security. For water-scarce countries, it would also be wise to do the reverse: study the possible implications of national water scarcity on trade. In water-scarce countries a trade-off is to be made between (over)exploitation of the domestic water resources in order to increase water self-sufficiency (the apparent strategy of Egypt) or virtual water import at the cost of becoming water dependent (Jordan).

---

* To whom correspondence should be addressed. e-mail: A.Y.Hoekstra@ctw.utwente.nl

Keywords: Water security, virtual water trade, water footprint, national water dependency, globalisation of trade

1. Introduction

The relation between international trade and water security is generally not something that government officials think about. The reason is that water is hardly traded internationally, due to its bulky properties. Besides, there is no private ownership of water so that it can also not be traded as in a market. It is often forgotten, however, that water is traded in virtual form, i.e. in the form of agricultural and industrial commodities. Although invisible, import of 'virtual water' can be an effective means for water-scarce countries to preserve their domestic water resources (Allan, 2001a). Import of water-intensive commodities can thus relieve the pressure on the domestic water resources. This is a mechanism that makes many countries in the Middle East survive. These countries meet their demand for food and save their scarcely available water resources through food imports from overseas. Mediterranean countries will expectedly experience increased water scarcity due to climate change, forcing them into the direction of increased import of water-intensive products.

Where food import can alleviate national water scarcity, export of water-intensive goods will increase the use and thus the scarcity of water in the exporting countries. Many water problems bear an international trade component (Hoekstra and Hung, 2005; Hoekstra and Chapagain, 2008). Subsidized water in Uzbekistan is overused to produce cotton for export; Thailand experiences water problems due to irrigation of rice for export; Kenya depletes its water resources around Lake Naivasha to produce flowers for export to the Netherlands; Chinese rivers get heavily polluted through waste flows from factories that produce cheap commodities for the European market.

International trade magnifies water scarcity in some places and relieves water scarcity in other places. Another likely effect of international trade is that it increases interdependences between nations, with as a net effect that net importers of water-intensive commodities are dependent on the net exporters. Water is like oil in this respect. Many countries that do not have oil depend on countries that do have oil. Similarly, countries that do not have much water and therefore import water-intensive commodities depend on countries that have more water and apply that for producing water-intensive export goods.

This paper reviews current knowledge with respect to the three issues raised above:

- How relevant is international trade in magnifying national water scarcity?
- How relevant is international trade in alleviating national water scarcity?
- To which extent does international trade contribute to national water dependency?

The chapter will be concluded with a discussion of risks and opportunities associated with the intensification of international trade in water-intensive commodities.

2. International trade amplifying national water scarcity

In the period 1997–2001, 16% of the water use in the world was not for producing products for domestic consumption but for making products for export (Hoekstra and Chapagain, 2007, 2008). The nations with the largest net annual water use for producing export products were the USA (92 billion cubic meters), Australia (57 billion cubic meters), Argentina (47 billion cubic meters), Canada (43 billion cubic meters), Brazil (36 billion cubic meters), and Thailand (26 billion cubic meters). The main products behind the national water use for export from the USA were oil-bearing crops and cereal crops. These products are grown partly rain-fed and partly irrigated. In Australia and Canada, the water use for export was mainly related to the production of cereals and livestock products. In Argentina and Brazil, water use for export was primarily for producing oil-bearing crops. The national water use for export in Thailand was mainly the result of export of rice. Much of the rice cultivation in Thailand is done during the rainy season, but irrigation is widespread, to achieve two harvests per year. In the period 1997–2001, Thailand used 27.8 billion cubic meters per year of water (sum of rainwater and irrigation water) to produce rice for export, mostly grown in the central and northern regions (Maclean et al., 2002). The monetary equivalent of the rice export was US$1,556 million per year (ITC, 2004). Hence, Thailand generated a foreign exchange of 0.06 US$/m^3.

Let me repeat that currently 16% of the water use in the world is not for producing products for domestic consumption but for making products for export and let us assume that, on average, agricultural production for export does not cause significantly more or fewer water-related problems (such as water depletion or pollution) than production for domestic consumption. That means that roughly one sixth of the water problems in the world can be traced back to production for export. Consumers do not see the effects of their consumption behaviour due to the tele-connection between areas of consumption and areas of production. The benefits are at the consumption side, but since water is generally grossly under-priced, the costs remain at the production side. From a water-resources point of view it would be wise for the exporting countries in the world to review their water use for export and see to which extent this is good policy given the fact that the foreign income associated with the exports generally does not cover most of the costs associated with the use of water. The construction of dams and irrigation schemes and even operation and maintenance costs are often covered by the national or state government. Negative effects downstream and the social and environmental costs involved are not included in the price of the export products as well.

3. International trade alleviating national water scarcity

An obvious effect of international trade in water-intensive commodities is that it generates water savings in the countries that import those commodities. This effect has been discussed since the mid-1990s (Allan, 2001b; Hoekstra, 2003; Oki and Kanae, 2004; De Fraiture et al., 2004; Chapagain et al., 2006; Yang et al., 2006). The national water saving associated with import can be estimated by multiplying the imported product volume by the volume of water that would have been required to produce the product domestically.

In many countries international trade in agricultural products effectively reduces domestic water demand (Table 1). These countries import commodities that are relatively water-intensive while they export commodities that are less water-intensive. In the period 1997–2001, Japan, the largest (net) importer of water-intensive goods in the world, annually saved 94 billion cubic meters from its domestic water resources. This volume of water would have been required, in addition to its current water use, if Japan had produced all imported products domestically. In a similar way, Mexico annually saved 65 billion cubic meters, Italy 59 billion cubic meters, China 56 billion cubic meters, and Algeria 45 billion cubic meters (Chapagain et al., 2006).

One of the water-scarce countries that most heavily depend on imports of water-intensive commodities is Jordan. It imports 5 to 7 billion cubic meters of water in virtual form per year, which is in sharp contrast with the 1 billion cubic meters of water withdrawn annually from domestic water sources (Haddadin, 2003; Hoekstra and Chapagain, 2007, 2008). People in Jordan thus survive owing to the fact that their 'water footprint' has largely been externalized to other parts of the world, for example the USA. Intelligent trade largely covers up Jordan's water shortage: export of goods and services that require little water and import of products that need a lot of water. The good side of Jordan's trade balance is that it preserves the scarce domestic water resources; the downside is that the people are heavily water dependent.

For countries that depend on the import of water-intensive products, it is important to know whether the water thus saved has higher marginal benefits than the additional cost involved in importing these products. Let us consider the example of Egypt, a country with a very low rainfall – the mean rainfall is only 18 mm/year – and with most of its agriculture being irrigated. The import of wheat in Egypt implies a saving of their domestic water resources of 3.6 billion cubic meters per year, which is about 7% of the total volume of water Egypt is entitled to according to the 1959 agreement on the use of the Nile River. The national saving is made with the investment of foreign exchange of US$593 million per year (ITC,

2004), so that the cost of the virtual water is 0.16 US$/m³ at most. In fact, the cost will be much lower, because the costs of the imported wheat cover not only the cost of water, but also the costs of other input factors such as land, fertilizer, and labour. In Egypt, fertile land is also a major scarce resource. The import of wheat not only releases the pressure on the disputed Nile water, but also reduces pressure to increase the area of land under agriculture. Greenaway et al. (1994) and Wichelns (2001) have shown that in the international context Egypt has a comparative disadvantage in the production of wheat, so that the import of wheat into Egypt implies not only physical water saving, but also an economic saving.

TABLE 1. Examples of nations with net water saving as a result of international trade in agricultural products. Period 1997–2001

Country	Total use of domestic water resources in agricultural sector[a] (10^9 m³/year)	Water saving as a result of import of agricultural products[b] (10^9 m³/year)	Water loss as a result of export of agricultural products[b] (10^9 m³/year)	Net water saving due to trade in agricultural products[b] (10^9 m³/year)	Ratio of water saving to water use
China	733	79	23	56	8%
Mexico	94	83	18	65	69%
Morocco	37	29	1.6	27	73%
Italy	60	87	28	59	98%
Algeria	23	46	0.5	45	196%
Japan	21	96	1.9	94	448%

[a]Hoekstra and Chapagain (2008).
[b]Chapagain et al. (2006a). Agricultural products include both crop and livestock products.

4. How international trade can enhance national water dependency

Nations can be 'water dependent' in two different ways. Most writings about national water dependency are concerned with the dependency of downstream nations on the inflow from water from upstream basins or the mutual dependency of nations sharing a border river. This type of water dependency is sometimes quantified by considering the ratio of external to total renewable water resources of a country. FAO (2007) defines the 'external renewable water resources' of a country as that part of the country's renewable water resources which is not generated in the country. It includes inflows from upstream countries (groundwater and surface water) and part of the water of border lakes or rivers. The 'internal renewable water resources' of a country concern the average annual flow of rivers and recharge of aquifers generated by endogenous precipitation. The total renewable water resources of a country are the sum of internal and external renewable water resources. Table 2 shows the 'external water resources dependency' for a number of selected downstream countries. For a country like Egypt the

dependency is extremely high, because the country receives hardly any precipitation and thus mostly depends on the inflowing Nile water. Similarly,
but to a lesser extent, Pakistan strongly depends on the water of the Indus,
Cambodia on the water of the Mekong and Iraq on the Tigris and Euphrates.
In all these cases water is an important geopolitical resource, affecting
power relations between the countries that share a common river basin. In a
country like the Netherlands external water resources dependency is high
but less important, because water is less scarce than in the previous cases.
Nevertheless, here too there is a dependency, since activities within the upstream countries definitely affect downstream low flows, peak flows and
water quality.

   The political relevance of 'external water resources dependency' of
nations makes water a regional geopolitical resource in some river basins.
The other type of water dependency, virtual water import dependency,
makes water a global geopolitical resource. The fundamental reason is the
combination of increasing scarcity of water, its unique character that pre-
vents substitution and its uneven distribution throughout the world. Where
water-abundant regions did not fully exploit their potential in the past, they
now increasingly do so by exporting water in virtual form or even in real
form. The other side of the coin is the increasing dependency of water-scarce
nations on the supply of food or water, which can be exploited politically
by those nations that control the water.

TABLE 2. Dependency on incoming river flows for selected countries

Country	Internal renewable water resources[a] (10^9 m^3/year)	External (actual) renewable water resources[a] (10^9 m^3/year)	External water resources dependency[b] (%)
Iraq	35	40	53
Cambodia	121	356	75
Pakistan	52	170	77
Netherlands	1.1	80	88
Egypt	1.8	56.5	97

[a] FAO (2007).
[b] Defined as the ratio of the external to the total renewable water resources.

   From a water resources point of view one might expect a positive
relationship between water scarcity and virtual water import dependency,
particularly in the ranges of great water scarcity. Virtual water import
dependency is defined as the ratio of the external water footprint of a
country to its total water footprint. As Hoekstra and Chapagain (2008) show,
countries with a very high degree of water scarcity – e.g. Kuwait, Qatar,
Saudi Arabia, Bahrain, Jordan, Israel, Oman, Lebanon and Malta – indeed

TABLE 3. Virtual water import dependency of some selected countries. Period: 1997–2001

Country	Internal water footprint[a] $(10^9 \, m^3/year)$	External water footprint[a] $(10^9 \, m^3/year)$	Water self-sufficiency[b] (%)	Virtual water import dependency[c] (%)
Indonesia	242	28	90	10
Egypt	56	13	81	19
S. Africa	31	9	78	22
Mexico	98	42	70	30
Spain	60	34	64	36
Italy	66	69	49	51
Germany	60	67	47	53
Japan	52	94	36	64
U.K.	22	51	30	70
Jordan	1.7	4.6	27	73
Netherlands	4	16	18	82

[a]Hoekstra and Chapagain (2008).
[b]Defined as the ratio of the internal to the total water footprint.
[c]Defined as the ratio of the external to the total water footprint.

have a very high virtual water import dependency (>50%). The water footprints of these countries have largely been externalised. Jordan annually imports a virtual water quantity that is five times its own yearly renewable water resources. Although saving its domestic water resources, it makes Jordan heavily dependent on other nations, for instance the United States. Other water-scarce countries with high virtual water import dependency (25–50%) are for instance Greece, Italy, Portugal, Spain, Algeria, Libya, Yemen and Mexico. Table 3 presents the data for a few selected countries. Even European countries that do not have an image of being water-scarce, such as the UK, Belgium, the Netherlands, Germany, Switzerland and Denmark, have a high virtual water import dependency. In those cases where large virtual water imports go together with national water abundance, the import is obviously not related to water scarcity but must be explained from other factors.

In most water-scarce countries the choice is either (over)exploitation of the domestic water resources in order to increase water self-sufficiency (the apparent strategy of Egypt) or virtual water import at the cost of becoming water dependent (Jordan). The two largest countries in the world, China and India, still have a very high degree of national water self-sufficiency (93% and 98% respectively). However, the two countries have relatively low water footprints per capita (China 700 m^3/capita/year and India 980 m^3/capita/year). If the consumption pattern in these countries changes to that of the USA or some Western European countries, they will be facing a severe water scarcity in the future and will probably be unable to sustain their high degree of water self-sufficiency. A relevant question is how China and India are going to feed themselves in the future. If they were to decide to partially obtain food security through food imports, this would put enormous demands on the land and water resources in the rest of the world.

5. Discussion

International transfers of water in virtual form are substantial and likely to increase with continued global trade liberalization (Ramirez-Vallejo and Rogers, 2004). Intensified trade in water-intensive countries offers both opportunities and risks. The most obvious opportunity of reduced trade barriers is that virtual water can be regarded as a possibly cheap alternative source of water in areas where freshwater is relatively scarce. Virtual-water import can be used by national governments as a tool to release the pressure on their domestic water resources. This import of virtual water (as opposed to real water, which is generally too expensive) will relieve the pressure on the nation's own water resources. For water-abundant countries an argument can be made for export of virtual water. Trade can physically save water if products are traded from countries with high to countries with low water productivity. For example, Mexico imports wheat, maize, and sorghum from the USA, which requires 7.1 billion cubic meters of water per year in the USA. If Mexico were to produce the imported crops domestically, it would require 15.6 billion cubic meters of water per year. Thus, from a global perspective, the trade in cereals from the USA to Mexico saves 8.5 billion cubic meters per year. Although there are also examples where water-intensive commodities flow in the opposite direction, from countries with low to countries with high water productivity, the available studies indicate that the resultant of all international trade flows works in a positive direction. Hoekstra and Chapagain (2008) show that international trade in agricultural commodities reduces global water use in agriculture by 5%. Liberalization of trade seems to offer new opportunities to contribute to a further increase of efficiency in the use of the world's water resources.

A serious drawback of trade is that the indirect effects of consumption are externalized to other countries. While water in agriculture is still priced far below its real cost in most countries, an increasing volume of water is used for processing export products. The costs associated with water use in the exporting country are not included in the price of the products consumed in the importing country. Consumers are generally not aware of – and do not pay for – the water problems in the overseas countries where their goods are being produced. According to economic theory, a precondition for trade to be efficient and fair is that consumers bear the full cost of production and impacts. Another downside of intensive international virtual-water transfers is that many countries increasingly depend on the import of water-intensive commodities from other countries. Jordan annually imports a virtual-water volume that is five times its own annual renewable water resources. Other countries in the Middle East, but also various European countries, have a similar high water import dependency. The increasing lack of self-sufficiency has made various individual countries, but also larger regions, very vulnerable. If for whatever reason food supplies cease – be it due to war or a natural disaster in an important export region – the importing regions will suffer severely. A key question is to what extent nations are willing to take this risk. The risk can be avoided only by promoting national self-sufficiency in water and food supply (as Egypt and China do). The risk can be reduced by importing food from a wide range of trade partners. The current worldwide trend, however, facilitated by the World Trade Organization, is toward reducing trade barriers and encouraging free international trade, and decreasing interference by national governments.

The current global trade pattern significantly influences water use in most countries of the world, either by reducing domestic water use or by enhancing it. Future national and regional water policy studies should therefore include an assessment of the effects of trade on water policy. For water-scarce countries, it would also be wise to do the reverse: study the possible implications of national water scarcity on trade. In short, strategic analysis for water policy making should include an analysis of expected or desirable trends in international or inter-regional virtual-water flows.

References

Allan, J.A., 2001a: Virtual water – Economically invisible and politically silent – A way to solve strategic water problems. *International Water and Irrigation*, Nov: 39–41.

Allan, J.A., 2001b: The Middle East water question: Hydropolitics and the global economy. I.B. Tauris, London.

Chapagain, A.K., Hoekstra, A.Y. and Savenije, H.H.G., 2006: Water saving through international trade of agricultural products. *Hydrology and Earth System Sciences*, **10(3)**: 455–468.

De Fraiture, C., Cai, X., Amarasinghe, U., Rosegrant, M. and Molden, D., 2004: Does international cereal trade save water? The impact of virtual water trade on global water use, comprehensive assessment, *Research Report 4* IWMI, Colombo.

FAO, 2007: AQUASTAT. Online database: www.fao.org/nr/water/aquastat/main/index.stm.

Greenaway, F., Hassan, R. and Reed, G.V. (1994) An empirical-analysis of comparative advantage in Egyptian agriculture. *Applied Economics*, **26(6)**: 649–657.

Haddadin, M.J., 2003: Exogenous water: A conduit to globalization of water resources. In: A.Y. Hoekstra (Ed.), Virtual water trade: Proceedings of the International Expert Meeting on Virtual Water Trade, Value of Water Research Report Series No. 12, UNESCO-IHE, Delft, pp. 159–169.

Hoekstra, A.Y. (Ed.), 2003: Virtual water trade: Proceedings of the International Expert Meeting on Virtual Water Trade, Delft, The Netherlands, 12–13 December 2002, Value of Water Research Report Series No. 12, UNESCO-IHE, Delft.

Hoekstra, A.Y. and Chapagain, A.K., 2007: Water footprints of nations: Water use by people as a function of their consumption pattern. *Water Resources Management*, **21(1)**: 35–48.

Hoekstra, A.Y. and Chapagain, A.K., 2008: *Globalization of water: Sharing the planet's freshwater resources*. Blackwell, Oxford.

Hoekstra, A.Y., and Hung, P.Q., 2005: Globalisation of water resources: International virtual water flows in relation to crop trade. *Global Environmental Change*, **15(1)**: 45–56.

ITC, 2004: PC-TAS version 1997–2001, Harmonized System. CD-ROM, International Trade Centre, Geneva.

Maclean, J.L., Dawe, D.C., Hardy, B. and Hettel, G.P., 2002: Rice almanac: Source book for the most important economic activity on earth, International Rice Research Institute, Los Baños, Philippines.

Oki, T. and Kanae, S., 2004: Virtual water trade and world water resources. *Water Science and Technology*, **49(7)**: 203–209.

Ramirez-Vallejo, J. and Rogers, P., 2004: Virtual water flows and trade liberalization. *Water Science and Technology*, **49(7)**: 25–32.

Wichelns, D., 2001: The role of 'virtual water' in efforts to achieve food security and other national goals, with an example from Egypt. *Agricultural Water Management*, **49(2)**: 131–151.

Yang, H., Wang, L., Abbaspour, K.C. and Zehnder, A.J.B., 2006: Virtual water trade: An assessment of water use efficiency in the international food trade. *Hydrology and Earth System Sciences*, **10**: 443–454.

REDUCING THE RISK: DROUGHT MITIGATION AND THE ECONOMY OF IRRIGATION

S. Szalai[1*] and L. Cselőtei[2]

[1] Hungarian Meteorological Service, 38 POB, 1525 Budapest, Hungary
[2] Hungarian Academy of Sciences, 7 Nádor u, 1051 Budapest, Hungary

Abstract. Hungary is situated in the Carpathian basin and its climate has similar tendencies to the South European region. In particular, temperature shows increasing tendencies in all seasons and precipitation decreasing tendencies throughout the year except summer. The warming is faster than the global average, but the year-to-year variability of precipitation still has larger impacts than the decreasing trend. Furthermore, the reducing snow/rain ratio and the growing intensity of precipitation worsen the situation of the surface water balance. Therefore, the country is very sensitive to the climate change, especially to drought. Irrigation is the tool most used to reduce drought damage. Because of the water problems in the country, irrigation is not possible everywhere due to the natural and economical problems. Irrigation is quite expensive, and the agricultural production has to return a profit, i.e. agricultural production has to be profitable. Different cultivation methods require different investments, need different degrees of human effort and produce different income. Therefore, the economy of irrigation can be judge only in a complex way, using natural, economical and social factors.

Keywords: Climate of Hungary, climate change, drying, irrigation, agricultural cultivation methods

1. Introduction

Hungary is situated in the bottom of the Carpathian basin. The main types of climate are maritime, Mediterranean and continental. According to this, the climate of the country shows large variability, especially the precipitation. The annual temperature average is about 10.5°C, the annual precipitation amount about 600 mm. Monthly precipitation can exceed 100 mm or sometimes even 200 mm in any month, but months without any precipitation may occur any time of the year. The growing season (for most of the plants April–September) shows even larger variations regarding monthly precipitation sums. All seasons show decreasing trends except summer, where no significant trend is detected. The growing frequency of summer droughts can be explained by the increasing intensity of rainfall and the warming. Warming is occurring fastest in summer of all the seasons; it reached

---

* To whom correspondence should be addressed. e-mail: szalai.s@met.hu

1°/100 year in the 20th century, while the annual average is about 0.7°C. Flood and drought can happen in the same place and same year according to the changing climatic conditions and lower the quality of the hydro-physical characteristics of soils in the given area.

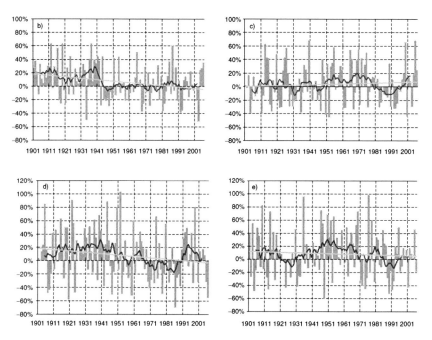

Figure 1. Country-wide seasonal precipitation time series (spring: upper left, summer upper right, autumn: lower left and winter: lower right). Columns are the seasonal deviations in percentage from the 1971–2000 average, the curve is a 10-year average (Szalai et al., 2005)

Days with precipitation of more than 1.0 mm show a significant decrease, and the average precipitation in these days slightly increases. More frequent heat waves and an increasing number of warm nights worsen the situation for plant water availability.

The water budget explains another source of vulnerability of the country: the difference between precipitation and actual evapotranspiration is about 6% of the surface water inflow from abroad (Table 1). Therefore, very accurate management of available water sources should be requested.

2. Land use

Hungary covers 9.3 million hectares. Agricultural production occurs in almost 8 million hectares. Figure 2 shows the different cultivation methods, the summarized income from them and the ratio of employments among them.

TABLE 1. The average yearly water budget of Hungary (1961–1990) (VITUKI, 2005)

	mm	M^3/s	km^3
Precipitation	598	–	55.6
Actual evapotranspiration	525	–	48.8
Difference	73	–	6.8
Surface water inflow	–	3,518	110.9
Surface water outflow		3,743	118.0

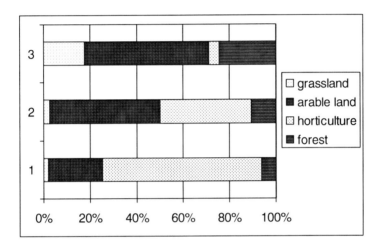

Figure 2. Different agricultural cultivation methods in Hungary. Row 3 shows the ratio of territory; row 2 is the ratio of incomes and row 1 is the ratio of number of employees. 100% means about 8 million hectares, 800 billion HUF (€1 is approximately 250 HUF) and 300,000 employees

Some very important features are not shown in the Fig. 2: these include natural, environmental, landscape, human comfort, etc. effects. The importance of these effects is undoubted, but their quantification is still not unambiguous. Usually, they are not taken into account in economic calculations, but they have to be replaced by similar employment after agricultural activity ceases in an area.

Horticulture occupies only 5% of the agricultural area, but produces about 40% of its overall income and provides work for about 200,000 people, about 2/3 of all people employed in agriculture. In contrast, almost 50% of the agricultural area is arable land, but about 20–25% of employees work there. Therefore, when we calculate the total agricultural benefits, we have to include financial, social and the non-quantified additional factors.

3. Variability of agricultural production

Rainfed agriculture has a large year-to-year variability according to the climate variability, primarily due to the drought events. The largest area equipped for irrigation was about half a million hectare, but practically, the largest irrigated area does not exceed 300,000 ha in dry years and 100,000 ha in wetter years. The changing economical and political conditions had large influence on the irrigated area. According to Table 2, Hungary has possibilities for irrigation on about 120,000 ha. Production on the 1–1.5% of the total agricultural land gives about 25% of the total income.

Glasshouses and ornamental plants have the most intensive production and irrigation. This type of production provides about 12–13% of the total and more than half of the income from irrigated agriculture in just 8,000 ha. On the other hand, half of the irrigated area, the irrigated arable land, produces about 15% of the total income from irrigated agriculture (Fig. 3). Therefore, irrigation is a basic condition in horticulture, and detailed calculation is required for arable irrigation to make it beneficial. Transpiration and evaporation are regulated differently in glasshouses and ornamental plant production. The temperature of the plants and the air, radiation and air humidity are individually controlled for these purposes.

Frost protection has to be set up for the irrigated arable land according to the climate of the country. According to research, the late spring and early autumn frosts are not related to global warming, and agriculture has to be prepared to them. Climate change can influence the frequency of natural disasters, which has to be taken into consideration.

TABLE 2. Comparison of the irrigated area in the countries of the region in the 1980s and in the first half of 1990s (Dirksen and Huppert, 2006)

Country	Total arable land million hectares	Former irrigated area, (second half 1980s) million hectares	Actual irrigated area (first half 1990s) million hectares
Bulgaria	4.805	1.250	0.03–0.04
Czech Republic	3.100	0.133	0.01
Germany	17.185	0.940/0.5	0.45/0.2
Hungary	4.600	0.3	0.1
Macedonia	0.612	0.1278	0.0274
Poland	14.050	0.3015	0.833
Russia	120.000	4.737	3.5061
Romania	9.416	3.2052	0.5–0.85
Slovenia	0.746	0.15	0.005
Ukraine	33.615	2.624	0.7–1.0

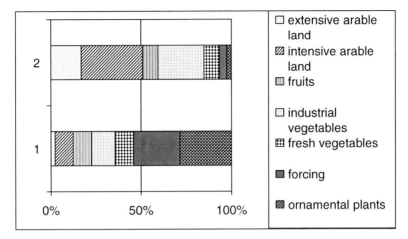

Figure 3. Territorial and financial distribution among the different irrigated cultivation methods. Upper row shows the ratio of area of different cultivation methods, and the lower row shows the appropriate incomes

4. Water resources irrigation demand

Total water consumption is about 60 km^3 in Hungary today. The largest part of this is evaporation and transpiration. Human water demand comprises the industrial and communal water (4.5 + 1 km^3), aquaculture and irrigation need about 0.5 km^3 water. Additionally, glasshouse and ornamental plant production requires 0.1–0.2 km^3 of irrigation water. Unfortunately, the use of wastewater is not satisfactory and not well measured. The use of sub-surface water by root systems and wells is important, but not very accurately estimated, because most wells operate without any permission.

Although it seems that Hungary has enough water for irrigation, there is a very strong competition for available water. According to the governmental regulations for dry periods, the first step is a marked reduction of irrigation water use. Water deficit plays a very important role in the biomass production (Nagy, 2007). Figure 4 shows that the water deficit and the reduction of yield have a nearly linear relationship.

The ratio of irrigated and non-irrigated agricultural productivity is at least two, but usually more. The problem is that the large variability of precipitation makes the operation of irrigation systems useless and sometimes under-designed.

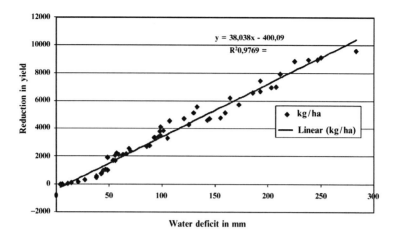

Figure 4. Connection between water deficit and yield reduction of maize (Nagy, 2007)

5. Legal and administrative basis of irrigation

Water use is regulated generally by the European Union and in detail by national rules. On a national level, some regulations relate to irrigation water use, its availability, reduction in the dry periods, etc. The latest documents support irrigation. Such documents are the New Hungary Development Plan and Rural Development Plan, which has already been accepted by the European Union. It is expected that the Hungarian government will accept the National Drought Strategy and the National Climate Change Strategy this year. The last document is deeply involved with adaptation measures, among others irrigation and estimates the present adaptive capacity of the country. Agricultural adaptation is strongly connected to the water management development in Hungary (Ligetvári, 2006). Large flood protection works were begun in the 19th century with the philosophy of taking the surplus water of floods out of country as soon as possible. Due to climate change, a different philosophy is developing today: how can we keep the surplus water of the winter floods for the summer droughts? The design of large water management projects in emergency reservoirs is going on.

On the other hand, the question about the dry farming versus irrigation is still not solved. Some opinions say that agriculture has to follow the direction of climate change, and produce appropriate products in an appropriate way. This is the green direction and they suggest dry farming. The other direction suggests that irrigation be used more as an adaptation method of agriculture to climate change, keeping the economical, social and landscape benefits of present agriculture. Whatever the final answer, water-saving technologies have to be supported and distributed not only in irrigation, but in every industrial and communal activity (Camp et al., 1990).

The EU regulations provide the background for the national activities, first of all the Water Framework Directive, which makes the complex problem of irrigation economy even more sophisticated.

6. Conclusions

The benefits of irrigation are usually calculated in a very simple way: water and electricity prices versus yield growth. The question is much more complicated: the social factors (number of employees, income tax against unemployment subsidy), landscape, and other factors have to be involved in the assessment. The benefit can be very different depending what, where and when is irrigated. The most profitable irrigated agricultural cultivation methods need large investments, which have to be subsidized, especially in the less developed countries. All these activities have to be strongly connected to the economic and rural development plans of the country, with special regards to the possible climate change and its impacts and the possible growing frequency of natural disasters.

Conversely, the legal and administrative framework has to be fit for the changing environmental, economic and social conditions, and reviewed regularly both at national and EU level.

References

Camp, C.R., Sadler, E.J., Sneed, R.E., Hook, J.E. and Ligetvári, F., 1990: Irrigation for humid areas In: *Management of farm irrigation systems*, St. Joseph: American Society of Agricultural Engineers, 549–578.

Dirksen, W. and Huppert, W. (Eds.), 2006: *Irrigation sector reform in Central and Eastern European countries.* Deutsche Gesellschaft für Technische Zusammenarbeit (GTZ), Eschborn, Germany.

Ligetvári, F. (Ed.), 2006: *Country report on irrigation sector reform in Hungary.* ICID Hungarian National Committee.

Nagy, J. (Ed.), 2007: *Kukoricatermesztés* (Maize production). Akadémiai Kiadó (In Hungarian).

Szalai, S., Bihari, Z., Lakatos, M. and Szentimrey, T., 2005: *Some characteristics of climate of Hungary since 1901.* Hungarian Meteorological Service.

VITUKI, 2005: *Country report on the water budget of Hungary.* VITUKI.

WORKING GROUP I: RISK ASSESSMENT AND WATER GOVERNANCE

Chair and rapporteur: **D. Brooks**
Friends of the Earth Canada (with additions by Tony Jones)

Although the initial remit given to this Working Group was to discuss the possible effects of changing management of water resources from State to private companies, most of the group consisting of delegates from former Soviet countries found that they preferred to discuss more regional issues and to focus on risk assessment. The two are not entirely separate issues, as the facilities of the State for providing early warning of hydrological extremes and for planning emergency response are arguably greater than those of privatised industry. There is also a strong argument for close State regulation of water companies, both in terms of environmental impacts and in terms of costs, profit-making and services to the consumer. Record fines imposed by the UK regulator Ofwat on Thames Water and SevernTrent Water during the first half of 2008 for presenting misleading information and poor customer services emphasise the point. Similarly, the UK environmental regulators, the Environment Agency and the Scottish Environmental Protection Agency, have fined privatised water companies as well as farmers and industries for numerous river pollution incidents. Some of the worst cases of pollution over recent years in the UK have come from privatised water companies.

The chapter by Jones (this volume) discusses further issues of water governance, especially of failures by private companies in Africa and South America and of emerging fears relating to the modes of finance being used by private companies. There are two highly important issues here:

1. States may lose control of their resources – their pricing and equal accessability for all the population – to foreign profit-making companies. For many poorer States in the developing world, the financial muscle of multinational companies may be greater than those of the State, or greater than the State could afford to risk in any legal challenges.
2. Importing investment from abroad has provided money for developing water infrastructure and "modern" management systems, but it can also expose infrastructure and management systems to less desirable aspects of global finance. These include:
   (a) Possible political pressure from foreign interests, such as "sovereign wealth funds" like the China State Investment Corporation (Jones, loc. cit.) or

(b) Exposure to the vagaries of the world financial market. The latter is of particular concern since the credit crisis of 2007. Companies that develop and maintain their systems using highly leveraged loans are now very much more constrained in their operations than before-hand. Again, it is a matter of 'wait and see' to appreciate the full impacts of the downturn in the money markets on privatised water companies and, potentially, on their contractual commitments to national governments.

1. Risk assessment issues in the wider Caucasus region

1. The Working Group identified a wide range of risks, some of which are more readily assessed by current methods than others:

   • Risk of ongoing water scarcity for current populations and for economic growth
   • Increased risk of floods and of droughts, along with increased scale of resulting damages
   • Risk of water resources contamination during times of water surplus or water deficit
   • Sources of and options for coping with natural and technological disasters
   • Risks associated with anticipated effects of climate change, especially in relationship to regional ecological features and to ongoing processes of economic globalization

2. Better information is needed in the region on water resources, water demand management, and predictive modeling. This requires better data collection networks and data exchange systems. The WG identified the following priorities for research to improve capacity for risk assessment and risk reduction:

   • Establishment of well-funded institutions for water resources surveys
   • Development, enhancement and application of systems for continuous monitoring of regional water resources and uses at scales from local to regional
   • Development of models of water ecosystems, as well as models that show the transportation of sediment and chemical pollutants through those ecosystems
   • Methods to predict anomalous events and to identify dangerous situations before the create disasters
   • Region-wide implementation of contemporary GIS technologies

In particular, members of the Working Group believe that the regions of the Caucasus and Central Asia would achieve strong benefits through further development of a system for transforming the multi-spectral satellite information of high resolution, with a specific focus on regional water basins and ecosystems, and with wide dissemination of the related to tools to the regional environmental services, universities, and scientific institutions.

4. The Working Group noted the necessity of establishing broad regional and international cooperation. Where institutions already exist to support cooperation, their work should be extended. In either case, the main emphasis should be placed on issues of transboundary waters. International cooperation should not focus exclusively on technical cooperation, but should also include joint policy development and, where possible, joint river basin management.

   The case of the Aral Sea was reviewed in some detail, with attention to ways in which risk management had been neglected and how the current disaster could have been avoided had research results and monitoring as recommended above had been in place. The WG concluded that there is no short-term way to resolve the ecological, economic and social impacts on the Aral Sea basin, but that gradual restoration of river flows and modification of economic planning would be of some help.

5. There was strong support for integrated river basin management despite the recognition that full development of the concept would always be more of a general goal than a specific accomplishment. The participants noted that, in recent years, the need for considerable water to be left *in situ* to provide ecosystem services, and that economic development planning should recognize the physical and biological risks from over-exploitation of water resources. Demand management to reduce the extraction of water resources should be in place for all regions and communities.

6. Participants recognized that, despite enormous gains in recent years, risk assessment can never be perfect, and that at all levels from local to regional plans should be in place to cope with floods and droughts, and that mock trials of disaster scenarios should be ongoing parts of risk management strategies. As well, social and economic alternatives should be sought for communities that are particularly at risk from particular combinations of regional ecology and local livelihoods.

7. Participants emphasized that risk assessment is not an exact science and that social and economic as well as technical and engineering concepts should be used in assessing and managing risks.

8. The Working Group recognises and urges attention to the need for parallel development of all forms of education in the field of water resources, including both the scientific and engineering components as well as the economic and policy components.

9. They also urge that better information on the nature of regional water systems and the possible risks from excessive use or pollution be conveyed to their citizens. Whenever possible the participation of local communities should be sought when reviewing options for river basin management, risk management, and other decisions affecting their lives and livelihoods.

2. Conclusions

2.1. GENERAL OBSERVATIONS OF THE WORKING GROUP

1. First, regardless of the degree or nature of the risk, wider public participation was needed to define appropriate ways to cope with and to define acceptable levels of risk, in order to ensure that officials and agencies could cope effectively with threats.
2. Environmental resources have been undervalued, particularly during the Soviet period.
3. Special environmental resources, e.g. the Aral Sea, deserve to be kept in an ecologically healthy state and their productive use should not impinge upon their ecology.
4. Nations upstream or upflow (in the case of groundwater) have the responsibility to ensure that water that flows across (or under) their border should not be significantly degraded in quality.

2.2. PRIORITIES IDENTIFIED

1. Need to establish/enhance international cooperation on regional transboundary waters.
2. Improved monitoring and modelling of water resources, quality and ecology, and forecasting for extreme events, including greater use of satellite and GIS systems.
3. Development of education on water and community participation in decision making.

PART II:

WATER QUALITY AND TERRORISM

DRINKING WATER SECURITY IN CRISIS SITUATIONS FROM A MEDICAL PERSPECTIVE

A.E. Gurzau[1*], **C. Borzan**[2], **I.R. Lupsa**[3], **L.O. Sfetcu**[4], **A.L. Ivan**[5] **and S. Gurzau**[5]

[1] *Environmental Health Center, Cluj-Napoca, Romania*
[2] *University of Medicine and Pharmacy Iuliu Hatieganu, Cluj-Napoca, Romania*
[3] *Public Health Authority of Hunedoara County, Deva, Romania*
[4] *Institute of Public Health, Timisoara, Romania*
[5] *Institute of International Studies, Cluj-Napoca, Romania*
[6] *Regional Center for Environmental and Health Assessment, Cluj-Napoca, Romania*

Abstract. The present paper deals with the drinking water security problems in large rural areas of Romania during the 2005 floods in the Timis, Bega and Mures River basins. It evaluates the intervention of public health authorities to prevent a major health risk from an epidemic or endemic waterborne diseases, as well as the health promoting education in this regard. From the medical perspective, the interventions in flood affected region showed once again that the main public health priority was to provide a basic secure water supply to the affected population. Along with the water supply, the prophylactic immunisation for waterborne diseases, like hepatitis A, resulted in crisis management without significantly affecting the health status of the population. Analysis of the events in recent years has shown the persistence of the communities' vulnerability in emergency situations, even though the capacity of the specialized institutions for action has increased.

Keywords: Drinking water, water security, floods, public health, medical intervention

1. Introduction

Romania is part of the Black Sea Region, an important region nowadays in modeling international relations not only politically but also economically, as well as socially. Also, the Black Sea Region has a strong connection with the Caspian Sea Region, this connection being much more complex than an economic one. Looking towards the future, international relations are going to be ever more based on cooperation, interdependency and knowledge sharing. The aspiration to work more on projects in these regions is increasing

---

[*] To whom correspondence should be addressed. e-mail: ancagurzau@ehc.ro

and this will promote more stability, because working on joint projects draws together a wide range of interests: political, economical, social and last but not least ecological.

The international debate on the concept of Human Security began in 1994 when it was first mentioned in the Human Development Report of the General Assembly of the United Nations Organization, stipulating that the solution for the worldwide security lies in development and not in arming.

The Human Development Report describes human security as being the state of safety from the constant threats of hunger, disease, crime and repression. It also means protection from sudden disruptions in the pattern of our daily lives. At the same time four domains that may affect the human security were stated as follows:

- Poor economic development
- Lack of food
- Poor health of population
- Degradation of the environment and natural disasters

The concept of Human Security unites the domains of security and development and consolidates the role of public health as part of worldwide security. Each nation's water resources must be protected, conserved, developed, managed, used, and controlled in ways which ensure efficient, sustainable and beneficial use of water in the public interest.

Global warming and climatic change may also influence the spread of diseases through the potential growth in occurrence and extent of natural disasters like landslides, earthquakes, tsunamis, hurricanes and floods. These events, like wars and violent conflicts, inevitably affect the state of health and the medical substructure, as well as massive transferring of people into shelters and over-crowded camps (Brower and Chalk, 2003). It is possible that such events have direct effects upon the public health by transforming the disaster areas into possible epidemiological bombs with delayed effect.

Natural disasters can have rapid or slow onset, with serious health, social, and economic consequences. During the past 2 decades, natural disasters killed millions of people, adversely affected the lives of at least 1 billion more people, and resulted in substantial economic damage (Watson et al., 2007).

The six catastrophic events of floods which affected Romania during 2005, showed that this country faces all these as manifestations of the global climatic changes. These consequences, expressed in deaths and substantial economic damage, estimated at more 1.5 billion euros, show that a rapid change in view and strategy in the fight against these natural disasters are necesary, as well as in the management of climate change.

2. Material and methods

We analysed the surveillance data of the drinking water security problems in large rural areas from Romania during the 2005 floods in the Timis, Bega and Mures river basins. During April 2005 in the Banat region of Romania, a huge freshet on Timis River breached the dam in the Crai Nou area, releasing about 320 billion cubic metres of water and creating the so-called "Banat Sea" – the most devastating floods in the last 40 years. This episode was followed by other flood events in other areas of the Timis County, which were insignificant compared to the above-mentioned one.

In Hunedoara County, the situation was different, most localities being flooded as a consequence of torrents and freshets on small rivers. From the medical perspective, the epidemiological issues associated with flooding are (Watson et al., 2007):

- Drowning (from rushing water)
- Injuries (trees, electrocution, debris, clean-up, power blackout)
- Water safety (flood water/sewage contamination)
- Food safety
- Vector-borne diseases
- Wild animal displacement/rabies
- Outbreaks of communicable diseases

Our work concerned only two epidemiological issues: water safety and outbreaks of communicable diseases.

3. Results and discussions

The official statistics reported in 2005 in Timis County showed the following: 34 affected localities with 5,375 households, 1,051 wells and 92,732 ha of land flooded and almost 3,000 people displaced. In Hunedoara County: 37 localities, 830 households, 1,687 wells and 5,477 ha of land were also flooded. Disruptions in drinking water and sewage networks were reported in both counties.

3.1. WATER SECURITY AND SAFETY

Water supply problems arise in all phases of the disaster-management cycle. As with all other elements of emergency management, water supplies can be designed and maintained in ways that help to reduce the health impact of disasters. Rural communities are usually less vulnerable than urban communities to disruption of water supplies in disasters, as their supplies are generally decentralized and based on simple technology, and alternative sources are frequently available. Certain hazards, such as floods and droughts, may have a greater impact in rural areas than in urban areas (AWWA 1999; Logue, 2006; WADEM, 1990; WHO, 1997, 2004). Emergency response strategy must include:

3.1.1. Priorities

The first priority is to provide an adequate quantity of water, even if its quality is poor, and to protect water sources from contamination. In all situations, a successful emergency response in the water supply sector depends on improvisation and gradual improvement of water supplies, progressing from basic services during the emergency and recovery phases, to more sustainable services in the long term. If the usual water source is not damaged or contaminated and is still safely accessible to the population, it is necessary only to monitor the source. If the accustomed source is contaminated, which is commonly the case after floods, an alternative source should be sought or water should be chlorinated before consumption until the source can be disinfected and protected. In any case, preventive disinfection will help reduce the health risks associated with contaminated water. Emergency repairs to damaged waters supply installations are necessary.

3.1.2. Assessment, monitoring and review

The most effective emergency water supply measures are ensured through a process of assessment, monitoring and review. Assessment is required to identify needs, damage and resources, so as to be able to respond appropriately and with maximum impact; monitoring of activities and the context is essential to ensure that the water supply activities are carried out as planned, with timely indications of problems and unsatisfied needs; periodic reviews of the situation and the response are essential to ensure that the response remains relevant to the needs and resources of the communities affected by the disaster (Logue, 1996; Watson et al., 2007).

In the most affected areas of Timis County, the drinking water was provided from three existing deep-drilled wells treated by mobile water treatment stations and delivered to the population in plastic bags and tanks. At the same time, other water sources were inspected in order to be rehabilitated as alternative drinking water sources and to reduce the distributed quantity of still and sparkling mineral water. Also, the private wells were investigated and the population were given recommendations for disinfection. In Hunedoara County, the situation was very different and not so dramatic. Most of the flooded settlements had functional central drinking water system supply and private wells, so the drinking water was provided from tanks. Almost 1,700 wells were disinfected, the population being supplied with drinking water from cisterns and with bottled water.

In both counties, the quality of drinking water was carefully monitored. It must be mentioned that the monitoring of drinking water quality in some localities was possible only after the flood water decreased because of the lack of access routes. The bacteriological and chemical analyses performed on 133 water samples showed that 48.8% of them were not in accordance with the quality standards, and therefore unfit for use as drinking water (e.g. only 16 out of the 47 private wells investigated, generally about 8 m deep, were not chemically and/or bacteriologically contaminated). In Hunedoara County, the situation with the flooded drinking water sources monitoring quality was similar for the private wells, but the hyper-chlorination of water for a short period of time in central systems could assure a good quality of drinking water for the population.

3.1.3. Hygiene promotion and participation

The emergency water supply response should be carried out with, or as part of, a hygiene promotion programme that works with the affected population to respond to disasters in order to reduce risk, increase resilience and miti-gate the impact of disasters on health.

In Hunedoara County, the local Public Health Authority published a guide of preventive behaviour addressed to the population affected by floods. The main message was "Your community may be affected or even devastated by the floods, and because of this you should remember a few basic rules to protect your health". Thirty-eight companies concerned with hygiene promotion and health education made recommendations related to:

- The hygienic conditions of water sources, its disinfection and food safety
- The supplying with drinking water only from known and authorized safe sources, still and sparkling mineral water
- The use of water for household purposes
- The drilling of deeper wells, over 100 m, as alternative water sources
- The cleaning and disinfection of water tanks
- The disinfection of wells with chlorine

3.2. OUTBREAKS OF COMMUNICABLE DISEASES: WATERBORNE AND WATER-RELATED COMMUNICABLE DISEASES

The risk for communicable disease transmission after disasters is associated primarily with the size and characteristics of the population displaced, specifically the proximity of safe water and functioning latrines, the nutri-tional status of the displaced population, the level of immunity to vaccine-preventable diseases and the access to healthcare services. Although outbreaks after flooding have been better documented than those after earthquakes, volcanic eruptions, or tsunamis, natural disasters, regardless of type, that do not result in population displacement are rarely associated with outbreaks (Noji, 1991; Craun et al., 2006).

The greatest water-borne risk to health in most emergencies is the transmission of faecal pathogens, due to inadequate sanitation, hygiene and protection of water sources. Waterborne infectious diseases include diarrhoea, typhoid fever, cholera, dysentery and infectious hepatitis. Access to safe water can be jeopardized by a natural disaster. Diarrhoeal disease outbreaks can occur after drinking water has been contaminated and have been reported after flooding and related to displacement. The risk for diarrhoeal disease outbreaks following natural disasters is higher in developing countries than in industrialized countries (Noji, 1992). Hepatitis A and E are also transmitted by the faecal-oral route, in association with lack of access to safe water and sanitation. Whatever the source and type of contamination, decisions on acceptable water quality in emergencies involve balancing short-term and long-term risks and benefits to health. At the same time, ensuring access to sufficient quantities of water is vital for health protection.

Natural disasters, particularly meteorological events such as cyclones, hurricanes, and flooding, can affect vector-breeding sites and vector-borne disease transmission (water-related diseases). The crowding of infected and susceptible hosts, a weakened public health infrastructure, and interruptions of ongoing control programs are all risk factors for vector-borne disease transmission (Noji, 1991; Western, 1982). The transmission is not directly associated with flooding. Such events may coincide with periods of high risk for transmission and may be exacerbated by increased availability of the vector's breeding sites (mostly artificial containers) caused by disruption of basic water supply and solid waste disposal services. The risk for out-breaks can be influenced by other complicating factors, such as changes in human behavior or changes in the habitat that promote mosquito breeding.

Crowding is common in populations displaced by natural disasters and can facilitate the transmission of communicable diseases. For example, measles and the risk for transmission after a natural disaster is dependent on baseline immunization coverage among the affected population, and in particular among children under 15 years of age (Noji, 1991; Watson et al., 2007; Western, 1982).

At this point, we present some aspects of communicable diseases prevention actions in flooded areas from Hunedoara County. During the March–September 2005 floods in Hunedoara County 3,873 people from 37 localities or villages were affected. The persons at higher risk from the communicable diseases point of view are the children younger than 14 years old (1,077), and respectively the people older than 65 years old (767). The prevention of communicable diseases includes the following actions:

- Cholera: prophylactic medication to all people who drink water from local sources in the areas where the disease has been declared in the last 12 months. The anti-choleric immunization is not recommended anymore.
- Dysentery: no prophylactic medication and no immunization.
- Typhoid fever: no prophylactic medication; the immunization of 5–55 years old people at higher risk to infection.
- Hepatitis A: immunization of people at high risk-persons aged 6 months to 15 years old with no medical history of hepatitis A and potentially exposed workers in water treatment plants and residual water treatment stations. In the case of a hepatitis A outbreak, vaccination of contacts is recommended.

In 2005, the Hunedoara County floods affected areas where the population totalizing 2,321 persons was vaccinated against hepatitis A (1,182 people), typhoid fever (32 people) and tetanus (1,107 people). Tetanus is not transmitted from person to person, but it is caused by a toxin released by the anaerobic tetanus bacillus, *Clostridium tetani*. Contaminated wounds, particularly in populations where vaccination coverage levels are low, are associated with illness and death from tetanus.

As we have mentioned, the first and the most important step in preventing waterborne diseases after flooding is the consumption of secure water, chlorination being compulsory for all water sources. The safe water distributed to the population and the rapid medical actions in both Hunedoara and Timis counties resulted in no waterborne outbreaks, and more than that no higher incidence of these diseases in 2005 compared to the previous 5 years. At national level, the incidence of waterborne diseases (acute diarrhoea disease, dysentery and acute hepatitis) in the 2000–2005 period has actually shown a decreasing trend.

4. Conclusions

Drinking water is an essential element for ensuring health and the quality of life. From the medical perspective, the interventions in flood-affected regions showed once again that the main public health priority was to provide a basic secure water supply to the affected population. The most effective emergency water supply measures were ensured through a process of assessment, monitoring and review. Along with the water supply, prophylactic immunisation for waterborne diseases like hepatitis A, resulted in crisis management without affecting significantly the health status of the population. Risk assessment may be the most important step in the risk management process, and may also be the most difficult and prone to error. In the interest of public health, the risks versus benefits of the possible alternatives must be carefully considered.

The analysis of the events in the last few years has shown the persistence of the communities' vulnerability in emergency situations, even though the capacity of specialized institutions for action has increased.

References

AWWA, 1999: Water quality and treatment. A handbook of community water supplies, 5th ed. McGraw-Hill, New York.

Brower, J. and Chalk, P., 2003: The Global Threat of New and Reemerging Infectious Diseases: Reconciling U.S.National Security and Public Health Policy, RAND, Santa Monica, CA.

Craun, G.F., Calderon, R.L., and Wade, T.J., 2006: Assessing waterborne risks: An introduction. *Journal of Water and Health*, **4** (Suppl. 2), 3–18.

Logue, J.N., 1996: Disasters, the environment, and public health: improving our response. *American Journal of Public Health*, **86(9)**: 1207–1210.

Noji, E.K., 1991: Natural disasters. *Critical Care Clinics*, **7**: 271–292.

Noji, E.K., 1992: Disaster epidemiology: Challenges for public health action. *Journal of Public Health Policy*, **13**: 332–340.

Watson, J.T., Gayer, M., and Connolly, M.A., 2007: Epidemics after natural disasters. *Emerging Infectious Diseases*, **13(1)**: 1–5; Centers for Disease Control and Prevention (CDC).

Western, K.A., 1982: Epidemiologic surveillance after natural disaster. Washington, DC: Pan American Health Organization, pp. 1–94; Scientific Publication no. 420.

World Association for Disaster and Emergency Medicine (WADEM), 1990: Position Paper – Priorities in medical responses to disasters. *Prehospital and Disaster Medicine*, **5(1)**: 64–66.

World Health Organization, 1997: *Water Pollution Control. A guide to the use of water quality management principles*. Published on behalf of WHO by F. & F.N. Spon, London.

World Health Organization, 2004. Water, Sanitation and Health. Technical notes for emergecies. Available at: http://www.who.int/water_sanitation_health

BIOSENSORS IN A SYSTEM OF INSTRUMENTAL TOOLS TO PREVENT EFFECTS OF BIOTERRORISM AND AUTOMOTIVE CONTROL OF WATER PROCESS PURIFICATION

N.F. Starodub[*]

Palladin Institute of Biochemistry of National Academy of Sciences of Ukraine, 9 Leontovicha St., 01601 Kiev, Ukraine

Abstract. Biosensors for the express control of drinking water, water resources, food and feeds are presented. Instrumental feedback control of water purification processes and ways for the decontamination of polluted environmental materials from different types of low molecular weight toxins are given.

Keywords: Water resources, water purification, control, biosensors

1. Introduction

Despite the 1972 Biological Weapons Convention, it appears worldwide that some nations are continuing to develop and produce biological warfare agents (BWA). Today the problem of environment protection from BWA which are applied not only for solving local conflicts but also for psychological terror aimed at arousing panic among people of stable countries and to form permanent terror in the word is very current. The some information about countries which worked on the program of biological weapons was given earlier by Roberts and Moodie (2002). The investigations within framework of this program were directed at the induction of acute diseases among people and animals, destroying sexual status, prevention of antibiotic effects and others. BWAs are divided into three main groups: toxins (the so called mid-spectrum agents); viruses; and bacteria. Toxins are biochemicals of various origin and chemical structure. Mycotoxins may serve as a characteristic example of BWAs. These substances form a large group which includes more than 300 individual toxins produced by different fungal strains (Mirocha et al., 1983). T2, aflatoxins, searelenone, patulin and others attract great interest, since they are widespread and characterized by high level toxicity. T2 mycotoxin has more toxic effect (400 times more) than mustard gas and lewisite. It is well-known that mycotoxin T2 was packed into rockets, bombs, cisterns, some explosive cylinders, and hand-grenade and they were applied in Laos and Afghanistan as yellow rain (Rosen and Rosen, 1982; Morris and Clifford, 1985).

---

[*] To whom correspondence should be addressed. e-mail: nikstarodub@yahoo.com

This mycotoxin may be prepared in a very simple way. Both the simplicity of obtaining it and its high level of toxicity create a very serious problem, since these toxic elements may be used by bioterrorists. It is necessary to mention that the use of toxins in general among others groups of BWA by terrorists is the most probable, since viruses and bacteria present a significant danger not only for the intended victims, but also for the executors of the terrorist acts. Providing appropriate services based on simple, very selective and sensitive methods for rapid revealing of BWA components in the environment is a pre-eminent approach among all others directed towards preventing serious consequences. Unfortunately, the analytical methodologies for analysis of mycotoxins as well as other low molecular weight toxins, include such instrumental analyses as high-performance liquid or gas chromatography with mass spectroscopy or liquid chromatography with mass spectroscopy. Due to the extremely high complication and cost of analysis using these methods, the development of innovative approaches, such as immune analysis and particular chemo- and biosensors, is very urgent (Rittenburg, 1990; Iqbal et al., 2000; Nabok et al., 2005; Starodub et al., 2006a–c).

In this review, the main attention will be paid to some mycotoxins and other chemical substances related to a group named as endocrine disrupting agents, for example, heavy metal ions (HMI), nonylphenol (NPh), surface active substances (SAS) and some pesticides. To start with, the proposed instrumental analytical methods based on the biosensorics principles and intended for the rapid revealing of the above numerated toxins in environmental materials will be discussed. This aspect will be analyzed from two points of view: estimation of total toxicity level at the screening analysis of environmental materials and then determination of the concentration of individual toxins in these materials.

In the analysis of the instrumental analytical device intended for the control of total toxicity of environmental substances and for the determination of the concentration of individual substances in these, the main attention will be paid modern approaches based on the principles of biosensorics. First, the efficiency of the developed biosensor will be analysed in the case of work with the standard solution. Then the obtained results will be compared with the results from application of the biosensors in real conditions. The type of biosensors, their design peculiarities and the details of the analytical methods will be presented in the appropriate section of this article.

Our approach focuses on three aspects: fundamental research, creation of working prototypes and development of some elements of technology. A fundamental aspect includes the selection of types of transducer (physical surface) and physical–chemical signals for monitoring the interaction of biological molecules, as well as choosing sensitive biological material and development of effective methods for the oriented immobilization of this material. Working prototypes of biosensors are created for application in the area of human and veterinary medicine, ecology and biotechnology. After examination of the biosensor work in real conditions, some elements of technology of biosensor production are worked out.

Our investigations in the field of development of biosensors and their application include the following steps: (1) control of the total toxicity of environmental materials by the use of bioluminescent bacteria with optical signal registration and determination of the intensity of *Daphnia* exometabolites by the chemiluminescent method; (2) revealing groups of toxic elements by the electrochemical and thermo-metrical biosensors with the sensitive elements presented by special micro-organisms and enzymes; (3) determination of the specific toxic elements by the electrochemical and optical biosensors, based on the enzyme and immune components as the selective structures on the one hand and the different types of transducers including surface plasmon resonance, porous silicon, electrolyte-insulator structures on the other.

2. Results and discussion

2.1. INSTRUMENTAL ANALYTICAL APPROACHES FOR DETERMINATION OF TOTAL TOXICITY

Usually different living organisms (*Crustacea*, fish, algae, fungi, some vegetables and others) are used to control total toxicity of environmental materials. An International Standard exists as the basis of the determination of some indexes of *Daphnia* immobilization (ISO 6341:1996(E)). Unfortunately, it is a very routine procedure. Other approaches which are used in practical applications are based on control of oxygen consumption by micro-organisms or determination of their luminescence.

2.1.1. *Method with the use of Daphnia as the sensitive object*

We propose a principal new approach based on the determination of the chemiluminescence (ChL) level of a live *Daphnia* medium. The differences of measuring cell signals before and after introducing *Daphnia* into the solution to be analyzed were recorded. In the experiments, *Daphnia magna* St. (*Cladocera*) was used, which was kept in the medium according to the International Standard rules (ISO 6341:1996(E)). In preliminary experiments, it was shown that only 1–5 *Daphnia* are sufficient for these experiments (Ivashkevich et al., 2002). The excited ChL of the medium was recorded in the presence of luminol and H_2O_2. The optimal concentrations of the above numerated chemicals were preliminary established in the special experiments (Ivashkevich et al., 2002). Stationary, semi-portable and portable devices supplied by optrods, high sensitive photomultiplier, or photo resistors were created for the determination of the intensity of ChL.

Depending on source of toxic substances we have obtained deviations of ChL values from the initial levels, which were commensurate with the intensity of the toxic effect. Potassium biochromate was used as the standard chemical solution and its toxicity was checked by the generally accepted method (according to the index of *Daphnia* immobilization) and using the biosensor based on the determination of ChL level of the live *Daphnia* medium. It was found that the generally accepted method allows use of just 0.1 mg/l of potassium biochromate as the minimum level. At the same time, the sensitivity of the proposed biosensor approach was almost two orders higher (Levkovetz et al., 2002). It is necessary to mention that the overall time of analysis made a big difference in both cases (about 24 h and 30 min for the generally accepted and biosensor methods, respectively).

The sensitivity of *Daphnia* to mycotoxins T2 and patulin was demonstrated by Gojster et al. (2003) and Pilipenko et al. (2007). Diapason of the measurements of T2 mycotoxin by the generally accepted method was in the range of concentration of 0.01–0.1 mg/l. At the same time, the biosensor method had a range from 0.001 to 1 mg/l. As for patulin, it was possible to make quantitative determinations by the biosensor method in the range of 0.001–1 mg/l.

2.1.2. Method with the use of bioluminescent bacteria as the sensitive entity

In the investigations, *Photobacterium phosphoreum* K3 (IMB B-7071), *Vibrio fischeri* F1 (IMB B-7070) and *Vibrio fischeri* Sh1 purified from the Black Sea and the Sea of Azov were used. The level of bioluminescence (BL) was measured by the developed devices. The level of toxicity was presented as the concentration which caused a 50% decrease of the intensity of BL (EC_{50}). For all cases of signal measurements, the value of EC_{50} oscillated in the range of 7–19 mg/l depending on the time of incubation of the bacteria in the T2 mycotoxin solution. It is necessary to emphasise that the sensitivity of *V. fischeri* F1 to micotoxin T1 is much higher in comparison with the sensitivity of *Ph.phosphoreum* Sq3 (Katzev et al., 2003).

Increasing patulin concentration from 0.63 to 40 mg/l caused sufficient decrease in the BL intensity at the influence on *Ph.phosphoreum Sq3* during 12–60 min. The value of EC_{50} for patulin was in the range 0.63–1.25 mg/l (Katzev et al., 2003). The dose-effect of patulin at the low concentration (as low as 1 mg/l) may be confidently registered in the case of three repeated measurements for each point. If this methodological approach is followed in the analysis, the toxic effect of patulin to bioluminescent bacteria may be revealed at concentrations of less than 0.15 mg/l. Moreover, extending time of influence up to 90 min, the toxic effect of patulin increased and value of EC_{50} was in the range 0.15–0.63 mg/l. Decreasing the pH of the medium to the lower physiological limit (5–5.5), the sensitivity increased up to one order. The value of EC_{50} has analogy with the semi-lethal dose established for animals and it correlates with other indexes of toxicity (cytotoxicity, irritation of mucouses, etc.) (Elnabarawy et al., 1988). It is necessary to mention that the intestinal barrier in animals is destroyed at patulin concentrations of about 1 mg/l (Manfoud et al., 2002). Taking this fact into consideration, the above indicated data testify that the proposed biosensor analysis with the use of bioluminescent bacteria may be effective for screening samples of water, juice, foods and other environmental substances.

In the study of the influence of different types of SAS on the intensity of bioluminescence of bacteria (*Ph. phosphoreum* K3 (IMB B-7071), *V. fischeri* F1 (IMB B-7070) and *V. fischeri* Sh1), it was revealed that the most of the investigated substances are inhibitors of this process. At first, the cationic and anionic SAS had similar kinetics of inhibition. Second, nonionic SAS, have an additional stage in which the inhibition is absent or some activation of bioluminescence is observed. Therefore, for revealing the toxicity of this group of SAS, it is necessary to incubate these substances with bacteria for a long time.

2.2. BIOSENSOR DETERMINATION OF GROUPS OF TOXIC ELEMENTS

To determine group specific toxic substances, for example, phosphororganics, chlororganics, cyanides and others, we have developed a multi-biosensor based on electrolyte-insulator-semiconductor (EIS) structures (Starodub and Starodub, 2000).

In these experiments, the simazine conjugates and antiserum to simazine were obtained according to Yazynina et al. (1999) and presented by Professor B. Dzantiev of the A.N. Bach Institute of Biochemistry, Moscow, Russia. Conjugates of 2,4-D with proteins and enzymes were obtained with the help of Fenton's reagent. The antiserum to simazine cross-reacted with atrazine (89%), terbutylazine (80%), and propazine (10%). Other analytes demonstrated cross-reaction in the range of 0.7–6.2%. The antiserum to 2,4-D did not have a cross-reaction with simazine.

The principles of the design and operation of biosensors were presented by Starodub et al. (1999). Specific antibodies to herbicides were immobilised through the staphylococcal protein A. The analysis was performed using the sequential saturation method where antibodies are left unbound after their exposure to the native herbicide in the investigated sample, then have interacted with the labelled herbicide. The sensitivity of the EIS structures based sensor to simazine, when the HRP-conjugates were used, was approximately 5 µg/l. The linear plot of the sensor response lay in the range of the concentrations from 5 to 150 µg/l. This sensitivity of the EIS structures based sensor towards both herbicides was lower than is needed in practice. We tried to elucidate the main reasons for such a situation. One of them may be

connected with difficulties to record sensor output due to the formation of
air bulbs, which appear as a result of high activity of the HRP. Use of high
concentrations of ascorbic acid may be another reason for the lower sensitivity of this sensor. We changed the HRP label to the GOD one and obtained a
sensitivity of the analysis approximately five times higher. The linear plots
for simazine and 2,4-D were in the range of 1.0–150 and 0.25–150 µg/l,
respectively (Starodub and Starodub, 1999b; Starodub et al., 2000).

An immune biosensor based on the EIS structures attracts attention
because of the simple analytical procedure and possibility of undertaking
multi-parametrical control of the environment. For repeated analyses, replace-
able membranes are very suitable. The overall time of the analysis is about
40 min. Therefore, the EIS structures based immune sensor may be used for
wide screening of the environment for the presence of herbicides. It offers
the possibility of carrying out analysis of eight to ten samples simultaneously.
It is suitable for a wide screening of not only herbicides but also other types
of toxicants. Other types of biosensors may be used for the verification of
the analytical results, for example, based on the ISFETs, the sensitivity of
which in the determination of the above mentioned herbicides is at the level
of 0.1 µg/l or less (up to 0.05 µg/l) which corresponds to practice demands
(Starodub and Starodub, 1999a, 2001; Starodub et al., 2000). We believe
that the sensitivity of the EIS structures based immune sensor can be increased
still further. One of the possible ways of doing this could be the development of special suitable membranes. It is necessary to provide a very high
density of the immobilised specific antibodies on the membrane surface.
Moreover, it would be very effective if these antibodies were immobilized
not only on the membrane surface but also in its large-scale pores, which
would be accessible for large molecules of conjugates of herbicides with
enzymes. In our opinion, synthetic biologically compatible polymers, which
can be prepared in a simple way with different levels of density and porosity, can serve as a prospective material for such membranes (Shirshov
et al., 1997; Rebrijev et al., 2002). Of course, to increase the sensitivity of the
analysis, it would also be very efficient to use monoclonal antibodies with a
high level of affinity to analytes, to choose enzyme labels with a high turn-
over of activity and to provide preservation of the enzyme activity during
preparation of the conjugate. If the membranes were prepared in advance,
the duration of the analysis may be shortened up to 10 min. Membranes are
simple to prepare, they are very cheap and they can be stored for a long
time in a refrigerator.

Since a number of enzymes which have serine residuum in the active
centre (first of all butyrylcholine esterase – BChE, acetylcholine esterase –
AChE and total choline esterase – ChE) are very sensitive to phosphororganic
pesticides (PhOrPe) and other ones (urease) with the thiol groups react with
HMI, there is the possibility of simultaneous determination of these classes
of toxic elements (Starodub et al., 1998; Rebrijev and Starodub, 2001).

The sensitivity of HMI and PhOrPe determination depends essentially on the incubation time of the enzyme membranes in the environment of these analytes. Two different approaches were tested: (1) registration of the sensor output signal in the mixture of a substrate and analytes; (2) separation of the inhibition reaction from the subsequent measurement of the residual enzyme activity. In the latter case, the threshold sensitivity of toxin analysis was about ten times higher. The time of incubation was chosen experimentally and it was 15 min. The concentration of HMI that could be determined by the urease channel of the sensor array lay within the range from 10^{-4} to 10^{-7} M, depending on the type of the metal used. The range of linear detection covered two to three orders of the concentration change. The effects of both pesticides are very similar. The limit of detection of pesticides indicated above was 10^{-7} M. The range of the linear response was from 10^{-5} to 10^{-7} M. At the same time, the sensitivity of BChE to HMI was substantially lower than that of urease. The maximum sensitivity of BChE to HMI was for concentrations of more than 10^{-4} M. The activity of GOD depends on the presence of HMI for concentrations above 10^{-4} M. GOD was used as the reference enzyme which has a minimal reaction in respect of both types of groups of toxins.

2.3. BIOSENSOR DETERMINATION OF INDIVIDUAL TOXINS

For this purpose, we apply SPR, TRIE and calorimetric based biosensors.

2.3.1. Analysis by SPR and TIRE based optical immune biosensors

The principles of construction of SPR biosensor and the main algorithm of analysis with its help were described by Starodub et al. (1997, 1999).

As a rule, we have analysed in detail three main variants of approaches: (a) specific antibodies from an antiserum were immobilised on the gold surface of an SPR transducer through the intermediate layer from *Staphylococcal* protein A or some lectin, and free analyte was in the solution to be analysed (direct method of analysis); (b) conjugate (NP, simazine, 2,4-D or T2 mycotoxine with some protein – BSA, or STI, or Ova) was directly immobilised on the gold surface of the SPR transducer, and free analyte with an appropriate antiserum was in solution (competitive approach with the immobilized conjugate); (c) the specific antibodies from the antiserum was immobilised as in "a", and free analyte and its conjugate with some protein were in the solution to be analysed (competitive method with the immobilised antibodies); and (d) immobilised and oriented antibodies as in "c" react with free analyte and then with the appropriate conjugate (approach to saturation of active binding sites on the surface). It was observed that orientation of the antibodies on the surface is more effective with the help of protein A than with the use of lectins. Maybe, this is connected with the possibility of presence of some carbohydrates not only in the F_c-fragment of the anti-bodies but also in the F_{ab}-fragments.

It was found that the sensitivity of 2,4-D analysis by the direct method is about 5–10 µg/l, which is not high. Such a low level of sensitivity with the direct method of analysis is observed in case of the determination of other low weight substances, for example, T2 mycotoxin and NPh. Methods "b" to "d" above were much more sensitive. A biosensor based on the TIRE allows us to identify mycotoxin T2 up to 0.15 ng/ml (Starodub and Starodub, 1999; Starodub et al., 2006). Both optical immune biosensors can provide the sensitivity of analysis which is needed for practice. The overall time of analysis is about 5–10 min, if the transducer surface is prepared before-hand. It is necessary to mention that the immune biosensor based on the SPR is simpler then the TIRE biosensor. In addition, to former may be developed as a portable device.

2.3.2. Analysis by calorimetric immune biosensor

We will demonstrate the efficiency of an immune biosensor for the deter-mination of low molecular substances using the results obtained in the experi-ments with NPh. For successful development of the calorimetrical biosensor, it was necessary to first set the optimal concentration of antiserum (for example, antiserum to NPh). For this purpose, 150 µl of antiserum in different con-centrations was brought in a measuring cell and incubated for 15 min to establish a baseline (for this time the temperature in a barn was set at an optimum level). Then, 50 µl solution of NPh in concentrations of 1, 5 and 10 µg/ml were brought into the cell. Thus, it was set that the optimum concentration of antiserum was about 5 mg of protein in 1 ml.

For the determination of NPh in solutions with the help of a thermal biosensor, it was necessary to construct a corresponding calibration curve. For this purpose, 150 µl of antiserum (in a concentration of 5 µg of protein in 1 ml) was brought into a measuring cell and then 50 µl of NPh in a range of concentration from 0.5 up to 10 µg/ml was pumped into the measuring cell. This demonstrated the opportunity of "direct" detection of NPh by a calorimetrical biosensor with the sensitivity of about 1 µg/ml. The overall time of analysis is about 20–30 min.

Certainly, the sensitivity of the determination of NPh by thermal immune biosensor is much less than in the case of application of the SPR or TIRE biosensor, but it is necessary to mention the simplicity of carrying out the measurement. Maybe, a thermal biosensor could be used for the screening of toxic elements in environmental materials with subsequent verification of the results of analysis by optical immune biosensors.

2.4. OPTICAL AND OTHER TYPES OF BIOSENSORS IN THE SYSTEM OF FEEDBACK CONTROL OF THE WATER PURIFICATION PROCESS

The creation of cheap and effective technology for the removal of toxic pollutants from water is an urgent need in modern environmental protection. The first and most important problem is modifying and changing technology, aiming towards energy savings and reaching the minimal emission levels in the hydrosphere. This problem can be resolved by an optimal combination of chemical–technological methods with biological ones (Klimenko et al., 2002). Treatment expenses depend on the degree of purification needed. There are certain purification limits determined by economy, under which the enterprise becomes non-profitable. The role of a combination of natural biodegradation processes with chemical–technical methods in this context is most important. The toxicity of pollutants entering the environment and their transformation as a result of wastewater treatment must to be taken into account during technology creation. Recent experiences indicate that the efficiency for the purification of wastewater from toxic xenobiotics can be improved by combining adsorption and oxidation methods with biological methods. However, effective automatic control of water treatment is needed for the optimization of the clean-up process. Biosensors offer unique possibilities of obtaining cheap, fast and sensitive control units that can be incorporated as sensors for water toxicity control and regulation at different stages of the treatment process (Goncharuk et al., 2007; Klimenko et al., 2007). In this report, the results about the application of optical and other types of biosensors for feedback control of water purification process will be presented.

The technology was accomplished by successive pre-photooxidation steps (O_3 and UV-irradiation), followed by simultaneous biological degradation in a biosorber with activated carbon (AC) with immobilized degradable bacteria. From the start, we proposed using a *Daphnia* chemiluminescent test for the express determination of total toxicity during water purification. It was established that this biotest is very sensitive to the presence of toxic pollutants, such as nonylphenolethoxylates, for example. Quantitatively, their control is possible at concentrations of about 20 mg/l and less. It was demonstrated that: (a) the optimal regime of photoozonation may be achieved if the concentration of the initial substances decreased by 50% from the initial; (b) the use of activated carbon after photo-ozonation caused a sharp decrease in general toxicity of the treated water due to adsorption of some organic radicals and semi-decay of pollutants. The stability of the biosorber work with AC is provided by simultaneous biodegradation of the AC during solution filtration through the AC layer. Special methods were developed to increase the efficiency of AC biodegradation to 95–98% from its

initial capacity. To provide stability of biosorber operation, special requirements were developed on the sorbents, first as regards their operating conditions, and second as regards their porosity. The pilot scale system aimed
at the complete treatment process based on a joint action of physical–
chemical and biochemical processes with biosensor control has been
developed and assembled. The general scheme of this system is presented
below (Fig. 1).

For the control of total toxicity of the initial water and for the determination of individual toxic substances (for example of NphEO), it is necessary to use the chemilminescent and bioluminescent tests and one of the
immune biosensors, based on electrolyte semiconductor structures or surface
plasmon resonance (SPR). The thermal cell biosensors with *Sacchromices
cerevisia* and bioluminescent test with *Vibrio fischeri* as the sensitive biological elements were non-effective at the low concentrations of such toxic
subsrances as detergents. It is the same situation in the intermediate stage of
water purification, after its treatment by ozone and UV-radiation. At the
final stage of water purification, it is more suitable to use a *Daphnia* chemiluminescent biosensor and immune biosensor based on ion-sensitive field
effect transistors (ISFETs), as this is more sensitive in comparison with the
other immune biosensors.

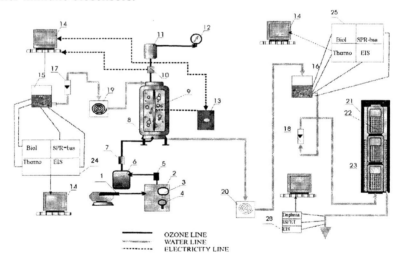

OZONE LINE
WATER LINE
ELECTRICITY LINE

Figure 1. Overall scheme of pilot scale system with biosensor feedback control for water
purification from surfactants. Where: 1-compressor; 2-ozone generator; 3-electronic mano-
meter; 4-manometer; 5-gas rotameter; 6-water trap; 7-back valve; 8-reactor for catalytic
photoozonation; 9-UV-lamp of low pressure; 10-photometric gas cuvette; 11-ozone trap
(KI solution); 12-gas clock; 13-energy source for UV-lamp of low pressure; 14-computer;
15, 16-tanks, containing surfactant solution; 17, 18-water rotameters; 19, 20-peristaltic
pumps; 21-bioreactor; 22-AC-active carbon; 23-AC with immobilized microorganisms;
24-biosensors for initial solution; 25-biosensors for solution after photoozonation;
26-biosensors for purified water

During the process of treatment, the following different integral parameters of water quality were analyzed: chemical oxygen demand, per-
manganate oxidizability, biological oxygen demand, total organic carbon,
residual concentration of ozone, dissolved peroxides, total toxicity and
concentration of individual toxic elements. The technology is suitable for
the textile industry (band textiles, silken, knitted, artificial technical fabric
and cotton factories as well as wool-spinning mills), and for water quality
control in galvanizing and oil extraction processes. The using of this
technology leads to recycling of the processing water within an industry.
The results might be used for wastewater purification from toxic pollutants
(surfactants, phenols, pesticides, oil products) and production of drinking
water from polluted sources of water supply.

References

Elnabarawy, M.T., Robideau, R.R. and Beach, S.A. 1988: Comparison of three rapid
toxicity test procedures: Microtox, Polytox and activated sludge respiration
inhibition. *Toxicity Assess*, **3**: 361–370.

Gojster, O.S., Starodub, N.F. and Chmel'nitskij, G.A., 2003: Determination of T2
mycotoxin by chemiluminescent method with the use of *Daphnia. Hydrobiol J.*,
5: 85–91.

Goncharuk, V., Klimenko, N., Starodub, N., et al., 2007: Control of technical parameters of waste water purification from surfactants by biosensors. In: *Proceedings
of Investigations on Sensor Systems and Technologies*, National Academy of
Science of Ukraine, Kiev, 340–350.

ISO, 1996: Water quality – Determination of the mobility of *Daphnia magna Straus*
(Cladocera, Crustacea) – Acute toxicity test, ISO 6341:1996(E).

Ivashkevich, S.P., Levkovetz, S.P., Nazarenko, V.I. and Starodub, N.F., 2002:
Chemiluminescence of medium *Daphnia* cultivation and optimization of conditions of it determination. *Ukr Biochem J*, 74: 93–97.

Iqbal, S.S., Mayo, M.W., Bruno, J.G., et al., 2000: A review of molecular recognition
technologies for detection of biological threat agents. *Biosensors Bioelectronics*,
15: 549–578.

Katzev, A.M., Gojster, O.S. and Starodub, N.F., 2003: Influence of T2 mycotoxin
on the intensity of bacterial bioluminescence. *Ukr Bioch J.*, **75**: 99–103.

Klimenko, N., Winter-Nielsen, M., Smolin, S., et al., 2002: Role of physico-
chemical factors in the purification process of water from surface-active matter
by biosorption. *Water Res.*, **36**: 5132–5140.

Klimenko, N., Starodub, N. and Nevynna, L., 2007: Influence of chemical and
biochemical products of destruction of non-ionic surfactants on water solution
toxicity and efficiency of biofiltration through activated carbon. *Water Chem.
Technol.*, **29**: 207–225.

Levkovetz, I.A., Ivashkevich, S.P., Nazarenko, V.I. and Starodub, N.F., 2002:
Application of chemiluminescent method for the determination of sensitivity
Daphnia magna to different toxic substances. *Ukr. Biochem. J.*, **74**: 120–124.

Manfoud, R., Maresca, M., Garmy, N. and Fantini, J., 2002: The mycotoxin patulin alters the barrier function of the intestinal epithelium: mechanism of action of the toxin and protective effects of glutathione. *Toxicol. Appl. Pharmacol.*, **181**: 209–218.

Mirocha, C.J., Pawlosky, R.A., Chatterjee, K., et al., 1983: Analysis for Fusarium toxins in various samples implicated in biological warfare in Southeast Asia. *J. Assoc. Anal. Chem.*, **66**: 1485–1499.

Morris, B.A. and Clifford, M.N. (eds.), 1985: *Immunoassays for Food Analysis.* London/New York, Elsevier.

Nabok, A.V., Tsargorodskaya, A., Hassan, A.K. and Starodub, N.F., 2005: Total internal reflection ellipsometry and SPR detection of low molecular weight environmental toxins. *Appl. Surface Sci.*, **246**: 381–386.

Pilipenko, L.N., Egorova, A.V., Pilipenko, I.V., et al., 2007: Investigation of toxic effect of patulin with the help of biosensorics systems. *Food Sci. Technol.*, **1**: 35–38.

Rebrijev, A.V. and Starodub, N.F., 2001: Photopolymers as immobilization matrix in biosensors. *Ukr Biochem J.*, **73**: 5–16.

Rebrijev, A.V., Starodub, N.F. and Masljuk, A.F., 2002: Optimization of conditions of immobilization of enzymes in a photopolymeric membrane. *Ukr. Biochem. J.*, **74**: 82–87.

Rittenburg, J.H. (Ed.), 1990: *Development and Application of Immunoassay for Food Analysis.* London/New York, Elsevier.

Roberts, B. and Moodie, M., 2002: Toward a threat reduction strategy. National Defense University. *http://www.ndu.edti/iass/pms/ndttithfi.html.*

Rosen, R.T. and Rosen, J.D., 1982: Presence of four Fusarium mycotoxins and synthetic material in yellow rain. Evidence for the use of chemical weapons in Laos. *Biomed Mass Spectrom.*, **9**: 443–450.

Shirshov, Y.M., Starodub, N.F., Kukla, A.L., et al., 1997: Creation of the multi-enzymatic sensor for simultaneous determination of phosphororganic pesticides and heavy metal ions in solutions. In: *Proceedings of the 11th European Conference on Solid-State Transducers. Eurosensors XI*, September 21–24, 1997, Warsaw, Poland, 727–730.

Starodub, N.F. and Starodub V.M., 2000: Immune sensors: origins, achievements and perspectives. *Ukr Biochem. J.* **72**, 147–163.

Starodub, N.F. and Starodub V.M. 2001: Biosensors and control of pesticides in water and foods. *Chem. Technol. Water* **6**, 612–638.

Starodub, N.F., Dibrova, T.L., Shirshov, Y.M., Kostioukevich, K.V., 1997: Development of sensor based on the surface plasmon resonance for control of biospecific interaction. In: *Proceedings of the 11th European Conference on Solid-State Transducers. Eurosensors XI*, September 21–24, 1997, Warsaw, Poland, 1429–1432.

Starodub, N.F., Torbicz, W., Pijanowska, D., et al., 1998a: Optimization methods of enzyme integration with transducers for analysis of irreversible inhibitors. In: *Proceedings of the XII European Conference on Solid-State Transducers and the IX UK Conference on Sensors and their Applications*, September 13–16, 1998, Southampton, UK. Edited by White, N.M., Bristol Institute of Physics 1: 837–840.

Starodub, N.F., Kanjuk, M.I., Kukla, A.L., Shirshov, Y.M., 1998b: Multi-enzymatic electrochemical sensor: field measurements and their optimization. *Anal. Chim. Acta* **385**: 461–466.

Starodub, N.F., Starodub, V.M., Kanjuk, N.I., et al., 1999: Biological sensor array for express determination of a number of biochemical quantities. In: *Proceedings of 2nd EUREL Workshop "European Advanced Robotic System Development. Medical Robotics"*, 23–24 September 1999, Piza, Italy, 57–64.

Starodub, N.F., Dzantiev B.B., Starodub V.M., Zherdev A.V., 2000: Immunosensor for the determination of the herbicide simazine based on an ion-selective field effect transistor. *Anal. Chim. Acta* **424**, 37–43.

Starodub, N.F., Mel'nik V.G., Shmireva O.M., 2006a: Instrumental approaches and peculiarities of design of stationary and portable analytical devices for determination of bio- and chemi-luminescence. In: *Proceedings of the Society of Automotive Engineering (SAE)*, Norfolk, VA, USA.

Starodub, N.F., Nabok A.V., Tsargorodskaya A., et al., 2006b: Control of T2 mycotoxin in solutions and foods by biosensors based on SPR and TIRE. In: *Proceedings of the International Conference on "Sensor+Test 2006"*, Nuremberg, Germany, 87–92.

Starodub, N.F., Nazarenko V.I., Ivashkevich S.P., et al., 2006c: Principles of express instrumental control of total toxicity of environmental objects and their realization in space conditions. In: *Proceedings of the Society of Automotive Engineering (SAE)*, Norfolk VA, USA.

Starodub, V.M. and Starodub, N.F., 1999a: Optical immune sensors for the monitoring protein substances in the air. In: *Eurosensor XII. The 13th European conference on solid-state transducers*, September 12–15 1999. The Hague, The Netherlands, 181–184.

Starodub, V.M. and Starodub, N.F., 1999b: Electrochemical immune sensor based on the ion-selective field effect transistor for the determination of the level of myoglobin. In: *Proceedings of the 13th European Conference on Solid-State Transducers*, September 12–15, 1999, The Hague, The Netherlands, 185–188.

Yazynina, E.V., Zherdev A.V., Dzantiev B.B., et al., 1999: Immunoassay techniques for determination of the herbicide simazine based on use of oppositely charged water-soluble polyelectrolytes. *Anal. Chem.* **71**, 3538–4353.

A HIGH SENSITIVITY NUCLEAR METHOD FOR REAL-TIME DETECTION OF ELEMENTS AND COMPOUNDS IN DRINKING WATER AND SOIL

P. Buckup[*]
Bohrlochmessungen-Dr. Buckup, Magdeburg, Germany

Abstract. Experiments using a neutron probe in lakes, wells, rivers and water collectors show that the instrument is capable of analyzing soil and water for contaminants in the ppm-range. A two-channel-neutron-neutron-probe is used; the neutrons are produced by a deuterium/tritium-source. The distribution of thermal neutrons is mainly influenced by the composition of the medium and therefore a direct recognition of the components disturbing the neutrons is possible. Measurements may be performed in time-dependent (monitoring) or depth/length-dependent (logs or profiles) mode. The monitoring may be used for permanent control of critical spots in water reservoirs and on soil (bomb detection, explosive indication). The logs and profiles may be run for depth-related information or area studies. The field studies have shown a high sensitivity. Concentration values may be obtained from the shape of the decay curve or by pulse distribution in the measuring windows. Comparisons with lab analysis have shown regression coefficients over 0.9 for different elements like Fe, As, Cl, Cd, P etc., but also for chemical compounds like BTEX or trichlorbezen. It was established experimentally, that a 25 kg bomb may be detected on a distance of 4.5 m. Thus, the pulse neutron technique may be applied for risk studies and planning measures for risk assessment in real time under in-situ-conditions.

Keywords: Water quality detection, nuclear methods, neutron pulse

1. Introduction

Nuclear measurements are widely used for elemental detection (Buckup, P. and Buckup, K., 2002). One of the methods is the neutron–neutron modification. The response of the neutron–neutron method to elements and compounds results from the intensity of neutron reactions with the environment. Emitting neutrons by a deuterium/tritium-reaction (D/T) with a neutron-flux of over 10^8 neutrons/s at a fixed energy of 14 MeV, offers opportunities for a continuous pulse neutron monitoring (PNM) under various conditions and especially for water reservoirs. Sufficient hardware has been developed

---

[*] To whom correspondence should be addressed. e-mail: philipp.buckup@ufz.de

(Fig. 1) to record thermal neutrons in their time distribution, which in physical terms is a neutron lifetime measurement.

Figure 1. Pulse-neutron-neutron-monitoring in operation

The neutron probe is built around an accelerator tube (Fig. 2), which operates with an optimal bursting frequency of 20 Hz.

Figure 2. Low frequency minitron

By the fusion reaction

$$D + T \longrightarrow 4\,He + n$$

fast neutrons with an energy $En = 14$ MeV will be produced inside the tube. These neutrons are thermalized and recorded by He3-counters in two channels at an optimized spacing, when they reach the thermal level 0.0248 eV. This process is described by an exponential decay curve:

$$N = N_o\,e^{-\tau}$$

where N_o – initial neutron output, and τ – neutron lifetime.

The neutron thermalisation process dictates the neutron lifetime and consequently the population of thermal neutrons (Hansen, 1997). This process is mainly influenced by the capture probability for thermal neutrons of the medium (Table 1). The neutron lifetime is measured in a time-range over 2 ms, usually sufficient in fluids to catch the majority of present neutrons.

The recording time of 2 ms is divided into windows of optimized length, depending on technical factors and each of these windows may be approached individually. This allows qualitative study of the presence of different materials, analysing the whole decay curve. There are two ways to obtain the information about the water body under investigation: by the shape of the decay curve (Fig. 3) or the continuity of the curve (Fig. 4). In view of the large diversity of practical cases with different contents of elements, it is practically impossible to model the response of the method on different elements and compounds. Under different practical conditions, it was possible to establish the sensitivity of the NMP response limit as shown in Table 2.

Recording decay curves offers the possibility of distinguishing between elements (Fig. 3).

TABLE 1. Capture probability Σ for selected elements and compounds (Gerhart Industries, 1980)

Quartz SiO_2	Calcite $CaCO_3$	Anhydrite $CaCO_4$	Fe	NaCl	B
4.26	7.07	12.5	193	753	99
Cd	H_2O	Oil	Gas/air	Barites	Glauconite
113	22	17–22	5	6.8	21

TABLE 2. Empirically established response levels of NMP for selected materials (in ppm)

Au	Cd	Cl	Hg	Pb	Fe	B	Organics
0.5	0.1	0.1	1.0	1.0	0.5	0.5	0.1

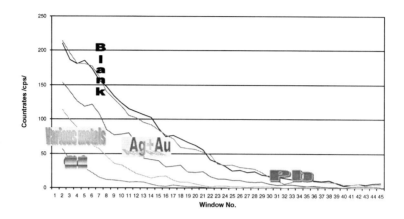

Figure 3. Decay curves for a polymetallic sand

The decay curves on Fig. 3, recorded for the same conditions, vary significantly; the highest amplitudes may be recorded for the clean (blank) sand sample and by contents of different metals at ppm-level the amplitude decreases. This effect is used for the Pulse-Neutron-Neutron continuous monitoring; the emitted neutrons were recorded by He3-detectors as thermal neutrons on a certain distance from the controlled neutron source. After theoretical research, hardware and software were developed to allow the application of the neutron technique under real conditions. The equipment is powered by a car battery, connected to a control panel. The software delivers an output, which may be converted to an EXCEL-file, processed consequently by standard mathematics. The software allows analysis of the decay curves in detail and tests were performed under real life conditions. The standard processing of the Excel file allows the quality of the recorded data to be assessed. Using standard mathematics, the recorded values are filtered and statistical deviations deleted.

The cleaned data are utilized to build the decay curves as per Fig. 4. Using hardware and the adequate software, different cases were studied. Historically, the first continuous neutron monitoring was performed as a pilot project on the Crimean peninsula and in Western Siberia; the first real monitoring test was performed in the River Rhine, close to the city of Koblenz, Germany (Buckup, P. and Buckup, K., 2005).

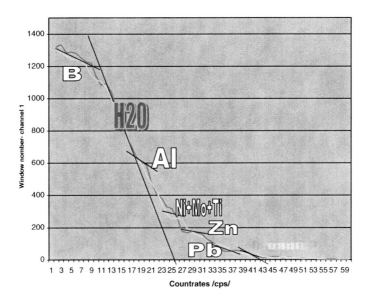

Figure 4. Appearance of different elements on the underground of a decay curve

2. Continuous neutron monitoring in the River Rhine

The Pulse-Neutron-Neutron continuous monitoring system was tested in the River Rhine where the technique could be compared with measured standard values, like temperature, conductivity, oxygen content etc. The rig-up of the instrument is shown on Fig. 5; it was placed on the river bed and kept in operation for 3 days on a recording cycle of 1 h.

The monitoring curve over the whole period is shown on Fig. 6 and indicates only small changes of water status, which is confirmed by the conductivity record; the neutron and conductivity curves show correspondence. In Fig. 6, all available time windows of the decay curve are shown individually accessed; the fluctuation of the recorded neutrons is small, but bigger than the conductivity curve.

Figure 5. Probe rig-up in the River Rhine

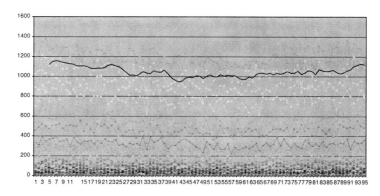

Figure 6. Twenty-four hours neutron monitoring in the River Rhine (black line-conductivity)

3. Long-term monitoring in a drinking water facility

One of the most challenging tasks for continuous monitoring is the direct control of drinking water quality in order to warning immediately that permitted concentration values are exceeded. For the "Gallus Quelle" water supply with a fractured aquifer system, it is crucial to have continuous information on the water status.

The tool was mounted directly opposite an open fracture and the neutron density continuously recorded (Fig. 7). The monitoring curve corresponds with the simultanously measured conductivity values over the monitoring period of 14 days at 1 h data sampling intervals. With the help of water samples, the neutron response was calibrated (Fig. 8) and the Cl-concentration over the monitoring time calculated. The lowest values of 15.7 mg/l were found on November 4th, the highest ones of 20.6 mg/l on the 2nd (Fig. 7). The extremal values are not reflected in the conductivity curve. The neutron monitoring curve shows the situation in the water body in more detail compared with the conductivity curve; the small range in the Cl-concentration has little influence on the conductivity.

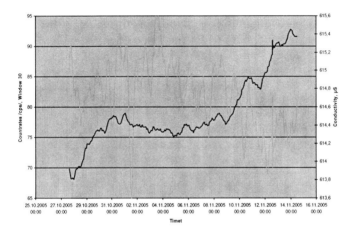

Figure 7. Monitoring a drinking water source conductivity values

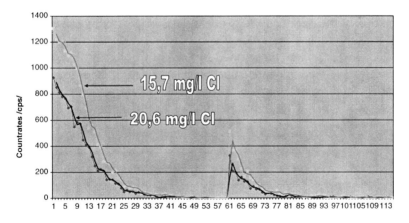

Figure 8. Calibration of decay curves for the Gallus Quelle

4. Pulse neutron monitoring on the River Meuse at Eijsden

A field trial in the River Meuse was carried out over a time period of 3 weeks with data sampling each hour (Fig. 9). The probe was lowered to the river bottom and Fig. 9 shows the general situation for both channels over the whole monitoring period. The average lines for the neutron intensity distribution in both channels are very similar and in view of their different depth of investigation, it may be concluded that the situation in the bottom sediments and in the water is comparable. In the background of Fig. 9 are presented the recorded countrates. The deviation of primary data from the trend line indicates a rapidly changing environment, caused by the transported material, which is of a different kind to that shown in Fig. 4. There are various compounds in small quantities of either organic or inorganic nature detected, but the dominant role is played by Cd.

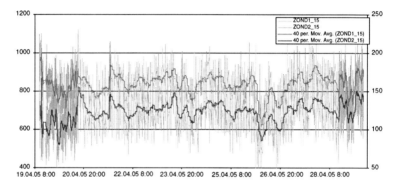

Figure 9. Continuous pulse-neutron-neutron monitoring in the River Meuse near Ejisden

Using chemical analysis from samples on April 19th and 20th as reference the position of the decay curve was calibrated in Cd-values and a maximum of Cd-concentration extrapolated, which was calculated for an event on April 26, where Cd was transported by the river water in higher quantities (Fig. 10). Organic compounds appeared periodically. The most remarkable anomalies for organic presence (suspected pesticides) were measured in the evening of April 20th, in the morning of April 22nd and on April 28th.

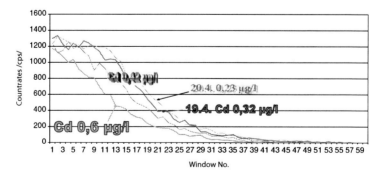

Figure 10. Influence of Cd-content on the decay curves at selected monitoring times

5. Pulse neutron monitoring for organic pollution control in water wells

The Pulse-Neutron-Neutron monitoring technique may also be used in water wells. On an industrially polluted site control wells were drilled to control the actual status. A logging unit was attached and the probe lowered to different depths to obtain data. Having preliminary information that the pollutant concentration is varying along the profile of the well, different spots were selected for monitoring. The pollution was identified as organic material around monochlorobenzene. For the neutron population therefore the Cl-content was the main factor and by increasing Cl-content a decrease in the neutron intensity was forecasted. The decay curves for the monitoring intervals at different depths are shown on Fig. 11 and confirmed the forecasted behaviour of the neutron count rates. In the processed version of Fig. 12, the average decay curves for the five monitored intervals are changing their amplitudes depending on the monochlorobenzene content.

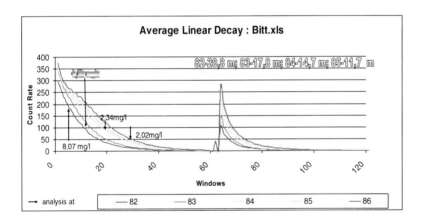

Figure 11. Decay curves for the monitoring intervals

The dependence is very strong and the regression coefficient R^2 for the function $C = f (I)$ (C – concentration; I – count rate in a selected window) in a logarithmic approximation is in the range of 0.9. In Fig. 12, the regression curves are shown for a selected window in the first and the second channel. The curve of the second channel has a higher regression coefficient, which is caused by the smaller influence of the water presence in the well, less influencing the readings in the second channel. For practical purposes, a linear approximation may be sufficient, as regression coefficients over 0.9 were calculated for both channels.

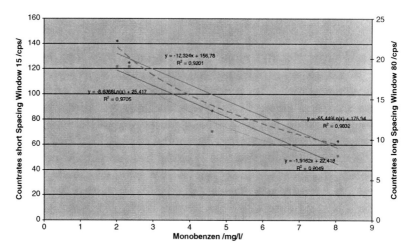

Figure 12. Correlation between INN-count rates and monochlorobenzene-content

6. Summary

The Pulse-Neutron-Neutron method was run as continuous monitoring under various conditions with different tasks, like pollution control for organic and inorganic compounds. Regression equations have been established to enable quantitative estimations to be made for a variety of elements. No matter in what type of water body the neutron monitoring is performed, a representative result may always be achieved.

References

Buckup, P. and Buckup, K., 2002: Multipurpose Application of the Pulse Neutron Method (PNM). Presentation VIII, Krajowa Konferencja Naukowa-Techniczna, Szymbark, Poland.

Buckup, P. and Buckup, K., 2005: Report on a PNN-demo for iron-detection in Deventer. Magdeburg, August 25th (unpublished).

Gearhart Industries, 1980: Formation Evaluation Data Handbook. Fort Worth, Gearhart Industries.

Hansen, S., 1997: Neutrons and Neutron Generators. *Bell Jar*, **6(3/4)**.

NATURAL DISASTERS AND SURFACE AND SUBSURFACE WATER POLLUTION RISK ASSESSMENT FOR SOME REGIONS OF GEORGIA

T. Davitashvili*

I.Vekua Institute of Applied Mathematics of Tbilisi State University, 2 University St., 0186 Tbilisi, Georgia

Abstract. In the present paper, some hydrological specifications of Georgian water resources are presented. The river Rioni's possible pollution by oil in a period of flooding is studied by numerical modelling. Some results of the investigation of pollution of Georgia's largest river, the River Kura, are also given. With the purpose of studying subsurface water pollution by oil in case of emergency spillage, the process of oil penetration into a soil with a flat surface containing pits are given and analyzed. Results of calculations have shown that the possibility of surface and subsurface water pollution due to accidents on oil pipelines or railway routes is high.

Keywords: Water pollution, oil pipeline, mathematical modeling

1. Introduction

The Transport Corridor Europe–Caucasus–Asia (TRACECA) using railways, highways and oil–gas-pipelines, is conveying oil, gas, coal and cotton, across Georgia from central Asia and Azerbaijan to other countries. At present, six main oil and gas pipelines are already functioning on the territory of Georgia. According to the experience of transit countries, the conveyance of oil and gas causes great damage to the ecology, thus counteracting the intended political and economical benefits. In addition to ordinary pollution of the environment, it is possible that non-ordinary situations like pipeline and railway accidents arise. As foreign experience with pipelines shows, the main reasons for crashes and spillages are the destruction of pipes as a result of corrosion, defects of welding and natural phenomena (floods, landslides, earthquakes, etc.). Also terrorist attacks and sabotage may occur. The TRACECA route crosses a multitude of surface water flows. Six major river crossings occur along the route on the territory of Georgia. Ground water along the route is also abundant and generally of high quality. Extra protective measures were carried out for the places where the oil-pipeline crosses the rivers. Thus, the pipeline is buried deeper in the soil under the rivers (3–4 m instead of normal 2.2 m). However, the mountain rivers are characterized by periodical

---

* To whom correspondence should be addressed. e-mail: tedavitashvili@gmail.com

floods. There are frequent cases in the Georgia when a flooded river under-
mines the bridges, bases and sometimes even washes away the bridges. So
eventually it can happen that the erosion of the soil reduces the protective
layer of the soil to a minimum, so that the broken stones carried by the
flooded river damage the protective covering of the pipeline and cause leakage of the oil. In addition, if we take into account that Georgia is located in a
seismically active zone, this increases the possibility of oil spillage on the
soil and into surface and subsurface waters and the rivers. As oil transportation by TRACECA and the pipelines goes through densely populated
areas, so the forecasting and modeling of possible emergency situations is
an extremely urgent subject for solving the problem of protecting the
population and the environment.

2. Some hydrological specifications of Georgian water resources

Georgia lies along the eastern coast of the Black Sea, to the south of the
main Caucasian Ridge. About 85% of the total land area is occupied by
mountain ranges. The Lesser Caucasus mountains occupy the southern part
of Georgia. These two mountain systems are linked by the centrally located
Surami mountain range, which bisects the country into the western and
eastern regions. To the west of the Surami Range, the relief becomes much
lower and elevations are generally less than 100 m along the river valleys
and the coast of the Black Sea. On the eastern side of the Surami Range, a
high plateau known as the Kartli Plain extends along the River Kura. The
two largest rivers in Georgia, the Kura and the Rioni, flow in opposite
directions: the Kura, which originates in Turkey, runs generally eastward
through Georgia and Azerbaijan into the Caspian Sea, while the Rioni runs
generally westward through the lower Rioni valley and drains into the
Black Sea. The average drainage density of the river network in Georgia is
0.6 km/km^2. The density of the river network is conditioned by the impact
of physical-geographical and climatic factors. The quantity of atmospheric
precipitation plays a specially critical role. Generally, the density of the
river network in Georgia decreases in parallel with the reduction of atom-
spheric precipitation from the west to the east. In particular, the density of
the river network in Western Georgia is about 1.07 km/km^2, while in
Eastern Georgia this figure is 0.68 km/km^2 and the density of the river
network in the driest regions Eastern of Georgia is very low at 0.1 km/km^2
(Svanidze et al., 1988). There is almost no river network in the driest
eastern regions of Georgia, where the annual quantity of precipitation is
very low at 100–250 mm, and the level of evaporation is extremely high.
About 99% of Georgian rivers are less than 25 km long, and only one river,
the River Kura, is longer than 500 km. Also there are 43 natural and
artificial lakes in Georgia, of which 35 are in East Georgia, for irrigation or

hydropower generation Surface water and groundwater resources include numerous thermal and mineral springs. Many snow- and glacier-fed rivers are present in the Greater Caucasus. Groundwater resources are abundant, especially in the lower slopes of the Greater Caucasus and in the lava plateaus of the Javakheti mountains (Svanidze et al., 1988).

In Georgia, about 1,600 water-suppliers provide a total of 620 million cubic meters of drinking water per year. From this quantity 90% is consumed by urban population and 10% by rural (Mindorashvili, 2002). The main source of drinking water is groundwater, accounting for about 90% of the total amount of water feeding the centralized water-supply networks. No special treatment of groundwater takes place before it is supplied to the users; the water is only chlorinated. When surface water is used as raw material, this water is specially treated – precipitated and chlorinated.

3. Georgian section of the transport corridor Europe–Caucasus–Asia and its environmental problems

The Caspian Sea region has abundant oil and gas reserves, but a limited accessible market for the products. Today, TRACECA is considered as a corridor, which is complementing other existing routes. The European Union considers Georgia as a partner in the development of the transport networks between the Black Sea and Central Asia because of its geopolitical position. Georgia participates in 19 of the 35 projects undertaken in the regional program TRACECA. The probability of failures in the oil transport pipelines increases with the age of the pipelines in service, and with the extent of their network. For example, 250 ruptures, which were accompanied by spillages of the transferred products, occurred every year from 1973 to 1983 in the US pipeline network, which has a total length of about 250,000 km. In West Europe it has been found that 10–15 leakages happen every year in a pipeline network of around 16,000 km length, resulting in a loss of 0.001% of transferred products (Davitashvili, 2004). At present, the Western Export Pipeline route (WEP) with diameter 530 mm is transporting up to 800 m^3 of oil per hour from the expanded Sangachal terminal near Baku, through Georgia to the Supsa terminal. In 1993, 80% of WEP was destroyed. All the oil contained in the oil pipeline was spilled onto the ground. During a single day, about four oil railway transports cross the territory of Georgia, each of them transporting about 1,000 t of oil products. According to official data from the Georgian Department of Railway Transport, between 1979 and 2000 there were six big accidents connected with the transportation of oil by rail (Davitashvili et al., 2005).

Data on accidents, received from the Georgian Railway Department, allows us to analyze some statistical characteristics of the space-time distribution of these phenomena. Investigations have shown that the probability of small accidents corresponding to 10 t and the probability of large disasters corresponding to more than 1,000 t are almost equal. But the probability of middle sized accidents of 100–500 t of spilled oil, is five to six times higher than for small and large accidents (Davitashvili, 2004).

4. Mathematical modelling of oil infiltration into soil for Assessment of undersurface water pollution

We have applied mathematical models describing oil penetration into soils as a result of accidents on railway and pipeline routes in the general case and for special conditions taking into account the specific properties of several types of soils found along the routes (Davitashvili and Komurjishvili, 2002; Davitashvili, 2004). The calculations were made for meadow-alluvial, sandy and sub-sandy soils. Numerical calculations were carried out during 180 days. Numerical calculations have shown that the process of oil filtration into soils from the pits varied for different period of the numerical integration. In particular, it was possible to distinguish three stages during the numerical calculations.

The first stage contained intensive processes of oil evaporation and oil filtration in the soil. This process continued for about 1.8 days until the ground surface was clear of the spilled oil, with the exception of the pit areas. For the first stage, the distribution of oil in the soil was almost the same both in the vertical and in the horizontal directions. Although we can note that oil penetration in the soil in the vertical direction was more intensive than in the horizontal direction in the areas around the lower border of the pit. During the second stage, two processes occurred: oil filtration and evaporation from the soil, and oil infiltration into the soil from the areas around the pit, until the pit was clear of the spilled oil. Numerical calculations have shown that this stage continued about 31 days. The second stage was mainly characterized by oil propagation into the soil from the edges of the pits, but we note that during the second stage we have used non-stationary boundary conditions at the lateral borders of the pit, taking into account oil filtration and evaporation processes simultaneously, until the pit clearing process was accomplished. Numerical calculations have shown that the pit clearing process was accomplished in 31 days and the maximum depth of oil penetration in soil reached 2.1 m. The third stage of oil infiltration into soil was less intensive than the first and second stages. We have kept the process of oil infiltration into the soil under our observation until the velocity of the front of oil distribution into the soil was infinitesimal, i.e. until the process became almost stable. Numerical calculations have shown that the third stage continued about 147 days. We note that this process in the vertical direction was more intensive than in the horizontal direction and in

180 days the line of the front of oil distribution in the soil was not very well distinguished, but had a sinusoidal character. The maximum depth of the front of oil distribution in the soil was observed on the depth 3.9 m, almost along the symmetric axes of the pit, and the maximum distance of oil penetration in the horizontal direction were observed to be about 2.8 m from the symmetric axes of the pit for the meadow-alluvial soils.

As underground water in flat areas adjacent to the River Kura with pseudo-clay, sandy and marshy wet soils comes close to the surface about 3–4 m below surface in East Georgia, and underground water comes close to the surface about 2–3 m below surface in hilly regions and about 1 m below in the flat areas in West Georgia, so possibility of underground water pollution owing to accidents along oil pipeline and railway routes is high.

5. The River Kura pollution

The most polluted rivers in Georgia are the Kura, Mashavera, Kvirila and Rioni. Because of financial and economical difficulties, the data from surface water monitoring has been very poor over the last 16 years. In most of these rivers, concentrations of phenols, hydrocarbons, copper, manganese, zinc and nitrogen are considerably higher than the national and international standards. The major sources of pollution are transport, the chemical industry and the energy sector. The largest polluter of surface water is municipal wastewater and oil products.

The River Kura passes large towns and also near the TRACECA route; that is why the Kura is polluted to such a high level. For example, processing the data for the Kura's water usage, namely water abstraction and water discharges into the Kura, for 1999–2001 shows that in 1999 the amount of water released into the rivers unpurified was 30%, but by 2001 the amount had reached 90%. In addition, the extent of contamination on the Mtkvari River has also become clear, and thus the importance of analyzing the current chemical-ecological state of the Mtkvari (Buachidze et al., 2002). Analysis of the data from 1997–2001 on the River Mtkvari revealed an increased amount of harmful substances – chlorines, sulfates and oil products – and a decreasing amount of organic substances.

It is also interesting to see how much the River Kura was polluted when it passed Tbilisi. The analysis of the data from February and March 2002, obtained for a number of locations along the River Kura in Tbilisi and for places situated above and about 1.5 km below Tbilisi, has shown that as the river passes through Tbilisi there are increases in the concentrations of chlorines (by 82.2%), sulphates (97%), hydrocarbons (38%), nitrates (5.4%) and nitrites (70.8%), which indicates that there was increased anthropogenic loading on the River Kura. Similarly, analysis of the data from the

Mtkvari shows that the intensity of smell increased by fourfold and water transparency deteriorated by sevenfold, which indicate that pollution in Tbilisi is extremely dangerous (Buachidze et al., 2002).

6. Mathematical modelling of the River Rioni pollution by oil products

In the event of petroleum getting into a river, the extent of its pollution is defined by the following factors: the mass of the petroleum poured into the river; the discharge, width and length of the river; the temperature of the water in the river and the speed of the wind blowing across the water surface (the intensiveness of evaporation is determined by these data), and finally by the type of petroleum pouring into the river.

By numerical modeling, we have obtained the concentration of petroleum in different moments of time. In addition to the area of petroleum covering dissolved in the water, the petroleum concentration and the quantity of the precipitated petroleum has been calculated. The quantity of the petroleum reaching the sea from the river has also been determined.

The distance between the place where the petroleum was spilled into the River Rioni and the Black sea coast was 5 km. The calculations showed that the concentration of petroleum products in such cases exceeds several hundred times the limits of permissible concentration. The maximum concentrations of petroleum products in the river Rioni at different periods of time are given in Table 1.

TABLE 1. Maximum concentration values of the crude oil, diesel fuel and boiler fuel for different moments in time after spillage (g/m^2) for the River Rioni

Time (s)	Crude oil	Diesel oil	Boiler oil
300	847.9	853.4	869.1
600	439.1	467.2	472.3
900	261.7	293.2	297.7
1,800	80.1	86.3	89.8
2,700	40.3	45.2	48.7
3,600	28.7	32.5	35.1
5,400	13.2	16.3	18.9
6,000	10.7	13.1	14.6
9,000	5.7	8.2	9.4

One can see that the concentration is much higher shortly after spillage. The average velocity in the River Rioni is 2 m/s and the maximal value of petroleum products reaching the sea occurs after about 50 min. At that time, the amount of vaporization and the amount of petroleum precipitating on the banks are insignificant. When the velocity of flow decreases to 0.6 m/s, the travel time of the petroleum slug increases and that causes an increase in the amount of petroleum precipitating on the banks, and that might act as a stationary source of pollution in the future. Results from the numerical calculations have also shown that the greater the amount of petroleum poured into the river, the higher is the degree of its pollution. As flows in Georgian rivers are mostly of the turbulent kind, the petroleum is sure to mix and spread across and downriver.

7. Conclusions

The pollution of surface and subsurface water by oil products impacts not only on the flora and fauna, but consequently also on human beings, as the surface and subsurface water is applied not only for industry, but as drinking water for people. It is therefore necessary to undertake special, additional treatment of this surface and subsurface waters before it is supplied to the users in Georgia. If the location of petroleum spillage into a river is not far from the sea, emergency activities can be hampered. That is why it must be considered necessary to erect purifying plants along the rivers. Also it is necessary to install doubly protected sections of pipe where the pipelines pass beneath the rivers.

References

Buachidze, N., Intskirveli, L., Kuchava, G. and Chachibaia, G., 2002: Chemical-ecological investigation of the River Mtkvari within the limits of Tbilisi. *Transactions of the Institute of Hydrometeorology*, **108**: 261–267.

Davitashvili, T., 2004: Numerical and theoretical investigation of spreading oil filtration in soils for Caucasus region. *Geography and Natural Resources*, Novosibirsk: Russian Academic Publishers, pp. 215–226.

Davitashvili, T. and Komurjishvili, O. 2002: Pollutants transfer in the environment with a new three-dimensional numerical scheme. *Reports of VIAM*, **28**: 25–30.

Davitashvili, T., Khantadze, A. and Samkharadze, I., 2005: Environmental and social-economical baseline of the Georgian Section of the Baku-Tbilisi-Ceyhan Pipeline. *Proceedings of Odlar Yurdu University*, Baku, Azerbaijan Academic Publishers, pp. 110–119.

Mindorashvili, A., 2002: Urgent problems for Georgia's population protection with drinking water and its quality monitoring. *Transactions of the Institute of Hydrometeorology*, **108**: 91–96.

Svanidze, G., Tsomaia, V. et al., 1988: *Water Resources of Transcaucasia*. Leningrad: Gidrometizdat. Press.

STORM SURGES ON THE SOUTHERN COAST OF GULF OF RIGA: CASE STUDY OF THE LIELUPE RIVER

T. Koltsova* and J. Belakova
*Latvian Environment, Geology and Meteorology Agency,
165, Maskavas St., LV-1019 Riga, Latvia*

Abstract. The Lielupe River is the second largest river in Latvia. Annual water runoff of the Lielupe River to the Gulf of Riga is 3.37 km³. After Jelgava city, the river starts forming a typical estuary with riverbanks, small islands and peninsulas, swamps etc. From here and down to the river mouth, the gradient of the river is 5–10 cm per kilometre, and on occasion the high water level in the Gulf of Riga has a damming effect on the flow of the river. The riverbed is much lower than the average Baltic Sea level over a length of 100 km upstream from the mouth. As a result, the river flows in the opposite direction from the sea in autumn and winter. In low flow periods in summer and winter, the water quality conditions become critical, because of decreasing oxygen content and increasing oxygen demand. The autumn of 2005 was characterised by very low flow and high water temperature. Hydrometeorological conditions plus the wastewater from the Jelgava sugar factory led to the formation of a dissolved oxygen deficit zone and fish losses as a result. Poor water management and water security were the main cause of that disaster. Storm surges facilitate the exchange of the water mass and thereby improve the quality of the water. Air temperature, wind direction and speed and water level data series were analysed with respect to climate change impacts on the water quality of the Lielupe River.

Keywords: Storm surges, backwaters, oxygen deficit zone, wind rose, wind gust index

1. Introduction

The Lielupe River basin measures approximately 17,600 km² of which 8,800 km² is in Latvia and 8,800 km² in Lithuania (Fig. 1). Lielupe River has an annual water runoff of 3,370 million cubic metres to the Gulf of Riga. The length of the main stream is approximatelt 285 km, of which 133 km is in Latvia. The Lielupe is the second largest river in Latvia.

The river has two major tributaries: the Musa River (165 km in length) and the Memele River (182 km in length). Thirteen kilometres from the border between Latvia and Lithuania, the Musa River merges with the Memele River at an elevation of 12 m above sea level to become the Lielupe River.

---

* To whom correspondence should be addressed. e-mail: tatjana.kolcova@lvgma.gov.lv

J.A.A. Jones et al. (eds.), *Threats to Global Water Security*,
© Springer Science + Business Media B.V. 2009

In this stretch, the river is 94 m wide and 0.5 m deep. The Lielupe River is navigable at 16 km downstream of this confluence. In Jelgava, the river is only 2.5 m above sea level; its width is about 200 m and its depth is more than 2.5 m. Downstream of Jelgava, the river forms a typical estuary with riverbanks, small islands and peninsulas, swamps etc. From here downstream, the slope of the river is 5–10 cm/km, and the high water levels in the Gulf of Riga have a damming effect on the flow of the river. The riverbed is much lower than the average Baltic Sea level over a length of 100 km upstream from the mouth. As a result, water flows upstream through backwaters in autumn and winter. In the highest backwater flooding event, seawater flows up the river mouth and sometimes reaches the Kalnciems region (c.50 km upstream). The entire water mass of Lielupe River flows downstream only during flood periods due to rainstorms and snow melt in spring (Shteinbah (ed.), 1969–1985). Also, periodic water level fluctuations should be considered. The water level rises over 12 h and falls during about 12.8 h (Pastors, 1965). The height of the tide is 10 cm on average during the open channel period, and about 9 cm during freezing (Fig. 2).

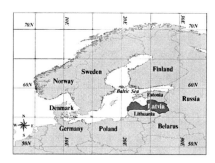 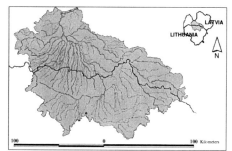

Figure 1. Study area and Lielupe River basin

This regime lasts 1–2 months and is normally interrupted by storm surges. For 3.5 h, when the water flow changes direction from inflow to outflow, its mean velocity is less than 1 cm/s. For these 3.5 h, the water travels only 63 m downstream and back upstream. So, the water does not move for 3.5 h twice per day and the pollution load increases significantly.

During low flow periods in summer and winter, the water quality conditions become critical because of the decreasing oxygen content (Fig. 3) and increasing oxygen demand.

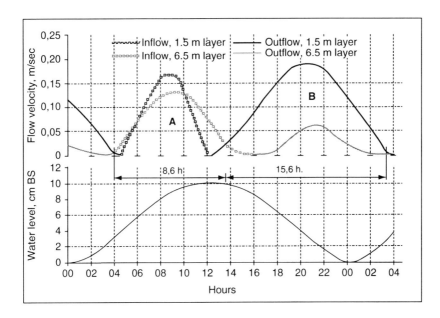

Figure 2. Flow velocity (above) at depth of 1.5 and 6.5 m: inflow (A), outflow (B); water level fluctuation (below) of the Lielupe River near the Priedaine bridge (Reference: A. Pastors, 1976)

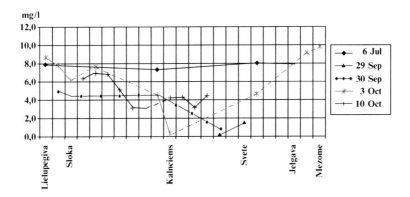

Figure 3. Dissolved oxygen concentration in Lielupe River, autumn 2005

The ecological situation in autumn 2005 is a striking example of this. Autumn 2005 was characterised by very low flow and high water temperature, 2.5°C above the long-term average. Hydrometeorological conditions, plus the wastewater from the Jelgava sugar factory led to formation of a dissolved oxygen deficit zone and as a result major fish losses (Fig. 4).

Figure 4. Lielupe River nearby Kalnciema (Photo from the homepage of Latvian State environmental service, 26.09.2005)

Poor water management and water security were the main cause of that disaster. Storm surges facilitate the exchange of the water masses and thereby improve the water quality.The aim of the present study is to evaluate the changes in water level data series of Lielupe River with respect to climate change impact on the water quality.

2. Data and methods

Surges in the Lielupe River are caused by the water mass exchange between the Baltic Sea and the Gulf of Riga, and the wind speed and direction. To estimate the changes in the peak values of surges, a comparison of the water level data series from three hydrological stations in the river mouth stretch was made (Kalnciems, Sloka and Lielupes Griva; 48, 27 and 5 km from the river mouth respectively). The nonparametric Sen's method (Helsel and Hirch, 2002) was employed to estimate the trend of storm wind durations for the period 1966–2006 and to evaluate changes in maximum and average water levels during the summer season for the period 1923–2006. For evaluating changes in the relationship between wind speed and storm water level, wind gust data of the Riga station and maximum water level data from the Lielupes Griva station were normalized with respect to the period 1961–1990, by subtracting the mean and dividing with the standard deviation.

3. Results and discussion

Considering the water quality problems of the Lielupe River during low flow periods, the average summer water levels were analysed (Fig. 5). Evidently, the water level at Sloka and Kalnciems in summer is not so much dependent on the sea water level. The Sen's slope of the summer season's average water level data series for the period 1922–2006 are: 0.3 cm/year for the Kalnciems and Sloka stations and 0.1 cm/year for the Lielupe Griva station.

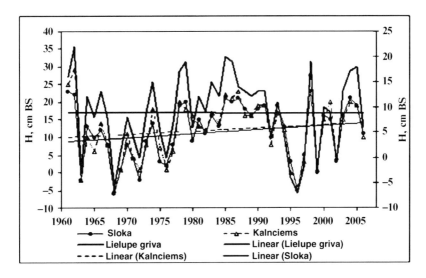

Figure 5. Average water level of summer low flow period (June–September)

Storm surges on the southern coast of the Gulf of Riga, especially in the mouth of the Lielupe River, are caused by storm winds and a water mass exchange between the Baltic Sea and the Gulf of Riga (Iljina and Pastors, 1976; Pastors, 1965). Mostly, northern to northwestern winds were recorded during surges. Actually, maximum water levels in storm surges strongly depend on wind gusts (Fig. 6).

During the recent two storm surges in 2005 and 2007, wind gusts with speeds of 30 and 28 m/s were recorded, respectively. A comparison of maximum water levels in Fig. 7 shows considerable differences between the trends in sea water level at the Lielupe Griva station and the river water levels at the Kalnciems and Sloka stations.

The process seems to be periodic with roughly a 40-year period in the river water level and an 80-year period in the sea water level. Moreover, only maximum sea water levels have increased; 48 km upstream from the river mouth, the maximum water level has decreased due to a decrease in spring flood levels.

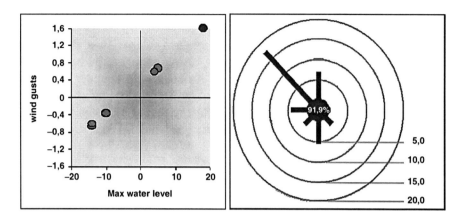

Figure 6. Anomalies from the 1961–1990 reference values (lines) of storm gusts (Y axis, standard deviations) versus maximum sea water level (X axis, standard deviations) for the decades during 1941–2000 (gray) and 2001–2007 (black). Wind rose of maximum wind (right)

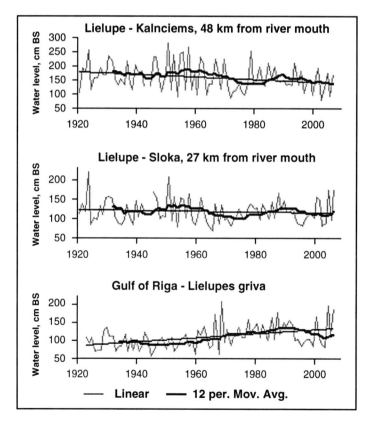

Figure 7. Maximum water level in the mouth stretch of Lielupe River for the period 1923–2006: thin line – regression line for the gauge observations, thick line – 12-year running mean of gauge observations

4. Conclusions

The average water level during the low flow period (June–September) is increasing at Sloka and Kalciems, but has not changed in the mouth. As a result, the pollution load in the Lielupe River mouth during the low flow period caused by tides is reduced, and the water quality is improving.

There is a linear correlation between the maximum water level in the mouth of the Lielupe River and storm gust speeds south of the Gulf of Riga. The wind gust index during the short period 2001–2007 is 1.6 (standard deviation) above the reference period 1961–1990.

Maximum water levels at Lielupe Griva station show a significant positive trend during the period 1923–2006, but the data series of Sloka station (27 km upstream from the mouth) show no trend at all, while the data series of Kalciems station (48 km upstream from the mouth) show a significant negative trend. So, the river stretch affected by backwater is becoming longer.

References

Hadonina, D. and Kadikis, N. (Eds.), 2001: Lielupe River basin management plan. Analysis of situation. Vilnus, *UAB "Baltijos kopija"*: 77–81.

Helsel, D.R. and Hirch, R.M., 2002: Statistical methods in water resources. Hydrological analysis and interpolation. In: *Techniques of Water Resources Investigations*, U.S. Geological Survey. Chapter A3, **4**: 266–274.

Iljina, J. and Pastors, A., 1976: Surges and ebbs in Gulf of Riga. Riga, *Collection of Scientific Papers of Riga Hydrometeorological Observatory*, **15**: 9–37.

Koltsova, T. and Belakova, J., 2007: Storm surges on the southern coast of Gulf of Riga. *BALTEX Newsletter*, **10**: 15–18.

Pastors, A., 1965: Surges and ebbs in downstretch of Lielupe River. Riga, *Collection of Scientific Papers of LATNIIG & M*, **3**: 279–308.

Pastors, A., 1976: Water periodical circulation on the mouth stretch of Lielupe River. Riga, *Collection of Scientific Papers of Riga Hydrometeorological Observatory*, **15**: 99–124.

Shteinbah, B. (Ed.), 1969–1985: *Annual reports of Riga Mouth Station*. Riga: Latvian Hydrometeorological Agency Press.

GROUND WATER VULNERABILITY ASSESSMENT OF THE APARAN AQUIFER, REPUBLIC OF ARMENIA, AND ITS REPRESENTATION IN A 3-D MODEL

A. Aghinian[*]

Yerevan State University, Geological Department, Alex Manoogian 1, Republic of Armenia

Abstract. The Aparan confined aquifer feeds the aqueduct supplying Yerevan's drinking water. Hydrogeologically, the terrain consists of an unconfined or water-table aquifer composed of recent alluvial-proluvial sediments, and a confined aquifer composed of Quaternary lacustrine-alluvial sediments. An assessment of the natural and bacteriological vulnerability of these aquifers was made, and a 3D hydrostratigraphic model was constructed. We have conducted the analyses according to depths to water level from land surface and the thickness of poorly permeable soils in the aeration zone. According to this method, the following types of soils have been distinguished in the cross-section of the aeration zone: (1) $K = 0.1–0.01$ m/day (loamy sand or fractured matrix rocks); (2) $K = 0.01–0.001$ m/day (loams or relatively confining matrix rocks); and (3) $K < 0.001$ m/day (clay or confining matrix rocks), where K is permeability (conductance). The quantitative assessment of the unconfined aquifer's vulnerability is being carried out when there is risk of bacteriological infection. This method is based on the downward time-of-travel to the groundwater, during which percolated bacteria or polluted water may reach the water table. The quantitative assessment of the confined aquifer's vulnerability that is nearest to the land surface is carried out on the basis of two parameters: total thickness of the overlaid soils that have slow infiltration rates (m_0) and the interrelation between level of above water-table (H_1) and established level of water of aquifer under study (H_2). A 3-D conceptual hydrostratigraphic model of the study area was constructed. As a result of evaluation, the unconfined aquifer was assessed as highly vulnerable, but the confined aquifer as protected by natural conditions.

Keywords: Aquifer, vulnerability, aeration zone, permeability, bacterial pollution

1. Introduction

The aquifer is located in an area of tremendous economic activity, affecting both the surface and subsurface. These activities, along with other natural exogenous processes, have a negative impact on hydrogeological conditions,

---

[*] To whom correspondence should be addressed. e-mail: gtorosyan@seua.am

J.A.A. Jones et al. (eds.), *Threats to Global Water Security*,
© Springer Science + Business Media B.V. 2009

mainly on the aquifers, which results in contamination and depletion of storage volumes. Currently, the contamination hazard of fresh water aquifers exceeds the danger of the storage depletion many times over. Hence, conservation of the ground water as a sustainable resource for drinking water supply is of paramount importance. Ground water is the only source for industrial and drinking water supply in the Republic of Armenia. The groundwater recharge areas are in the mountain ranges that naturally discharge water from the lower slopes and in the intermontane basins. Unconfined groundwater conditions occur on slopes. Aquifers in intermontane basins have both confined and unconfined nature.

Agriculture, including crop production and cattle-breeding, is developed in the mountain areas, 1,600 m above sea level and higher. Industrial manufacturing and agriculture, including horticulture, vegetable growing and growing of other industrial crops are developed at intermontane basins and foothills. All these activities are potential sources of groundwater contamination (Bochever, 1979). Ground water contamination is understood as changes in water quality under the impact of human activities compared with the quality of water in natural conditions, the "background level", that leads partially or entirely to the water being no use for human activities.

2. Characteristics of the study area

The Aparan basin extends over about 300 km^2 between the Aragats volcanic massif (4,095 m) in the west, the Pambak and Tsahkunyats Ridges on the north and east respectively, and the Aray volcanic massif to the southeast. The study area is located at the middle reaches of the Kasakh River at 1,300 m. above sea level.

The geological sequence mainly comprises Quaternary alluvial, lacustrine-postglacial sediments and volcanic lavas reaching up to 300 m in thickness. Water bearing rocks are represented by:

(1) Volcanic rocks of the Aragats massif (andesite, andesite-dacite), which outcrop westward from Aparan village as well as on the right bank of the Kasakh River. The lavas are intensely fractured, discontinuously water bearing, and a number of water springs occur in this hydrogeologic setting (yielding 100–600 l/s).

(2) Lacustrine-alluvial sediments, represented by boulder-pebbles and sandy-loamy sediments. Both unconfined and confined aquifers occur in these hydrogeologic settings. Ground water monitoring revealed mixing of very deep thermal waters into these aquifers. The temperature of the water in well #8 reaches 10°C, while the mean temperature of water in the adjacent wells is about 7–8°C.

The till loams provide confining for the aquifer units, and act as a regional aquitard. The Quaternary succession largely controls the confining conditions of the regional aquifer as well as its recharge, which occurs on topographic highs where bedrock either outcrops or is covered by a thin discontinuous till.

3. Methodology

Two types of ground water contamination are being distinguished, depending on the quantity of contaminating products:

(i) Concentration of contaminating products in water greater than under natural conditions, continues to grow, but at less than the maximum permissible concentration (MPC)
(ii) Concentration of contaminating products in water greater than the MPC (Goldberg, 1988)

The first type is considered the start of contamination, and the second type as already contaminated. It is important to determine the tendency of water impairment at the initial stages to prevent further contamination of ground water. Percolation or seepage of industrial and domestic liquids, pesticides, as well as natural substandard waters into operating aquifers may result in ground water contamination. Abandoned wells might also result in ground water contamination (Grozdova, 1979).

In the territory of the Republic of Armenia, we have investigated aquifers with various degrees of contamination caused by the possible contamination sources and seepage pathways mentioned above.

At the present time, both qualitative and quantitative evaluation of water table aquifers and compiling of respective maps are of highly practical importance.

3.1. VULNERABILITY OF UNCONFINED AQUIFERS

We have conducted the analyses according to depths to water level from the land surface and the thickness of poorly permeable soils in the aeration zone. According to this method in the cross-section of the aeration zone, the following types of soils with low infiltration rates have been distinguished:

- $K = 0.1–0.01$ m/day (loamy sand or fractured matrix rocks)
- $K = 0.01–0.001$ m/day (loams or relatively confining matrix rocks)
- $K < 0.001$ m/day (clay or confining matrix rocks), where K is permeability/ conductance

3.2. INTRINSIC VULNERABILITY OF WATER-TABLE AQUIFERS

Table 1 presents numerical scores assigned to depth to water table and to poorly permeable soils in aeration zone based on their thickness.

TABLE 1. Numerical scores assigned to depth to water level below land surface and to soils according their thickness and lithology

Water table depth below surface, m					Total thickness (m) and lithology of poorly permeable layer (a, b, c)											
					<2			2–4			4–6			6–8		
<5	10–20	20–30	30–40	>40	a	b	c	a	b	c	a	b	c	a	b	c
1	2	3	4	5	1	1	2	2	3	4	3	4	6	4	6	8

Continuation of Table 1

Total thickness (m) and lithology of poorly permeable layer (a, b, c)																				
8–10 m			10–12 m			12–14 m			14–16 m			16–18 m			18–20 m			>20 m		
a	b	c	a	b	c	a	b	c	a	b	c	a	b	c	a	b	c	a	b	c
5	7	10	6	9	12	7	10	14	8	12	16	9	13	18	10	15	20	12	18	21

As can be seen from the Table, high scores are assigned to poorly permeable soils in contrast to water level scores. The category of vulnerability for the unconfined aquifer is derived by summing the numerical scores in Table 1. The aquifer's vulnerability is assessed according to six categories, where the first category is assessed as highly vulnerable and the sixth category as highly protected, see Table 2.

TABLE 2. Classification of unconfined aquifer by degree of vulnerability

Category	Total numerical scores	Description
I	<5	Highly vulnerable
II	5–10	Vulnerable
III	10–15	Slowly protected
IV	15–20	Protected
V	20–25	Well protected
VI	>25	Highly protected

3.3. QUANTITATIVE ASSESSMENT OF WATER-TABLE

Aquifer vulnerability was analyzed for risk of bacteriological infection. This method is based on the downward time-of-travel (TOT) of water, during which percolated bacteria or polluted water may reach the water table. Percolation time is defined by formula $t_0 = \dfrac{Mo * \Pi o}{K_0}$ (1), where

t_0 – percolation time, days
Mo – thickness of aeration zone, m
Πo – porosity of soils in aeration zone, m
K_0 – permeability(conductance) of soils in aeration zone, m

The average period of vital functioning of bacteria under slow con-tamination is taken as 200 days, but under mass infection by natural or man-caused disaster or continuous contaminant spill conditions, it is taken as 400 days. Six categories of ground water vulnerability to bacteriological contamination have been distinguished (Table 3).

TABLE 3. Evaluation of unconfined aquifer vulnerability to bacteriological contamination

Category	Time of travel, days	Vulnerability
I	<10	Highly vulnerable
II	10–50	Vulnerable
III	50–100	Slowly protected
IV	100–200	Protected
V	200–400	Well protected
VI	>400	Highly protected

3.4. QUANTITATIVE ASSESSMENT OF CONFINED AQUIFER VULNERABILITY

Assessment of the vulnerability of the top of the aquifer is carried on the basis of two parameters: total thickness of overlain soils that have slow infiltration rates (m_0) and levels of water table (H_1) and the established water level in the confined aquifer (H_2).

Three categories of confined aquifer vulnerability have been singled out:

I   – protected, ($m_0 > 10$ m, $H_2 > H_1$)
II – relatively protected, ($m_0 \approx 5$–10 m, $H_2 > H_1$ or $m_0 > 10$ m; $H_2 \leq H_1$)
III – vulnerable, ($m_0 \leq 5$ m; $H_2 \leq H_1$)

3.5. THE 3D MODEL

Development was undertaken in three main stages: conceptualization, calibration and prediction. The model permits us to incorporate and analyze hydrogeological properties of the study area and to export the information to other software systems to meet the specific requirements of an application (Ross et al., 2004).

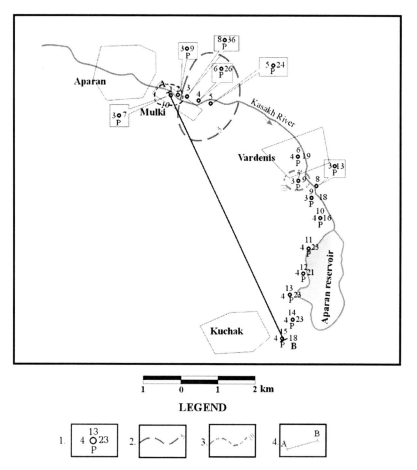

LEGEND

1. Borehole number above, below-vulnerability degree of confined aquifer: P – protected, right – bacteriological vulnerability of water-table aquifer in days, left – vulnerability of water-table aquifer in numerical scores
2. Vulnerability isolines in numerical scores
3. Vulnerability isolines of bacteriological contamination, days
4. Direction of cross-sections in Fig. 2

Figure 1. Schematic map of natural vulnerability of groundwater horizons

4. Results

The vulnerability of the unconfined aquifer in the study area was evaluated by the methods illustrated in Fig. 1. Contours in the map are drawn at intervals of five numerical scores. Two categories of intrinsic vulnerability are singled out, whereby category I is highly vulnerable and category II is vulnerable.

The vulnerability assessment data for bacteriological contamination in the study area given Fig. 1 represents mapping of bacteriological vulnerability, and the isolines are drawn in 10-day intervals. According to these isolines, the unconfined aquifer is evaluated as highly vulnerable (TOT < 10 days) and vulnerable (10–100 days).

Because of the considerable thickness of the overlying strata, the intrinsic vulnerability of the confined aquifer in the vicinity of the boreholes is eva-luated as protected and corresponds to category I vulnerability.

By constructing longitudinal and lateral cross-sections of the study area and by various interpolation methods, the conceptual hydrostratigraphic model has been constructed in Fig. 2.

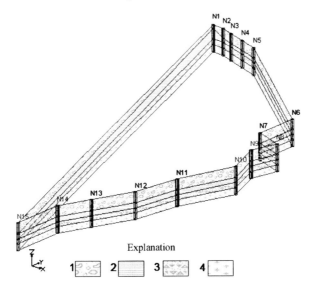

Explanation

1   2   3   4

1. Boulder-pebble sediments, 2. clay, loam, 3. pebble, sand, 4. compact tuffs

Figure 2. 3D cross-section of the study area

The principles and techniques of ground water vulnerability assessment discussed in this paper are not a universal solution and can be reviewed depending on specific hydrogeological conditions.

References

Bochever, F.M., 1979: Groundwater conservation from contamination. Nedra Press, Moscow, pp. 156–174.

Goldberg, V.M., 1988: Methodical recommendations for revealing and assessment of ground water contamination. VSEGINGEO Press, Moscow.

Grozdova, O.I., 1979: Mapping and regional predictions of man-caused modifications of underground hydrosphere, hydrogeology and engineering geology. VIEMS Press, Moscow.

Ross, M., et al., 2004: 3D geologic framework models for regional hydrogeology and land use management, *Hydrogeology Journal*, Springer.

WASTEWATER MODELING TO REDUCE DISASTER RISK FROM GROUNDWATER CONTAMINATION

N. Haruvy[1*] and S. Shalhevet[2]
[1] Netanya Academic College, Netanya, Israel
[2] SustainEcon, Brookline, MA, USA

Abstract. The disaster risk from groundwater contamination in Israel has been increasing due to continuous over-pumping as well as urban, industrial and agricultural activities. This risk can be reduced by making changes in water pumping policies and by higher level treatment of wastewater. However, these changes are expensive, and their specific effects on groundwater salinity depend on the local hydrological conditions. We have developed a water management model that enables optimal planning of water supply from different sources and treatment levels by water source. The model includes regional planning, hydrological, technological and economic considerations, and was applied on several different hydrological cells in Israel, with the aim of choosing the optimal policy that would minimize the disaster risk of groundwater contamination at the lowest economic cost possible.

The hydrological model predicts the level of groundwater salinity under different water supply combinations. The combined multi-disciplinary model shows the required salinity threshold level for freshwater supply from different sources and for irrigation with wastewater that is needed to remain below the threshold level of groundwater salinity, as well as the economic costs involved. These results were examined for various scenarios of salinity thresholds for urban water supply, irrigation, and groundwater recharge, under different water management policies.

Keywords: Groundwater, wastewater, modelling, hydrological model, economic model

1. Introduction

Israel is a semi-arid country with scarce water resources. A rapidly growing population places further stress on the freshwater sources, resulting in some areas in an increase in over-pumping water from the local aquifers. As a partial solution, seawater desalination plants are increasingly developed to supplement the water supply from renewable water sources for urban water use. While this practice can be a significant source of freshwater supply, seawater desalination is very expensive, and involves some environmental costs as well.

---

* To whom correspondence should be addressed. e-mail: navaharu@netvision.net.il

As most of the demand for water in Israel is for the agricultural sector, a partial solution has been found by using wastewater, treated to secondary or tertiary treatment level to decrease organic and inorganic hazards, for irrigation. This practice provides a reliable supply of water at a low cost as well as a partial solution for effluent disposal, and reduces the need for fertilizer use in agriculture (Haruvy et al., 1997). However, the use of wastewater can cause damage to crops and soils, and increases the risk of groundwater contamination. Wallach et al. (2005) found that long term irrigation with wastewater damages soils and increases the risk the ground-water quality in Israel. The effects of irrigation with wastewater on increasing the salinity level of a given hydrological cell was described by Yaron et al. (2000), and Haruvy et al. (2004) described the economic and policy impli-cations of these effects. The increased soil salinity causes a reduction in crop yields; the level of salinity can be decreased by additional desalination processes, but this increases the costs significantly. The risks of increased salinity require continuous follow up and control to reduce the risk of long run deterioration in aquifer water quality.

The disaster risk from groundwater contamination in Israel has been increasing due to continuous over-pumping as well as urban, industrial and agricultural activities. This has become such a problem that in some cases the authorities issued instructions to refrain from drinking tap water until the contamination was cleared up. And several towns have been forced to close down contaminated wells, stop relying on their own water supply, and to connect to the National Water Carrier instead.

This risk can be reduced by making changes in water pumping policies and by higher level treatment of wastewater. However, these changes are expensive, and their specific effects on groundwater salinity depend on the local hydrological conditions. Therefore, the optimal combination of water supply from National Carrier water, water pumped from local aquifers, treated wastewater and desalinated seawater should take into consideration the economic costs as well as the environmental implications, and should be designed to fit the local hydrological conditions in each area.

2. Methodology

We have developed a water management model that enables optimal plan-ning of water supply from different sources and treatment levels by water source, and applied the model to a case study of several different hydro-logical cells in Israel. The model includes regional planning, hydrological, technological and economic considerations. The model includes varying alternatives of water supply for drinking and for irrigation, as well as dif-ferent wastewater treatment alternatives. The technological options are based on the desalination alternatives of brackish groundwater, National Carrier water, wastewater and seawater. The goal was to identify the optimal treatment and reuse methods that that would minimize the disaster risk of groundwater contamination at the lowest economic cost possible, and ensure a safe, stable and sustainable supply of drinking water.

The general structure of the model is described in Fig. 1. The hydrological model predicts the groundwater salinity levels over time, based on alternative scenarios of water supply for drinking and for irrigation. Then, taking into account the timing and level of groundwater treatment, and the costs of the technological options available, the economic model enables to predict the financial costs of the treatment.

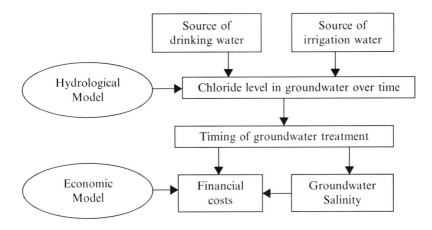

Figure 1. The water management model

The hydrological model, described in Fig. 2, predicts the level of groundwater salinity under different water supply combinations. The model simulates the contaminants flow through the unsaturated zone of the soil and the effect of irrigation with wastewater on chlorides in groundwater. The chlorine concentration in the groundwater increases until it reaches a predefined threshold. At that point the groundwater is treated and combined with freshwater sources, to reach the permitted concentration level.

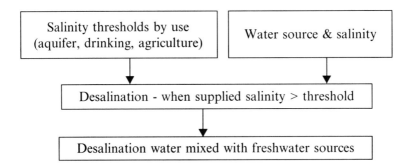

Figure 2. The hydrological model

The economic model, described in Fig. 3, calculates the total costs, including the cost of desalination, wastewater treatment, and water supply from pumped aquifer water and the national carrier water. The economic value of the environmental damage to groundwater is the cost of earlier desalination required due to increased salinity.

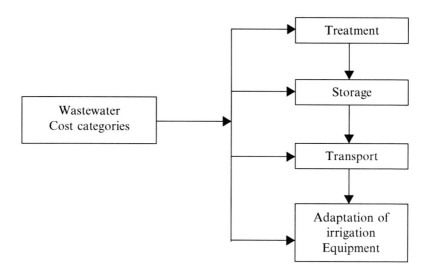

Figure 3. The economic model

The combined multi-disciplinary model shows the required salinity threshold level for freshwater supply from different sources and for irrigation with wastewater that is needed to remain below the threshold level of groundwater salinity, as well as the economic costs involved. The alternative water sources included local groundwater (well water), National Carrier water, treated wastewater and desalinated seawater. The model inputs included geographical data, based on mapping of the area's settlements; location and types of crops; data on available water sources; and hydrological (aquifer) data.

The model outputs included predictions of the water demand and supply; aquifer water levels and salinity; the quantity of desalinated water; water treatment methods; and the resulting financial costs. These results were examined for various scenarios of salinity thresholds for urban water supply, irrigation, and groundwater recharge, under different water management policies.

3. Case study

The model was applied on a case study of two areas in the coastal aquifer of Israel, whose characteristics are described in Table 1.

TABLE 1. Land and water use in the case study region[a]

	Coastal 1		Coastal 2		Total region	
	Land use	Water use	Land use	Water use	Land use	Water use
Agriculture: total	99%	**24.5**	51%	**34.9**	67%	**59.3**
National carrier		1%		34%		21%
Local aquifer		39%		56%		49%
Wastewater		61%	49%	10%	33%	31%
Urban: total	1%	**2.6**		**24.7**		**27.3**
National carrier		1%		88%		80%
Local aquifer		99%		12%		20%
Total	**8,354**	**27.0**	**16,164**	**59.6**	**24,518**	**86.6**

[a]Units: Land use – total hectares or percentage of total. Water use: total MCM or percentage of total.

The first area, referred to here as "Coastal 1", is mostly agricultural, with very few urban settlements. The area relies on wastewater supply for slightly more than one half of its total supply, with local aquifer water making up most of the rest of the supply. The second area, referred to here as "Coastal 2", is almost evenly divided between agricultural and urban areas. It relies on National Carrier water for more than half of its total water supply, with local aquifer water making up most of the rest; wastewater supply constitutes approximately 5% of the total water supply.

The basic hydrological characteristics are described in Table 2. Each one of the two areas is composed of four hydrological cells. The "Coastal 1" area has a higher level of groundwater salinity, over 230 mg/l Cl., while the level of groundwater salinity in the "Coastal 2" area is below 164 mg/l Cl.

TABLE 2. Hydrological cells in the case study region

	Coastal 1	Coastal 2
No. of hydraulic cells	4	4
Groundwater salinity	>230 mg/l Cl.	<164 mg/l Cl.
Other		Effluents from wastewater plant

The model examined eight alternative scenarios of chlorine thresholds for drinking water and for irrigation, varying between 50 to 250 mg/l Cl. The source of water for irrigation in each scenario was either freshwater or wastewater, at salinity levels of 250 or 350 mg/l Cl.

The major scenario characteristics are presented in Table 3.

TABLE 3. Threshold for water salinity in case study scenarios

Scenario	Town water mg/l Cl	Wastewater irrigation mg/l Cl	Priority of pumped water
1	250	250	Town
2	250	250	Town
3	150	250	Town
4	50	250	Town
5	250	Not included	Town
6	250	Not included	Agriculture
7	250	350	Agriculture
8	250	350	Town

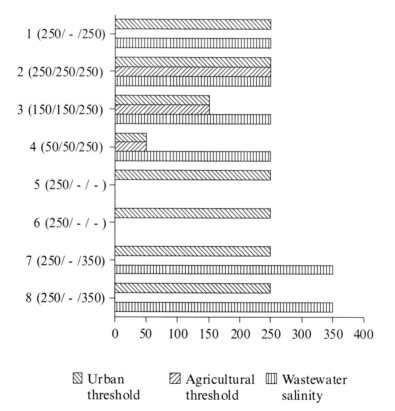

Figure 4. Scenarios of salinity thresholds (mg/l Cl)

In the first two scenarios, the predetermined salinity thresholds for urban and agricultural use are 250 mg/l Cl, and the urban water supply is first priority of the pumped water. Scenarios 2, 3 and 4 have increasingly strict salinity thresholds for urban water use. In the fifth and sixth scenario, irrigation with wastewater is not allowed; and in the seventh and eighth scenarios the salinity threshold for irrigation with wastewater is increased to 350 mg/l Cl. In the last two pairs of scenarios, we examined the impact of giving the priority of pumped water to the town or to agriculture, under the cases of no irrigation with wastewater or irrigation with higher salinity wastewater.

The scenarios are described graphically in Fig. 4. Each bar shows the salinity threshold for urban and agricultural water as use and the wastewater salinity level. The figure illustrates the increasingly strict thresholds in scenarios 1 to 4, and the lower salinity thresholds of the last two scenarios, as well as the absence of predetermined thresholds in scenarios 5 and 6, where irrigation with wastewater is not allowed.

4. Results

The predicted increase in aquifer salinity by scenario in Coastal 1 area is shown in Fig. 5. The salinity level is lower in the scenarios with greater restrictions, so that the fourth scenario results in the lowest line of salinity level over time, followed by the third scenario in the line above it, and the second scenario in the line above the third scenario. The scenarios with higher permitted salinity thresholds result in increasing groundwater salinity over time, with the highest line of salinity level under the last scenario. The differences in the groundwater salinity levels between the different scenarios increase gradually over time.

The economic implications of the increase in salinity levels between the different scenarios are presented in Fig. 6, for the case study of the Coastal 1 area. The two bars on the right side of the figure, presenting the cost calculations of the last two scenarios, show that Irrigation with wastewater increases desalination costs, but decreases the total cost of water supply. The three bars on the left side of the figure, representing scenarios 2–4, show that greater salinity restrictions increase costs significantly.

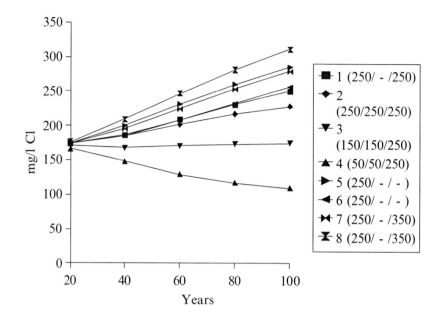

Figure 5. Predicted groundwater salinity in Coastal 1 area

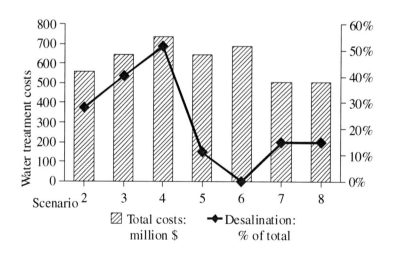

Figure 6. Water treatment and desalination costs in Coastal 1 area

The results for the Coastal 2 area showed similar trends in both salinity levels and cost comparisons between the different scenarios, but with different quantitative results. The economic results for the two areas are compared in Fig. 7. The second area, Coastal 2, with a higher salinity level and greater use of National Carrier water, resulted in a higher cost under every scenario.

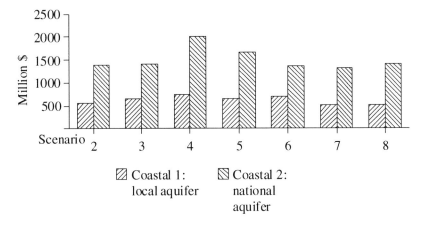

Figure 7. Water costs by water source and scenario

5. Recommendations

Analysis of the economic effects of desalination by water source showed that brackish (groundwater) desalination was the least expensive. Desalination of National Carrier water is effective only if done in large quantities. Desalination of wastewater is significant to maintain the salinity thresholds in irrigation. Finally, desalination of seawater is worthwhile only when its contribution to the water balance is significant. The main recommendation resulting from the analysis of the model's results is to allow wastewater reuse in agriculture in order to reduce costs of treatment and to insure a stable supply of water for irrigation. However, this requires careful monitoring of the groundwater salinity levels in order to reduce the disaster risk of groundwater contamination.

References

Haruvy, N., Hadas, A. and Hadas, A., 1997: Cost assessment of various means of averting environmental damage and groundwater contamination from nitrate seepage. *Agricultural Water Management*, **32**: 307–320.

Haruvy, N., Shalhevet, S. and Ravina, I., 2004: Irrigation with treated wastewater in Israel – financial and managerial analysis. *Journal of Financial Management and Analysis*, **17**: 93–102.

Wallach, R., Ben-Arie, O. and Graber, E.R., 2005: Soil water repellency induced by long-term irrigation with treated sewage effluent. *Journal of Environmental Quality*, **34**: 1910–1920.

Yaron, D., Haruvy, N. and Mishali, D., 2000: Economic considerations in the use of wastewater for irrigation. *Water and Irrigation*, **400**: 19–23.

SURVIVAL IN GROUNDWATER AND FT–IR CHARACTERIZATION OF SOME PATHOGENIC AND INDICATOR BACTERIA

Z. Filip and K. Demnerova[*]

Department of Biochemistry and Microbiology, Institute of Chemical Technology Prague (ICTP), Technicka 3-5, 166 28 Prague 6, Czech Republic

Abstract. Groundwater represents a capital resource of drinking water in many countries, and there is a growing public concern in regard to contamination of groundwater aquifers by health relevant bacteria. In microcosmos filled either with groundwater alone, or containing also sand from a deep groundwater aquifer, the survival of different health relevant bacteria was tested at 10°C ± 1°C. While *Bacillus megaterium* and *Staphylococcus aureus* died off within 30 days, all other bacteria survived up to 100 days or even longer. If a natural population of groundwater bacteria was present in the experimental system, however, an enhanced die off of introduced species or strains was observed. The FT-IR spectroscopic traits of some of the bacteria concerned varied with cell age and culture conditions, and thus their use in a routine identification of bacterial contaminants to groundwater remains questionable.

Keywords: Groundwater, bacterial contamination, survival of bacteria, FT-IR traits

1. Introduction

Groundwater represents less than 1% of all waters on the Earth, but it accounts for some 90% of fresh water reserves (Stetzenbach et al., 1986; Coates and Achenbach, 2002). In countries such as Germany, USA and others, up to 70% of drinking water originates from groundwater (Ward et al., 1985; Schleyer and Kerndorff, 1992). There is no doubt that for the human population drinking water should be available all day and every day both in a sufficient quantity and good quality. No diseases must be caused in those who drink or in any other way use potable water. Although technologies are available to improve water quality, they can not be used in some areas for different reasons and thus, raw groundwater has to be utilized. Because of this and because an intentional contamination of aquifers by health relevant microorganisms can not be excluded, the fate of such microorganisms in groundwater is of enduring interest. In Germany, a "50 days die off theory"

---

[*] To whom correspondence should be addressed. e-mail: katerina.demnerova@vscht.cz; zdenek.filip@vscht.cz

suggested by Knorr (1951) has been widely accepted, and this has been regularly used as an empirical measure in the sizing of groundwater protection zones. Most of our results do not support this hypothesis.

2. Bacterial survival in groundwater

From the curves shown in Fig. 1, the fate of different test bacteria in groundwater samples, which were sterilized by filtration and kept at an ambient temperature of $10°C \pm 1°C$, can be estimated.

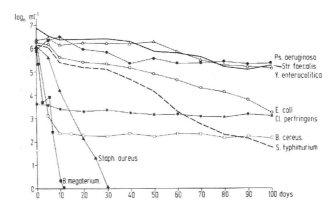

Figure 1. Survival of some health relevant bacteria in groundwater (Filip et al., 1988)

Apparently, no multiplication of the introduced bacteria occurred in the groundwater. Nevertheless, and despite of somewhat declining counts, most of them survived for an extended period of 100 days at quite high cell densities. Only a few bacteria, such as *Bacillus cereus*, *Bacillus megaterium*, and *Staphylococcus aureus* demonstrated strong decreases of 2 and up to 6 log already after 10 days. The survival of *B. megaterium* terminated after 12 days, while *S. aureus* could no longer be detected after 30 days. On the other hand, *B. cereus* the counts of which decreased by some 99.9% within the first 10 days of the experiment, demonstrated rather constant colony counts of 10^2 and 10^3 ml^{-1} up to 100 days. The members of the health important family of *Enterobacteriaceae*, such as *Escherichia coli* and *Salmonella typhimurium*, as well as the majority of other bacteria under testing remained clearly detectable in counts of $10^2–10^6$ ml^{-1} after 100 days.

The survival curve of *S. typhimurium* did not show any difference up to 50 days in experiments with groundwater alone or in those with the addition of sand which was obtained from a deep aquifer. Later on however, up to the 100th day, the number of survivors was about one log unit higher than in groundwater alone (Fig. 2).

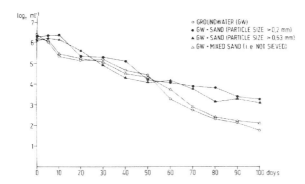

Figure 2. Survival of *S. typhimurium* in groundwater with or without addition of sand (Filip et al., 1988)

Undoubtedly, the survival of bacteria in a groundwater aquifer is affected by a number of physical, chemical, and biological factors. These environments are naturally inhabited by autochthonous bacteria which perform different metabolic activities in accordance with the ambient conditions and substrate available in the underground (Smith, 2002). Usually, these bacteria appear in the cell densities of 10^2–10^4 ml^{-1} in water, and 10^4–10^7 g^{-1} in aquifer sediments (Pelikan, 1983).

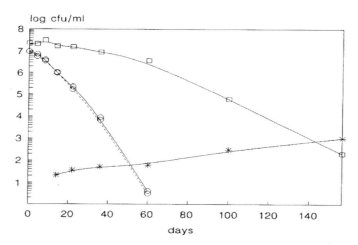

Figure 3. Effect of bacteria autochthonous to groundwater (*) on survival of *E. cloacae* (□), and *E. cloacae* RP4. (○) in presence (dotted line) or absence of different antibiotics (Claus et al., 1992)

According to results obtained in our laboratory experiments, the development of autochthonous and artificially introduced bacteria in groundwater might show a contrary motion, i.e., while the counts of autochthonous bacteria increase, those of the health relevant ones decline. In Fig. 3 this is documented for *Enterobacter cloacae* strains bearing or not a recombinant plasmid RP4. Even more, the plasmid bearing strain became no longer detectable, i.e., died off, after 60 days. These phenomena indicate that a competition might be expected between autochthonous and artificially introduced health relevant bacteria.

TABLE 1. Infrared absorption bands of *Escherichia coli* (Filip et al., 2008)

Band (cm^{-1})	Possible assignments[a]
3,300–3,000	NH$_2$ stretching, e.g., in adenine, cytosine, quanine; H-bonded OH groups
2,960–2,850	C–H stretching in aliphatics of cell walls (fatty acids, carbohydrates)
1,660–1,650	NH$_2$ bending, C=O, C=N stretching (amide I band in proteins)
1,540–1,535	Amide II band (in proteins)
1,460–1,455	C–H deformations of CH$_2$ or CH$_3$ groups in aliphatics
1,396–1,389	C–H bending, –CH$_3$ stretch in fatty acids
1,240–1,234	Amide III band (in proteins)
1,150–1,030	C=O, P=O, P–O–C (P–O–P) asymmetric stretching (glycopeptides, ribose, aliphatic esters)
650–480	C–O–C, P–O–C bonding (phospholipids, RNA, aromatics)

[a]Assignments after Parker (1971), Naumann (2000), and others.

3. FT-IR spectroscopic characterization of *Escherichia coli*

Fourier-Transform Infrared (FT-IR) spectroscopy is a widely utilized technique for chemical analysis (Johnston, 1991). In soil microbial biomass, IR spectroscopy appeared useful in establishing the spectral fingerprints of different cell constituents, such as proteins, lipids, carbohydrates and nucleic acid (Filip, 1978a, b). Time-consuming and difficult isolation of individual cell structural units can be omitted using this technique, when intact cells (biomass) are under investigation. Also, small amounts of biomass (1–2 mg) are sufficient for the analysis, and the spectra can be easily stored in the data files of most instruments. Naumann et al. (1990, 1991) documented the usefulness of FT-IR spectroscopy in bacterial diagnostics, including that of numerous *E. coli* strains. This potentially pathogenic bacterium inhabits mammalian intestine, and thus, it is widely used as an indicator of faecal water pollution. In our investigations (Filip et al., in press), the cell mass of *E. coli* delivered well-differentiated IR-bands (see Table 1).

As shown in Fig. 4, the FT-IR spectra of *E. coli* changed rather markedly with the age in hours of the culture. The same was true for biomass harvested from differently composed nutrient media (not shown in figure). From these results one can conclude that FT-IR spectra, seem only capable of delivering useful and reproducible information on bacteria under testing if the cell mass is obtained under well standardized cultural conditions.

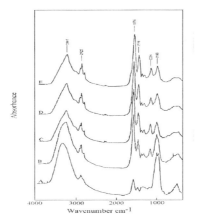

Figure 4. FT-IR spectra of *E. coli* cell mass harvested after (A) 3 h, (B) 16 h, (C) 41 h, and (D) 168 h from a minimum nutrient broth (Filip et al., in press)

References

Claus, H., Rötlich, H. and Filip, Z., 1992: Survival in groundwater of some bacteria with natural and recombinant plasmids. *Microbial Releases*, **1**: 103–110.

Coates, J.D. and Achenbach, L.A., 2002: The biogeochemistry of aquifer systems. In: Hurst, C.J. (Ed.), *Manual of Environmental Microbiology*, Washington, DC: ASM Press, pp. 719–727.

Filip, Z., 1978a: Infrared spectroscopy of two soils and their components. In: Krumbein, W. (Ed.), *Environmental Biogeochemistry and Geomicrobiology*, Ann Arbor, MI: Ann Arbor Sci. Publ., pp. 747–754.

Filip, Z., 1978b: Infrarotspektren der mikrobiellen Biomasse und der Huminsäure im Podzolboden. *Zeitschrift für Pflanzenernährung und Bodenkunde*, **141**: 711–715.

Filip, Z., Hermann, S. and Demnerova, K., 2008: FT-IR spectroscopic characteristics of differently cultivater *Escherichia coli*. Czech J. Food Sci. **26**: 458–463.

Filip, Z., Kaddu-Mulindwa, D. and Milde, G., 1988: Survival of some pathogenic and facultative pathogenic bacteria in groundwater. *Water Science and Technology*, **20**: 227–231.

Johnston, S.F., 1991: *Fourier Transform Infrared – A Constantly Evolving Technology*. New York: Ellis Horwood.

Knorr, M., 1951: Zur hygienischen Beurteilung der Ergänzung und des Schutzes großer Grundwasserverokommen. *Grundwasser F.*, **92**: 104–110.

Naumann, D., 2000: Infrared spectroscopy in microbiology. In: Meyers, R.A. (Ed.), *Encyclopedia of Analytical Chemistry*. Chichester: Wiley, pp. 102–131.

Naumann, D., Helm, D. and Labischinski, H., 1990: Einsatzmöglichkeiten der FT-IR Spektroskopie in Diagnostik und Epidemiologie. *Bundesgesundheitsblatt*, **33**: 387–393.

Naumann, D., Helm, D. and Labischinski, H., 1991: Microbiological characterizations by FT-IR spectroscopy. *Nature*, **351**: 81–82.

Parker, F.S., 1971: *Infrared Spectroscopy in Biochemistry, Biology and Medicine.* London: Hilger.

Pelikan, V., 1983: *Groundwater Protection.* Praha: SNTL (in Czech).

Schleyer, R. and Kerndorff, H., 1992: *Die Grundwasserqualität westdeutscher Trinkwasser-ressourcen.* Weinheim: VCH.

Smith, R.L., 2002: Determining the terminal electron-accepting reactions in the saturated subsurface. In: Hurst, C.J. (Ed.), *Manual of Environmental Microbiology*, Washington, DC: ASM Press, pp. 743–752.

Stetzenbach, L.D., Kelley, L.M. and Sinclair, N.A., 1986: Isolation, identification, and growth of well-water bacteria. *Ground Water*, **24**: 6–10.

Ward, C.H., Giger, W. and McCarty, P.L. (Eds.), 1985: *Ground Water Quality.* New York: Wiley.

WORKING GROUP II: THE THREAT FROM ARMED CONFLICT AND TERRORISM

Chairs and rapporteurs: **S. Arlosoroff[1] (terrorism) and J.A.A. Jones[2]**
[1] *Mekorot, The National Water Corporation of Israel, Tel-Aviv Israel*
[2] *Aberystwyth University and IGU Commission for Water Sustainability, Aberystwyth, UK*

Both terrorism and armed conflict can be threats to water security, either directly or indirectly. Both forms of violence have used disruption or poisoning of water supplies as a weapon, and both may cause collateral damage to water supplies. The distinction between war and terrorism can be arguable. Special Operations forces may use some tactics similar to terrorists, with similar repercussions for water systems, but the main practical distinction is between formal military actions and more informal, smaller scale guerrilla-style activities that may be directed more at civilian personnel and designed to engender fear and panic as much as specific damage.

1. Armed conflict

1.1. RISK ASSESSMENT

It is generally difficult to assess risk on anything but a short-term basis. But it is clearly possible to undertake some preventative measures. These will be similar in the main to those for terrorism, although heavy bombing and shelling disrupt water distribution and sanitation systems more and are likely to leave a more long-lasting "war footprint", specially in the environment, e.g. in pollution from explosives or materials released from the plant under attack. Examples might be depleted uranium from shells or PCBs from electricity stations, e.g. in Bosnia. Mining or bombing of dams, distribution pipelines and treatment plants, the disablement of management systems, whether human organisational or computer control systems, and general environmental pollution are the main risks.

1.2. EMERGENCY RESPONSE

This is the key area for action and improvement. The present situation seems rather inadequate and amateur, especially in leaving a large part of the work to NGOs like the Red Cross and Red Crescent. We recommend the following:

1. Creating special sections of the military or under military control that are more specialised in water and sanitation provision and restoration than the current military engineers, in order to take the strain off the NGOs and to enter the arena before it is safe for the NGOs. NGOs have a multitude of other interests and are funded on a non-professional basis, mainly by donations.
2. A pre-existing emergency plan should be developed wherever possible with community leaders organising the initial response.
3. The military should take overall control of restoration as soon as possible.
4. Finally, perhaps in the long-term the UN might develop a formal and specific set of rules for responsibilities concerning protection and restoration of water supply and sanitation in armed conflicts.

1.3. PRIORITIES FOR EMERGENCY SERVICES

The overall aim must be two-fold: (i) to provide adequate volumes of water for drinking and, most importantly, sanitation, and (ii) to prevent the outbreak of epidemics of disease and poisoning, e.g. from chemical, bacteriological, viral toxins or petroleum products in the water. We recommend the following order of priority:

1. *Provision of any water is the first priority* – quantity is more important than quality especially for use in sanitation rather than for drinking, as the most immediate danger is from disease caused by inadequate sanitation facilities.
2. *Continuity of supplies of both water and power is the second* most essential element. Restoring and securing electricity supplies is essential for pumps and control systems.
3. Provisions should be *extended to the whole of the population that has been affected*, whether by restoring public water supply networks or by mobile supplies.
4. *Attend to water quality* if the emergency supply was unsafe. This may involve importing freshwater in bottles or containers and provision of portable water treatment systems. But part of the community plan should be:
   (a) *Education of the people* in what to do in an emergency. Advise should be provided prior to or as early as possible in a conflict. This should comprise advise on (i) storage of water, including refrigeration where possible, (ii) purifying and disinfecting their own water by boiling (10 min at 100°C kills most biological vectors; pasteurisation at 60°C for 10 min removes living bacteria but not spores), other measures like exposure to ultraviolet light in sunlight or use of chlorine

or dilute bleach to kill pathogens, filtering for helminths and other parasites, and also for sediments as bacteria and viruses tend to adhere to surfaces in multiple layers that make some inaccessible to UV irradiation, and (iii) the dangers of specific diseases. The EU standard for safe water is 100 bacteria/ml, but safe levels for viruses vary: one virus may be enough if it spreads or invades bacteria which then spread their transformed cells. Water temperatures of 10–30°C are typical optimum temperatures for proliferation. The worst sources for emergency water supplies, especially in warm climates, are surface and shallow groundwater because of the danger of faecal pollution. Typical shallow groundwater in Africa, from a metre or two in depth is likely to be warm enough for bacterial proliferation, whereas in central Europe it is likely to be cooler than 10°C, especially in winter, so that *E. coli* (a good indicator species) cannot multiply. Somewhat deeper groundwater will be cooler and will have been filtered more effectively during a longer period of percolation.

(b) The *availability of cheap, basic equipment* with which to do this. Filtering is best done with a sub-2 μm membrane rather than simple filter paper. This will also remove many pathogens.

1.4. ENVIRONMENTAL REHABILITATION

This must generally await cessation of hostilities. It will begin with the long-term reconstruction of permanent storage, distribution and sewage networks and treatment plants. Even longer-term activities will involve cleanup, likely to be mainly chemical, of the remaining products of war in soils and water bodies. Again, we recommend that this should be made part of the formal responsibilities of combatants.

1.5. SUMMARY STATEMENT ON DEALING WITH IMPACTS OF ARMED CONFLICT

1. Protection and reconstruction of water supply systems is a priority and community plans should be in place prior to hostilities wherever possible. The military should have responsibility for securing water supplies before it is safe for aid agencies. Consideration should be given to establishing clear UN rules on responsibilities for the protection and reconstruction of supplies.

2. Instruct public in treatment methods and distribute emergency equipment at the earliest opportunity.

2. Terrorism

2.1. RISK ASSESSMENT

A terrorist attack can occur anywhere, any time, perpetrated by a small cell without any links to foreign commands or international financial transfers. Many attempts have been made to poison water resources, not just since 9/11. The main result of such an event is not just the cases of mortality or morbidity, but the panic caused. An extreme case of panic could cause mass infection by people evading quarantine spreading the infection nationally or even globally (cp. SARS and Avian flu).

There are three broad types of risk:

1. Disruption of the public water supply system – a quantity issue
2. Poisoning of the PWS – a quality issue
3. Disruption of sewage systems – not so effective and of lesser significance

Terrorists generally have less powerful mechanisms available to them than national armies. Fear and panic tend to be a major aim of terrorist actions; widespread death and illness are generally of secondary importance and often beyond their capability.

2.2. METHODS OF ATTACK DISCUSSED

These included:

1. Chemical, biological, radiological and physical. The first three fall broadly under the heading of poisoning. The last involves disruption and destruction of water supply and sewage networks, *including computer control systems*.
2. Many simple methods are available, but anthrax, smallpox or radioactive materials like polonium and caesium 137 are not so easily obtained.
3. Mixtures of agents will be more difficult to eliminate than single agents.

Chemical agents are the easiest to get and use. Biological agents like myco-toxins are harder to get and more expensive, and the effects may be delayed for weeks.

2.3. EMERGENCY RESPONSE

Devising effective counter-measures is complicated by the fact that the impacts may be long, medium or short term. However, chemical, biological and radiological events are generally similar to other sources of water pollution, so that many counter-measures will serve as protection for both normal risks and terrorist attacks. It is impossible to continually monitor every part of the system. There is especial danger from "backwater" introduction of agents. A small pump fitted to a domestic water tap can counteract the mains water pressure and allow agents to flow backwards into the system. Irish and Scottish terrorists threatened backwater introduction via street fire hydrants Jones (this volume) in 1999 and 2007, respectively.

2.4. RECOMMENDATIONS

1. Priority must be given to *intelligence information*, especially that derived from intercepting electronic traffic, and to *continuous online monitoring*. Further development of online monitoring systems is needed to increase effectiveness and the number of parameters covered, including DNA analysis, luminescence and live animals, and to reduce the currently very high costs ($100,000 per instrument). This has an important feedback to WG3 (this volume) in terms of the possible willingness or ability of private water companies to afford the equipment – who should control security measures?
2. Remove sensitive information on water systems from the public domain, e.g. from websites.
3. Training of medical community in identifying signs of an attack, e.g. increases in diseases and disorders (cp. American College of Preventative Medicine online guide, sponsored by the EPA and American Water Works Association).
4. Education of the public, raising their awareness of the risks, and their potential role at all stages from early warning to safe response. This is an area where Nato Science could play a unique role in an educational campaign. One suggestion might be to promote local or home storage of freshwater for emergency purposes in areas at risk.
5. Improving cyber security and vetting of employees – disablement of computer control systems by hackers is an increasing threat to automated systems.
6. Wider adoption of national agencies responsible for coordinating anti-terrorist activities, as in the USA and UK. The EPA works with other federal agencies, like the Centers for Disease Control and Prevention, the FBI, and Department of Defense, and with the water sector organizations, like the Water Environment Research Foundation, to develop water security technologies and protocols for detection, treatment and response, and to disseminate information to stakeholders. The EPA's Drinking Water Laboratory Response Preparedness Project aims to develop regional incident response plans. Israel also continues to make serious efforts to protect water supplies.

7. Improved physical protection of facilities – fencing, security cameras, etc.
8. Tightening of global money flow controls.
9. Developing improved and cheaper domestic-scale in-house safeguarding systems, like UV irradiation and carbon filters with pre-programmed automatic responses when overloaded or used up.

Online protection should include remote control for double dose chlorination, and diagnoses of swimming patterns of fish, especially *Gnatonemus Petersi* (Elephant fish), and *Daphnia*. Fluorescein can aid observation of bacteria and *Daphnia*. Expansion of real time DNA-series identification for early warning of biological agents is recommended, as funded by Nato and the US Homeland Security Office.

Poisoning by drinking polonium-200 or caesium-137 seems to be much less effective in large water systems because of the large dilution effect. But water contaminated with radioactive elements could be sprayed from a truck and cause panic as well as many casualties. Water soluble caesium chloride, obtainable in powder from hospital supplies, is very dangerous and there is currently very little awareness of the problem. There is a danger of inadvertently giving terrorists ideas.

Although it is not easily obtained, anthrax could cause widespread deaths and persist for ages. On the other hand, *Botulinum* bacteria can be found naturally and botulism is a powerful, short-term infection that could kill thousands.

2.5. SUMMARY STATEMENT ON TERRORISM

1. Awareness is the key to prevention of attacks, including prior intelligence and engendering awareness amongst the general population.
2. Avoiding panic, preparing the people for possible events and careful restriction of information to the public. Aim to avoid "waking sleeping dogs" by giving information to the enemy that may help them execute the attack.
3. Install systems for continuous monitoring of water quality.
4. Long term aim should be removing the causes of terrorism.

PART III:

**MANAGING EXTREME EVENTS
AND CLIMATE CHANGE**

CLIMATE CHANGE, GLACIER RETREAT, AND WATER AVAILABILITY IN THE CAUCASUS REGION

M. Shahgedanova[1*], **W. Hagg**[2], **D. Hassell**[3], **C.R. Stokes**[4]
and V. Popovnin[5]
[1] Department of Geography, The University of Reading, Whiteknights, Reading, UK
[2] Department of Geography, University of Munich, Munich, Germany
[3] The Hadley Centre, UK Met Office, Exeter, UK
[4] Department of Geography, The University of Durham, South Road, Durham, UK
[5] Faculty of Geography, Moscow State University, Leninskie Gory, Moscow, Russia

Abstract. The paper discusses the observed and projected warming in the Caucasus region and its implications for glacier melt, water availability and potential hazards. A strong positive trend in summer air temperatures of $0.05°C$ year^{-1} is observed in the high-altitude areas (above 2000 m) providing for a strong glacier melt. A widespread glacier retreat has also been reported between 1985 and 2000, with an average rate of 8 m year^{-1}. A warming of 5–7°C is projected for the summer months in the 2071–2100 period under the A2 emission group of scenarios, suggesting that enhanced glacier melt and a changing water balance can be expected.

Keywords: Glaciers, climate change, climate modeling, water resources, Caucasus

1. Introduction

The Caucasus region is a mountainous country divided into the Greater Caucasus, stretching for 1,300 km from the Black Sea in the west and the Caspian Sea in the east, and the region known as Transcaucasia, located further south. The Greater Caucasus spans the border between Russia and Georgia. Transcaucasia covers Georgia, Armenia, and Azerbaijan. Elevations in the Greater Caucasus exceed 5,000 m above mean sea level (a.m.s.l.) and about 1,600 km^2 of this area is glaciated (Stokes et al., 2006). Transcaucasia is a mosaic of mountains, plateaux, and lowlands with vastly different climates ranging from semi-arid, with annual precipitation about 200 mm, to very humid, with annual precipitation in excess of 3,000 mm (Volodicheva, 2002). Water availability varies immensely across the region: most of the Greater Caucasus and western Georgia have either sufficient or abundant

---

[*] To whom correspondence should be addressed. e-mail: m.shahgedanova@ reading.ac.uk

water resources while most of Armenia and Azerbajan are water deficient. However, diverse water resources across the Caucasus have one feature in common: most rivers originate in the mountains and glacier and snow melt are important sources of their nourishment.

It is widely acknowledged that glaciers are sensitive to climate change. Across the world, glaciers are retreating and perennial snow packs in the mountains are declining in response to the observed climatic warming (Oerlemans, 2005). The glaciated and snow-covered mountains are the 'water towers' of the region providing resources for irrigation, industrial, and domestic use. Therefore, changes in ice and snow cover are of great importance for regional water security. Potentially associated with rapid melt, are hazards such as outbursts of glacial lakes and mud flows and these are intensifying in many mountainous regions (Kääb et al., 2005).

This paper discusses the observed and projected changes in temperature and precipitation of the high-altitude areas of the Caucasus region and the likely impacts of these changes on glaciers and runoff. The following questions are addressed: (i) how did air temperature and precipitation change in the Caucasus Mountains between the 20th and early 21st centuries?; (ii) how are glaciers responding to the observed changes in climate?; (iii) what are the implications of glacier retreat with regard to hazards potentially associated with glacier melt?; (iv) how will climate of the region change in the future?; and (v) how will these changes affect regional water balance?

A variety of techniques, including field observations, remote sensing, regional climate modelling and hydrological modelling are employed in the attempt to resolve these questions. The paper begins with a brief analysis of the changes in air temperature, precipitation, glacier mass balance and the spatial extent of glaciers in the region. It proceeds with a discussion of regional climate change scenarios developed using regional climate model (RCM) PRECIS and presents the attempts at modelling runoff in glaciated regions of the Greater Caucasus.

2. Data and methods

The following observational data have been used in this study:

(i) Air temperature and precipitation records from two high-altitude stations were used for the analysis of the long-term trends, validation of the climate and calibration of the hydrological models: Terskol located in the glaciated region of the Greater Caucasus (43°15'N; 42°33'E; 2141 m a.m.s.l) and Aragats located further south (40°29'N; 44°11'E; 3227 m a.m.s.l).

(ii) Glacier mass balance data collected at Djankuat, a small valley glacier located at 43°12'N and 42°46'E on the northern slope of the central section of the Glavny Ridge, the most heavily glaciated area in the Greater Caucasus (Fig. 1). Its elevation lies between 2,700 m and 3,900 m a.m.s.l. and its meltwater eventually drains into the Caspian Sea via the Baksan and Terek Rivers. Typical of the Caucasus, Djankuat is a temperate glacier. In 2000, its surface area was 3.01 km^2. Mass balance measurements, reported as millimetres of water equivalent (mm w.e.), refer to the mass balance year beginning in October and ending in September of the following calendar year. Two components are measured: October–May accumulation and June–September ablation or melt (Shahgedanova et al., 2005). The measurements began in 1967 and are ongoing.

(iii) Discharge measurements for calibration of the hydrological model have been obtained from the Tyrnyauz gauging station located on the Baksan River, in its upper reaches (1,281 m a.m.s.l.). The Terskol and Tyrnauz stations and Djankuat glacier are located within approximately 20 km of each other.

Satellite imagery (Landsat TM and ETM+ scenes obtained at the end of summer in 1985, 1991 and 2000) has been used to evaluate glacier retreat in the Greater Caucasus. Detailed description of the methodology is given in Stokes et al. (2006).

Regional climate change scenarios have been developed using an RCM 'PRECIS' (PREdicting Climate for Impact Studies) developed by the UK Met Office (Jones et al., 2004). PRECIS has a horizontal resolution of 25 km. It derives lateral boundary conditions (LBC) from HadAM3P, a global atmosphere-only model with a resolution of 150 km, which is in turn forced by surface boundary conditions from the global circulation model HadCM3. PRECIS has been specifically developed to generate climate change scenarios for areas with complex terrain and has been used in high-altitude areas previously (Frei et al., 2003). Two integrations have been performed for two time slices (i) 1961–1990 providing 'baseline' climate and (ii) 2071–2100 providing the future climate change regional projections based on the aggressive A2 CO_2 emission change family of scenarios (SRES, 2001).

Runoff in the Baksan catchment has been simulated using a conceptual runoff model HBV-ETH originally developed for hydrological forecasting on Swedish lowlands (Bergström, 1976) and extended for the application in glaciated catchments (Braun and Renner, 1992). The model works on a daily time-step and has a very low data demand: daily means of precipitation and air temperature, and daily discharge for calibration. The distribution of area by altitude and aspect classes, for the total catchment and for glaciated parts, is needed and has been derived from a digital elevation model (DEM).

A detailed description of HBV-ETH is given in Hagg et al. (2006). The model shows a robust performance and was successfully used in the Alps (Braun et al., 2000), Tian Shan (Hagg and Braun, 2005), Altay (Hagg et al., 2007), and Himalayas (Konz et al., 2006).

3. The observed climate and glacier change in the Caucasus Mountains

There are about 2,000 glaciers in the Greater Caucasus located predominantly in its central part (Fig. 1). Glacier melt in the Caucasus is particularly strong in June, July and August (JJA) and variations in JJA temperatures are of particular importance. The long-term observations confirm that these are rising (Fig. 2). A strong increase in JJA temperatures has been observed at Terskol and Aragats since the late 1960s. At both sites, in the last 40 years, JJA temperatures have been increasing at a rate of 0.05°C per year and there is strong positive linear trend in the time series explaining 36% of the total variance. The Aragats record, dating back to 1929, confirms that the last 2 decades were the warmest in almost 80 years of observations. This warming is making a profound impact on glaciers. Glacier melt has intensified in the 1990s and its highest values have been recorded in the last two decades (Fig. 3). Thus, in the summers of 1998, 2000, and 2007 ablation exceeded the long-term average by two standard deviations from the long-term mean, reaching its highest value on record in 2007.

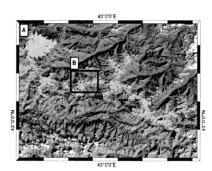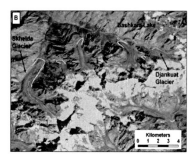

Figure 1. Landsat ETM+ satellite image of Main Caucasus Ridge (A) aligned north-west to south-east. This band combination (RGB, 5, 4 and 3) colours glaciers as bright blue and vegetation as green. Note the large glaciated area of Mt Elbrus in the top left. Clouds appear as white. The inset map in (B) shows six glaciers in the Adylsu Valley and their terminus positions in 1985 (red) and 2000 (yellow). All the glaciers have retreated and the maximum retreat rate is reported for the Skhelda Glacier (~350 m). Also note the large unvegetated proglacial area in front of the glaciers that delimits their most recent maximum at the LIA

Changes in the cold-season precipitation (Fig. 2) do not offset impacts of the observed summer warming. At Terskol, positive anomalies observed in the individual years do not compensate for a rapidly increasing melt.

As a result, the cumulative mass balance of Djankuat is declining and has reached its lowest value on record in the mass balance year of 2006/2007 (Fig. 3). At Aragats, the cold-season precipitation has been declining from

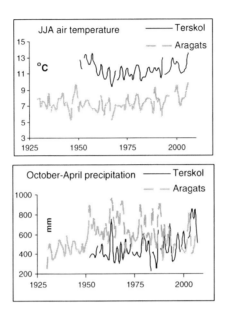

Figure 2. JJA air temperature and October–April precipitation at Terskol and Aragats

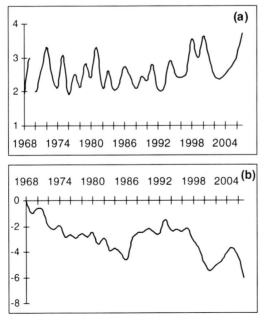

Figure 3. Seasonal ablation totals (**a**) and cumulative mass balance (**b**) of Djankuat glacier (m w. e.) between 1968 and 2007

its maximum recorded in the 1960s (Fig. 2). This trend has been especially strong in the last 2 decades possibly in response to the strongly positive phase of the North Atlantic Oscillation (NAO).

Analysis of changes in the extent of 113 glaciers in the central Greater Caucasus (Fig. 1) between 1985 and 2000 has shown that 94% of the glaciers have retreated (Stokes et al., 2006). The mean retreat rate was 8 m year^{-1} and maximum rates approached ~38 m year^{-1}. The termini of most glaciers have retreated by 50–150 m but the maximum retreat exceeded 500 m. Glaciers whose snouts are positioned at lower elevations exhibited stronger retreat; those at higher elevations were less sensitive to the warming climate. Fig. 1b illustrates changes in the extent of six neighbouring glaciers including Djankuat. The Skhelda glacier retreated by 350 m in just 15 years. The satellite imagery also reveals the inferred maximum limit of the Shkhelda Glacier at the end of the Little Ice Age (LIA). This limit is revealed by the 'fresh' unvegetated proglacial zone in front of the snout. Although the timing of the LIA limit is somewhat complex in the Caucasus Mountains (two possible maximal positions at AD 1300 and AD 1700) it can be seen that for the Skhelda, the recent retreat represents a significant proportion (~25%) of retreat since the most recent maximum. This is common for many other glaciers (Stokes et al., 2006) and suggests that the rate of retreat has increased in recent years.

Glacier retreat in the Caucasus is accompanied by the formation and expansion of glacial lakes (Stokes et al., 2007). Figure 1b shows that the Bashkara glacier terminates in a lake. It formed in 1995. Currently, it measures about 200 m across and its depth exceeds 15 m. The total number of lakes and their surface areas have increased dramatically in the Greater Caucasus from 16 in 1985 to 24 in 2000, representing a 57% increase in total lake area coverage (Stokes et al., 2007). Given that many of the glacial catchments feed directly into larger valleys, which house settlements and towns, this is a concern. Indeed, in August 2006 (the second warmest summer since 1951 in the central Greater Caucasus; Fig. 2), an outburst of a glacial lake, Birdjaly-Charan, located on the north-eastern slope of Mt. Elbrus, occurred in response to the enhance melting of the Chungurchat-Chiran glacier. It released 400,000 m^3 of water over 2 days (Chernomorets et al., 2007) leading to a substantial loss of property and infrastructure. This event emphasizes the vulnerability of human activities to glacier-related hazards, which are likely to intensify in response to the climatic warming.

4. Future climatic changes in the Caucasus region

Regional climate change scenarios have been developed using PRECIS to quantify changes in the future climate (2071–2100) relative to the climate of the baseline period (1961–1990). The projected changes in air temperature and precipitation in the Greater Caucasus and in a region encompassing Armenia and adjacent areas are shown in Fig. 4. These projections are for the A2 group of scenarios, which assume a rapid increase in the world population to 15.1 billion and an increase in mean CO_2 concentrations to 834 ppm by 2100, representing an aggressive increase in the emissions of greenhouse gases and global temperature (SRES, 2001). The model outputs have been validated against the observations from individual weather stations conducted during the baseline period of 1961–1990 (Fig. 5). The meteorological network is not dense in the high-altitude region of the Greater Caucasus and only eight stations were used in model validation. This spatial sampling allows one to judge the quality of air temperature simulations but evaluation of precipitation simulations requires higher density observations, particularly in the context of highly variable terrain (Frei et al., 2003). Armenia has a much denser network and 34 stations were used, insuring good spatial representativity. The model performed well with regard to air temperature across the modelling domain. The largest difference between the modelled and the observed data occurred in the spring–summer months and was about 2°. Much lower errors occurred in all other months. The quality of precipitation simulations varied across the region. The annual precipitation cycle has been reliably simulated for the semi-arid regions of Transcaucasia and in the areas where frontal precipitation maxima occur (e.g. the Black Sea coast). In the glaciated region, where a convective precipitation maximum is observed in summer, results were characterised by a greater bias. The winter precipitation rates have been overestimated and summer rates have been considerably underestimated by the model. An additional bias correction of simulated precipitation is required to make the data fully suitable for the use in hydrological modelling.

Our results show that air temperatures will increase across the Caucasus region. Most importantly for glacier and snow melt, the 30-year averages of air temperatures of the warm season (May–August) in 2071–2100 will exceed the 1961–1990 averages by 6–7°C resulting in the enhanced glacier and snow melt. The September–August air temperatures are estimated to increase by 3–5°C. By contrast, little or no change in winter precipitation is projected (Fig. 4). A 10% increase in the cold season precipitation is required to compensate for a 1°C warming if glaciers are to remain stable (Braithwaite et al., 2003). Our projections indicate that there will be no compensating change in winter precipitation under the A2 scenarios.

Two other important changes may facilitate the demise of the glaciers and reduction in perennial snow pack: (i) reduction in the length of the period with negative air temperatures and (ii) declining precipitation in April–June. The former will result in a shorter accumulation season and

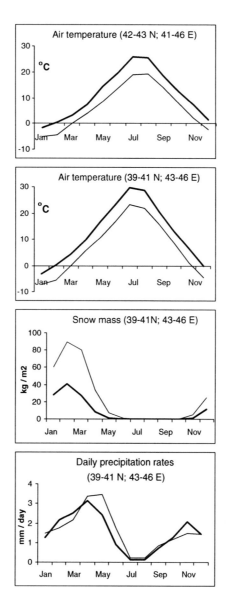

Figure 4. The 1961–1990 (thin line) and 2071–2100 (bold line) climatic averages for the Greater Caucasus (42–43°N; 41–46°E) and Armenia and adjacent regions (39–41°N; 43–46°E)

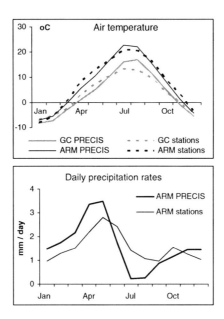

Figure 5. Modelled versus observed data for 1961–1990 for the glaciated region of the Greater Caucasus (GC; 42.5–43.3°N; 42–44.5°W) and Armenia and the adjacent regions (ARM; 39–41°N; 43–46°E)

lower accumulation totals. Although the bulk of snow mass is accumulated on the Caucasus glaciers in December–January (Shahgedanova et al., 2005), snowfalls in October, November, and March currently also make a significant contribution (snowfalls in these months are responsible for the recently observed peaks in Terskol precipitation; Fig. 2). A growing frequency of rainfall events will negatively affect glacier mass balance. Indeed, the modelling results show that snow mass will become markedly lower at the end of the 21st century (Fig. 4). A drier and warmer spring will result in longer and more intense ablation seasons. In addition to air temperature, precipitation is an important factor controlling melt through its effects on the albedo of glacier surface. On Djankuat, there is a close negative correlation ($r = -0.58$) between daily ablation rates and the albedo of the snow-covered surface (Shahgedanova et al., 2007). Low precipitation results in soiled snow and reduction of surface albedo forcing more intensive melt. It is a stronger melt of Djankuat in May 2007, which was the warmest May on record since 1951 and was preceded by a dry spring, that resulted in much faster transition from snow-covered glacier surface to bare ice (at the end of June instead of the mid-August) and reduction in the albedo from 0.4–0.9 to 0.2–0.3. As a result, ablation was the highest on record (Fig. 3) although the JJA temperatures were not record-breaking.

5. Application of a conceptual runoff model in the Baksan river basin

The HBV-ETH model has been set up to model components of water balance in the Baksan River basin. Model outputs are daily and mean values of all terms of the water balance (Table 1). Discharge measurements from Tyrnyauz and air temperature and precipitation data from Terskol, located within the modelled watershed, have been used for the 1991/1992–1999/2000 period for which both types of data were available. Topographic model input has been derived from a SRTM DEM with a ground resolution of 90 m. The size of the catchment has been determined by this DEM as 1,090 km². Glacier outlines have been taken from the Global Land Ice Monitoring from Space (GLIMS) database (Armstrong et al., 2005). Total glaciated area was 182 km² or 17% of the catchment. The distribution of the area by altitude and exposition classes (North, South, East-West-Horizontal) for the whole catchment and its glaciated parts were determined using GIS (Fig. 6a). Model calibration was executed by optimisation runs where free parameters were tuned by comparing modelled and measured discharge. Each run produced ten iterations for a pair of two parameters. For each combination of parameter values, the model efficiency criterion after Nash and Sutcliffe (R^2) was calculated as a numerical criterion for the goodness of fit. Step by step, the discrepancy between modelled and measured discharge

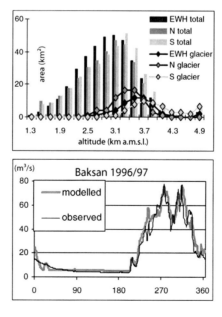

Figure 6. (**a**) Distribution of area by altitude and aspect classes (EWH: east-west-horizontal, N: north, S: south) for the whole catchment of the Tyrnyauz gauging station and its glaciated parts. (**b**) Example of hydrographs of the modelled and measured discharge

was minimized. Worldwide testing of conceptual runoff models (Rango, 1982) has shown that R^2 values exceeding 0.8 should be considered as 'good' in a high mountain terrain. For the 1991–2000 period, the mean R^2 value is 0.82 (Fig. 6b).

The HBV-ETH model calculates all terms of the hydrological cycle and the mean values for the terms of the water balance in the calibration decade (Table 1). These results are valuable estimates helping to understand hydrological processes in the Caucasus Mountains at present. As a following step, the results of the PRECIS climate model will be implemented in the hydrological simulations after bias correction in the modelled climate data and runoff scenarios for future climatic and glacial conditions will be generated.

TABLE 1. Mean terms of the water balance (1991/1992–1999/2000) as simulated by the HBV-ETH model and measured at Tyrnyauz gauging station (in mm)

Basin precipitation	815
Evapotranspiration	125
Glacier mass balance	–76
Snow storage changes	68
Ground storage changes	50
Modelled runoff	640
Measured runoff	616

6. Conclusions

A strong climatic warming is observed in the high altitude areas of the Caucasus and JJA temperatures are growing at a rate of 0.05°C year^{-1}. The last 20 summers were the warmest in 80 years in the high-altitude areas of Transcaucasia. At the same time, winter precipitation has remained the same (Greater Caucasus) or is declining (Transcaucasia). The observed climatic changes have already resulted in glacier retreat. Glacier mass balance is negative and is declining in response to the intensifying summer melt. This is indicated by the long-term mass balance record from Djankuat glacier and the widespread pattern of glacier retreat observed on satellite imagery. This retreat is accompanied by the expansion of glacial lakes. Lake outbursts have already occurred in response to the very high summer temperatures and melt rates in 2006. Future climate scenarios (based on the A2 group of CO_2 scenarios) indicate that summers will be 5–7°C warmer in 2071–2100 than in 1961–1990 and that winter precipitation will experience little or no increase. Solid precipitation is likely to be replaced by rainfall in the autumn and late spring. It is expected that glacier retreat will continue and small glaciers at lower elevations will be most vulnerable to complete disappearance by the end of the 21st century. The HBV-ETH hydrological model has been set up to quantify these changes and hydrological projections will be available following bias correction in modelled precipitation.

Acknowledgements This work was supported by the following grants: UK Royal Society (2005/R2JP); EU INTAS (06-1000017-8608); Russian Foundation of Basic Research (06-05-64094a); Programme on the Leading Scientific Schools of Russia (4861.2006.5). Meteorological data for Armenia have been kindly provided by Dr. H. Melkonyan of the Armenian Hydrometeorological Service.

References

Armstrong, R., Raup, B., Khalsa, S.J.S., Barry, R., Kargel, J., Helm, C., and Kieffer, H., 2005: *GLIMS glacier database*. Boulder, CO: National Snow and Ice Data Center.

Bergström, S., 1976: *Development and Application of a Conceptual Runoff Model for Scandinavian Catchments*. Department of Water Resources Engineering, University of Lund, Bulletin A/52, 134 S.

Braithwaite, R.J., Zhang Y. and Raper S.C.B., 2003: Temperature sensitivity of the mass balance of mountain glaciers and ice caps as a climatological characteristic. *Zeit. Gletscherkun. Glazialgeol.*, **38**: 35–61.

Braun, L.N. & Renner, C.B., 1992: Applications of a conceptual runoff model in different physiographic regions of Switzerland. *Hydrol. Sci. J.*, **73/3**: 217–231.

Braun, L. N., Weber, M. & Schulz, M., 2000: Consequences of climate change for runoff from Alpine regions. *Ann. Glaciol.*, **31**: 19–25.

Chernomorets, S., Petrakov, D., Tutubalina, O., Seinova, I. and Krylenko, I., 2007: Outburst of a glacial lake on the north-eastern slope of the Elbrus on 11 August 2006. *Data Glaciologic. Res.*, **102**: 219–224. In Russian.

Frei, C., Christensen, J.H., Deque, M., Jacob, D., Jones, R.G., and Vidale P.L., 2003: Daily precipitation rates in regional climate models: Evaluation and intercomparison for the European Alps, *J. Geophys. Res.*, **18(D3)**: D4124, doi: 10.1029/2002JD002287.

Hagg, W., Braun, L.N., 2005: The influence of glacier retreat on water yield from high mountain areas: Comparison of Alps and Central Asia. In: De Jong, C., Ranzi, R., Collins, D. (eds.) *Climate and Hydrology in Mountain Areas*, ISBN: 978-0-470-85814-1. Wiley: Chichester, 263–275.

Hagg, W., Braun, L.N., Weber, M., and Becht, M., 2006: Runoff modelling in glacierized Central Asian catchments for present-day and future climate. *Nord. Hydrol.*, **37/2**: 93–105.

Hagg, W., Braun, L.N., Kuhn, M., and Nesgaard, T.I., 2007: Modelling of hydrological response to climate change in glacierized Central Asian catchments. *J. Hydrol.*, **332**: 40–53.

Jones, R.G., Noguer, M., Hassell, D.C., Hudson, D., Wilson, S.S., Jenkins, G.J. and Mitchell, J.F.B., 2004: *Generating high resolution climate change scenarios using PRECIS*, Met Office Hadley Centre, Exeter, UK, 40pp.

Kääb, A., Huggel, C., Fischer, L., Guex, S., Paul, F., Roer, I., Salzmann, N., Schlaefli, S., Schmutz, K., Schneider, D., Strozzi, T., and Weidmann, Y., 2005: Remote sensing of glacier- and permafrost-related hazards in high mountains: An overview. *Nat. Haz. Earth Syst. Sci.*, **5**: 527–554.

Konz, M., Braun, L., Grabs, W., Shrestha, A. and Uhlenbrook, S., 2006: Runoff from Nepalese Headwater Catchments - Measurements and Modelling. *IHP/HWRP-Berichte, Heft 4, Koblenz*, ISSN 1614-1180, 160pp.

Oerlemans, J., 2005: Extracting climate signal from 169 glacier records. *Science*, **308**: 675–677.

Rango, A., 1992: Worldwide testing of the snowmelt runoff model with applications for predicting the effects of climate change. *Nord. Hydrol.*, **23**: 155–172.

Shahgedanova, M., Stokes, C.R., Gurney, S.D. and Popovnin, V.V., 2005: Interactions Between Mass Balance, Atmospheric Circulation and Recent Climate Change on the Djankuat Glacier, Caucasus Mountains, Russia. *J. Geophys. Res.*, **110(D4)**: D04108, doi: 10.1029/2004JD005213.

Shahgedanova, M., Popovnin, V., Aleynikov, A., Petrakov, D. and Stokes, C.R., 2007: Long-term Change, Inter-annual, and Intra-seasonal Variability in Climate and Glacier Mass Balance in the Central Greater Caucasus, Russia. *Ann. Glaciol.*, **46**: 355–361.

SRES, 2001: *IPCC Special Report on Emission Scenarios*. IPCC. Geneva.

Stokes, C.R., Gurney, S.D., Shahgedanova, M., Popovnin, V., 2006: Late-20th-century changes in glacier extent in the Caucasus Mountains, Russia/Georgia, *J. Glaciol.*, **52**: 99–109.

Stokes, C.R., Popovnin, V., A. Aleynikov, and Shahgedanova, M., 2007: *Recent Glacier Retreat in the Caucasus Mountains, Russia, and Associated Changes* in Supraglacial Debris Cover and Supra/proglacial Lake Development. *Ann. Glaciol.*, **46**: 196–203.

Volodicheva, N.A., 2002: The Caucasus. In: Shahgedanova, M. (Ed.), *The Physical Geography of Northern Eurasia*. Oxford University Press, Oxford, 284–313.

RISK MANAGEMENT AND MITIGATION IN HIGHLY URBANIZED RIVER BASINS

S. Pagliara[*]
Department of Civil Engineering, University of Pisa, Pisa, Italy

Abstract. Aim of the work is to present examples of flood risk management and mitigation in central Italy. Many studies have been conducted in this region in order to cope with flood risk. Different aspects are still open, particularly, for rivers and tributaries that pass through highly urbanized areas where the risk management and its mitigation could be complex. The work analyses interesting case studies of complex river basins in which the hazard maps are derived from many assumptions and in which a temporal evolution of the flood is connected with the presence of civil protection plans.

The case of the Camaiore River and its tributaries will be analyzed together with the case of the area enclosed between the Arno and the Bisenzio rivers. The last part of the paper will deal with the mitigation measures that can be foreseen in order to diminish the risk.

Keywords: Flood risk, mitigation, flood defense, flood management

1. Introduction

Flooding of urban areas is a frequent disaster (Yen, 1995). In Italy, a large amount of work has been done in order to cope with floods in the last 20 years. With the national law n.183/1989 the concept of river security at a basin level has been introduced. Further, the flood disasters in Tuscany with the Versilia River (1996), in Campania with the Sarno River (1998) and in Calabria at Soverato in the year 2000 have focussed attention on the necessity for developing specific master plans regarding the hydraulic risk for the whole nation. In this sense, many River Basin Authorities have approved a master plan of the structural and non-structural measures to reduce the hydraulic risk and the Hydro-geological Asset Plan (P.A.I.).

The P.A.I. today represents the synthesis of the knowledge sharing among all the stakeholders involved in the government and in land management issues. Actually, in Italy the acceptable risk level corresponds to a flood with a return period $T_r = 200$ years and the P.A.I. must contain the *flood risk maps* for different return period up to 500 years.

---

[*] To whom correspondence should be addressed. e-mail: s.pagliara@ing.unipi.it

J.A.A. Jones et al. (eds.), *Threats to Global Water Security*,

In Tuscany, all the River Authorities, both regional and national, have approved the P.A.I. with major consequences for urban planning and the urbanization process. In fact, the norms connected with the P.A.I. foresee considerable limitations of the urbanization process depending on the specific risk level of the area.

1.1. ZONING OF THE HYDRAULIC HAZARD

The hazard maps are different according to the different river basin authorities. In mapping the hydraulic hazard the Arno River Authority distinguishes between four levels:

(a) Very high hydraulic hazard (PI4): covering areas flooded with return period $T_r \leq 30$ years with water depths h ≥ 30 cm
(b) High hydraulic hazard (PI3) covering areas flooded with return period $T_r \leq 30$ years with water depths h <30 cm and areas flooded with return period $30 < T_r \leq 100$ years with water depths h ≥ 30 cm
(c) Medium hydraulic hazard (PI2) covering areas flooded with return period $30 < T_r \leq 100$ years with water depths h <30 cm and areas with flooded for $100 < T_r \leq 200$ years
(d) Low hydraulic hazard (PI1) covering areas flooded with events that have a return period of $200 < T_r \leq 500$ years

Regional river basin Authorities have chosen a different hazard classification, dividing the flood prone territory into three zones:

(a) Areas with very high hydraulic hazard level (PIME): corresponding to areas flooded by the flood with a return period of 30 years
(b) Areas with high hydraulic hazard level (PIE): corresponding to areas flooded by the flood with a return period of 200 years
(c) Areas with low hydraulic hazard level

The mapping of hazardous areas is the main output of the work and it is important because it involves laws and regulations that deeply interact with the life of the territory. In fact, each of the areas involved in the PAI is subjected to building restrictions and rules for flood protection.

The aim of the present work is to analyze selected cases of flood prone areas and mitigation measures in river basins located in the Tuscany region.

2. Material and methods

In order to build a flood hazard map, a huge amount of information is necessary. Actually, mathematical models are available for a considerable part of the whole Tuscany territory that are able to predict the flood prone areas. These models use hydrological and hydraulic data in order to propagate the flood in the river by means of 1D-models connected with 2D unsteady flow models able to route the floods outside the river channels in the floodplains (Pagliara, 2005).

One of the most suitable models is the one developed at the Public Works Research Institute and reported by Pagliara and Suetsugi (1997). In order to simulate the flow in the floodplain, the model uses the fully dynamic two-dimensional unsteady flow equations consisting of:

– Continuity equation,

$$\frac{\partial h}{\partial t} + \frac{\partial M}{\partial x} + \frac{\partial N}{\partial y} = 0 \tag{1}$$

– Dynamic equation in the x-direction,

$$\frac{\partial M}{\partial t} + \frac{\partial (uM)}{\partial x} + \frac{\partial (vM)}{\partial y} + gh\frac{\partial H}{\partial x} + \frac{1}{\rho}\tau_x = 0 \tag{2}$$

and a dynamic equation in the y-direction,

$$\frac{\partial N}{\partial t} + \frac{\partial (uN)}{\partial x} + \frac{\partial (vN)}{\partial y} + gh\frac{\partial H}{\partial y} + \frac{1}{\rho}\tau_y = 0 \tag{3}$$

with:

$$\tau_x = \frac{\rho g n^2 u\sqrt{u^2 + v^2}}{h^{1/3}} \tag{4}$$

$$\tau_y = \frac{\rho g n^2 v\sqrt{u^2 + v^2}}{h^{1/3}} \tag{5}$$

where, g = gravitational acceleration; ρ = density of water; M = u·h = flux (flow rate per unit width) in the x-direction; N = v·h = flux in the y-direction; h = water depth; H = water surface elevation; x, y = spatial coordinates; t = temporal coordinate; τ_x = shear stress at the bottom in the x-direction; τ_y = shear stress at the bottom in the y-direction. The numerical method used to solve the system of partial differential equations uses an explicit staggered finite difference scheme (Iwasa et al., 1980). Of course, different rainfall scenarios and return period floods need to be considered in order to obtain reliable maps. Different hypotheses must be formulated also in order to have homogeneous criteria between the various river authorities.

Actually, return period of 20, 30, 100, 200 and 500 years are considered and the relative hazard maps are available.

For practical applications, we can approximate the risk value R through the following "risk equation":

$$R=EVH_t \tag{6}$$

where H_t defines the hazard, that is the probability of observing the event in a period T_r and V and E respectively represent the Vulnerability and the Damage of the elements subjected to flood risk.

The hazard is linked to the return period T_r, which is the time interval in which the intensity of the event is exceeded on average once, by the equation:

$$H_t=1-(1-1/T_r)^t \tag{7}$$

where t is the reference period considered.

The actual flood maps in Italy are based on the hazard H_t and further work has to be done in order to obtain risk maps that consider the risk as defined by Equation 6.

The following are selected cases of flood studies undertaken in central Italy.

(a) Camaiore river basin

The Camaiore river is located in the north–west part of Tuscany (Fig. 1). The basin is highly urbanized and is located between the Tyrrhenian Sea and the Apuan Alps. Large flood events have occurred in recent decades. The regional river authority has developed a master plan in order to reduce the risk at an acceptable level.

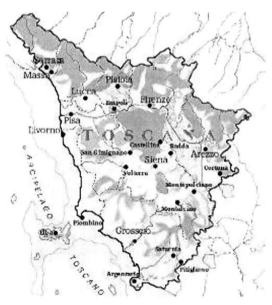

Figure 1. Tuscany region, central Italy

The estimated time to build the mitigation structures is foreseen as 15 years. Figure 2a shows a topographic map of the final, highly urbanized part of the Camaiore river basin. It is evident that the drainage network has some part with channels of very low gradients. Figure 2b represents the hazard map for a return period of 200 years. The map shows, for each point, the envelope of the maximum water depth reached by the flood during its routing. There are urbanized areas in which the water level reaches depths greater than 2 m with high risk for the people.

Figure 2. (**a**) Topography of the flood prone area; (**b**) 200-year return period hazard for the Camaiore River

The same area is represented in Fig. 3. Figure 3a shows the flood map for a return period of 30 years. Figure 3b shows the arrival time of the flood. The four figures represent the flood map at time steps of 1, 5, 9 and 14 which correspond to up to 7 h from the beginning of the event. It is evident that the railway embankment modifies the flood propagation and changes the hydraulic risk in the area. These maps are used for civil protection plans.

Figure 4 shows the final results for the regional river Authority that sub-divides the territory into PIME and PIE areas. In those areas, the P.A.I. suggests many restrictions for new building construction and many regulatory

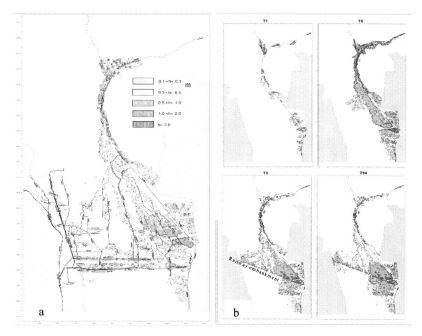

Figure 3. (**a**) Flooded areas for the 30-year return period hazard due to the Camaiore River and the minor drainage system; (**b**) arrival time of the 200-year return period flood

Figure 4. Hazard classification in the Camaiore river basin

rules. All these regulations remain valid until the realization of the hydraulic works that will reduce the hydraulic risk to the predefined acceptable level.

(b) Ombrone and Bisenzio river basins

Other areas where the hydraulic risk is high are the ones between the Ombrone and the Bisenzio rivers, located in north Tuscany north–east of the city of Florence. Figure 5 shows the results of a flood routing simulation with the envelope of the water depths due to a flood event with a return period of 200 years. In this case the flood comes from different tributaries. The routing pattern changes because of the presence of highways, railways and embankment of different rivers. Different areas suffer of more than 1 m water depth with high water velocities. It is evident that the flood risk mitigation must foresee different works in all the drainage basin with high costs for the regional community.

Figure 5. Example of flooding risk map in a highly urbanized environment (Pistoia–Arno)

(c) Mitigation measures

Each master plan foresees a large number of mitigation measures. The plan for the flood protection of the Arno River proposes works costing a total of about €3,000 million (estimated in 1992). In Tuscany, the main types of river protection works are levée systems and detention basins. Detention basins have been particularly used. In the Arno river basin, a total of 300 million cubic meters of storage are planned: this means about 200 new detention basins. Figure 6 shows a small part of the Arno river basin between the Ombrone and Bisenzio rivers with the planned structural measures. Clearly, a large area is needed in order to build the detention basins.

Actually, flood maps are available that account for the realization of a range of different structural measures. This is important in order to plan the urbanization in areas where the hydraulic risk will be removed by the hydraulic works.

Figure 6. Mitigation measures in the Arno river basin (clear areas represent detention basins). The Arno River flows from right to left in the lower part of the figure

The next report is a study conducted in the south of Tuscany close to the city of Follonica (Fig. 7). Different rivers create a high hydraulic risk. To cope with this, several hydraulic works have been planned.

Figure 8 shows the results, on the flow pattern, of the building of different structural measures for containing the flood. The actual flood hazard is represented in the frame F4; frame F4a shows the protection of a small tributary, frame F4b a bridge protection, F4c a levée system protection, F4d and F4e detention basins construction.

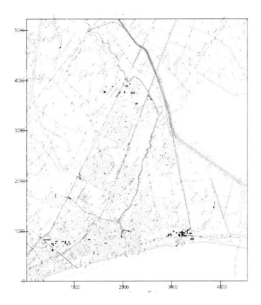

Figure 7. Location of the study area in the city of Follonica

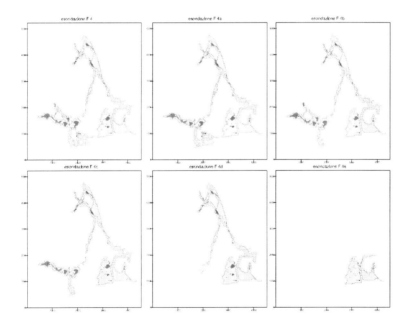

Figure 8. Flood hazard map in the city of Follonica due to the realization of different structural mitigation measures (see text for details)

3. Conclusions

The work analyzes approaches to flood risk in Italy, describing the methodology for developing flood maps. Different examples show the work
that has been done in the Tuscany region and the effect of structural
mitigation measures.

References

Yen B.C., 1995: Urban flood hazard and its mitigation. In: F.Y. Cheng and
    M.S. Sheu (eds), *Urban disaster mitigation*, Elsevier, Amsterdam.
Iwasa Y., Inoue K., Mizutori M., 1980: Hydraulic analysis of overland flows by
    means of numerical method. *Annals, Disaster Prevention Research Institute,
    Kyoto University*, 305–317 (in Japanese).
Pagliara S., 2005: Flood defence by means of complex structural measures.
    NATO-ARW – Baile Felix – Romania.
Pagliara S., Suetsugi T., 1997: Floodplain analysis by two-dimensional model.
    Technical Memorandum of PWRI No. 3520, Tsukuba, Japan.

SPATIAL DATA INTEGRATION FOR EMERGENCY SERVICES OF FLOOD MANAGEMENT

Gh. Stancalie[*], V. Craciunescu and A. Irimescu

National Meteorological Administration, 97 Soseaua Bucuresti-Ploiesti, 013686 Bucharest, Romania

Abstract. For Romania, as for many European countries, the most damaging geophysical events are floods, which cause significant damage every year over large areas, with the loss of human lives and economic consequences. Flood forecasting, warning and emergency response play important roles in reducing flood risk during the event. Options for flood risk management range from structural measures, such as raising dikes or channel enlargement, to non-structural measures to reduce exposure and vulnerability. The April 2000 floods in the western part of Romania (Crisuri River Basin) represented the starting point in creating a dedicated system for flood management using Earth Observation (EO) data. Starting from April 2005 this satellite-based surveillance system, connected to a dedicated GIS database, was extended to cover the whole Romanian territory. In order to obtain high-level thematic products, the data extracted from EO images were integrated with other spatial ancillary data (topographical, pedological, meteorological) and hydrological/hydraulic model outputs. A series of specific image processing operations were performed, using the ERDAS Imagine software: geometric correction and geo-referencing, image improvement, statistic analyses, etc. The optical and microwave satellite data supplied by the new European and American orbital platforms like the EOS-AM "Terra" and EOS–PM "Aqua", DMSP, Quikscat, SPOT, ERS, RADARSAT, Landsat7 have been used to identify and highlight flooded areas.

This approach is used in different phases of establishing the sensitive areas such as: the database management; the calculation of risk indices from morpho-hydrographical, meteorological and hydrological data; interfacing with the models in order to improve their compatibility with input data; recovery of results and the possibility of working out scenarios; presentation of results as synthesis maps easy to access and interpret, adequate to also be combined with other information layouts resulted from the GIS database. The products which are useful for flood risk analysis include: accurately updated maps of land cover/land-use, comprehensive thematic maps at various spatial scales with the extent of the flooded areas and the affected zones, maps of hazard-prone areas. The satellite-based products contributed to a preventive consideration of extreme flood events by developing

---

[*] To whom correspondence should be addressed. e-mail: gheorghe.stancalie@meteo.inmh.ro

plans for flood mitigation, building infrastructure in flood-prone areas and by optimizing the distribution of spatial flood-related information to local and river authorities.

Keywords: Floods, Romania, flood warning, flood risk maps, GIS, emergency response

1. Introduction

Floods are the most important disaster affecting many countries in the world year after year. From the Romanian perspective, floods are among the most hazardous natural disasters in terms of human suffering and economic losses. The large floods that occurred in spring and summer of 2005, the worst ones in more than 40 years, affected large regions of Romania: in the Timis county (floods of 19–23 April), over 1,300 homes were damaged or destroyed, 3,800 people were evacuated and about 30,000 ha of agricultural land flooded; in five counties situated in eastern Romania (floods of 10–14 July), 482 villages, towns and cities were flooded, 11,000 homes inundated, 8,600 people were evacuated, 53,000 ha farmland flooded, and 379 bridges damaged or destroyed.

On the Danube River, historic discharges greater than those of 1970 and 1981 were recorded in 2006 due to a combination of local runoff and flood wave propagation effects. This was the greatest flood event ever registered upstream of the Romanian sector of the Danube, registering 15,800 m^3/s at Bazias, as well as downstream in the Portile de Fier sector, in the last 100 years. It had the maximum discharge/stages, the longest duration of high water stage over the inundation thresholds.

Flood management evolves and changes as more knowledge and technology become available to the environmental community. Satellite imagery can be very effective for flood management in facilitating the detailed mapping that is required for the production of hazard assessment maps and for input to various types of hydrological models, as well as in monitoring land use/cover changes over the years to quantify prominent changes in land use/cover in general and the extent of impervious area in particular (Nirupama and Simonovic, 2002). Geographic Information System (GIS) can be used to extract some types of information, which are otherwise difficult to access by traditional methods, particularly for flood forecasting and floodwater movement. GIS is also considered a vital tool for making use of remotely sensed data for disaster mitigation. The GIS data handling capability also plays a major role in supporting the effectiveness of automated procedures developed for flood hazard control.

The decision process starts with the collection of observed data that supports the creation of information through modelling, the information evolves into knowledge through visualization and analysis, and finally the knowledge supports hydrological decisions. The decision makers are concerned with identifying the hazard events (which appear because of natural events and/or hydraulic and terrain configurations), determining their impact on the elements at risk and then adopting mitigation measures. Nowadays, it is accepted that the main component of the conceptual framework for flood management is a decision support system (DSS).

Recognizing the threat of floods and the need for further improvement of flood management the Romanian Meteorological Administration, under the framework project "Monitoring of Extreme Flood Events in Romania and Hungary Using Earth Observation Data" of the NATO Science for Peace (SfP) Programme, developed an efficient and powerful flood-monitoring tool, which is expected to significantly contribute to the improvement of the efficiency and effectiveness of the action plans for flood defence. This also includes the distribution of the graphic and cartographic products, derived using the GIS facilities and based on satellite data, maps and field surveys, to the interested authorities, media and public contributes to flood-preventive activities for land development and special planning in the flood-prone areas. (Brakenridge et al., 2003).

The paper describes the integration of the EO data into a warning system for emergency services of flood management and the main function of the system.

2. Romanian system for emergency response

The Romanian organisation for a flood emergency situation is based on procedures and information to assist the early warning and notification messages to responsible emergency management authorities. The responsible institutions for emergency situation management are presented in Fig. 1.

The roles of the main institutions are as following:

Ministry of Environment and Water Management represents the highest level of the system. Its role is to:

– Develop the national strategy for protection against floods, severe meteorological phenomena, accidents at hydrotechnical structures
– Coordinate and survey the construction of the hydrological structures with a defence role
– Coordinate flood protection activities at national level
– Cooperate with other specialized international organizations, on the basis of the conventions signed by Romania, concerning protection against floods, severe meteorological phenomena, and accidents at hydrotechnical structures
– Ensure the functioning of the Departmental Committee and the Operative Centre with permanent activity for emergency situations

The National Institute of Hydrology and Water Management has responsibilities for:

– Production of hydrological forecasts and warnings
– Establishment of critical thresholds and frequency of transmissions.

The National Administration "Romanian Waters" has to:

– Provide technical assistance for out of town, district and communal planning against floods
– Develop plans to warn and alert localities and industries downstream in the basins
– Ensure coordination of the Technical Support Groups for the management of risks, functioning within the District Committees

The National Meteorological Administration is involved in:

– Production of forecasts and warnings concerning severe meteorological phenomena and their transmission
– Production of instructions concerning the establishment of critical thresholds for severe meteorological phenomena
– Ensures the safe functioning of the national meteorological measurements network

The Inspectorate for Emergency Situations has the responsibility for:

– Public protection against flood disaster consequences
– Organizing the sheltering and evacuation of population
– Assessment of the post-disaster situation

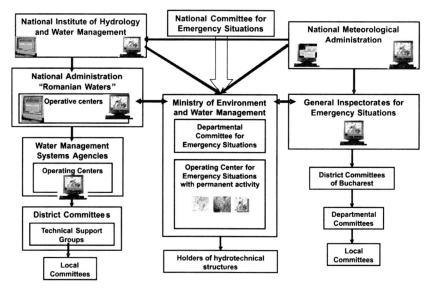

Figure 1. Flowchart of the system for the management of emergency situation generated by severe meteorological phenomena and floods

3. Methods for obtaining useful products for flood risk assessment

It is generally recognized that the management and mitigation of flood risk requires a holistic, structural set of activities, approached in practice on several fronts with appropriate institutional arrangements made to provide the agreed standards of services to the community at risk (Samuels, 2004). Flood management includes pre-flood, flood emergency and flood recovery activities.

The pre-flood activities are:

- Flood risk management for all flood causes
- Disaster contingency planning; building flood defence infrastructure and implementing forecasting and warning systems
- Maintenance of flood defence infrastructure
- Land-use planning and management within the whole basin;
- discouragement of inappropriate development within flood-prone plains
- Public communication and education on flood risk and actions to take in a flood emergency

Operational flood management includes:

- Detecting the likelihood of a flood
- Forecasting future river flow conditions from hydrological and meteorological observations
- Warnings issued to the authorities and the public on the extent, severity and timing of the flood
- Response to the emergency by the public and the authorities

Post-flood activities are:

- Relief for the immediate needs of those affected by disaster
- Reconstruction of damaged buildings, infrastructure and flood defence
- Recovery of environment activities in flooded areas
- Review of flood management activities to improve planning for future events in the area

Within the framework of flood surveying, optical and radar satellite images can provide up-to-date geographical information. Integrated within the GIS, flood derived and landscape descriptive information is helpful during their characteristic phases of the flood (Townsend and Walsh, 1998):

- Before flooding, the image enables the description of the land cover of the area under normal hydrological conditions.
- During flooding the image data set provide information on the inundated zones, flood map extent, flood's evolution.

- After flooding, the satellite image displays the flood's effects, showing the affected areas, flood deposits and debris, with no information about the initial land cover description unless a comparison is performed with a normal land cover description map or with pre-flood data.

In order to obtain high-level thematic products, the data extracted from EO images must be integrated with other non-space ancillary data (topographical, pedological, meteorological) and hydrological/hydraulic model outputs. This approach may be used in different phases of establishing the sensitive areas such as: the database management; the calculation of risk indices from morpho-hydrographical, meteorological and hydrological data; interfacing with the models in order to improve their compatibility with input data; recovery of results and the possibility to work out scenarios; presentation of results as synthesis maps easy to access and interpret, adequate to be in addition combined with other information layouts resulted from the GIS database (Fig. 2).

The products useful for flood risk analysis include: accurately updated maps of land cover/land-use, comprehensive thematic maps at various spatial scales with the extent of the flooded areas and the affected zones, maps of hazard-prone areas.

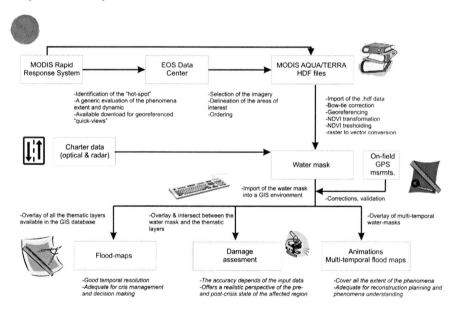

Figure 2. Methodology used to obtain the thematic products

3.1. THE LAND COVER/LAND-USE MAPPING

The main stages of the methodology for obtaining land cover/land use maps from medium- and high-resolution images (Stancalie and Craciunescu, 2005)

include: selecting medium- and high-resolution images, preliminary activities for data organizing and selection; computer-assisted interpretation and quality control of results; digitisation of the obtained maps; database validation on the level of the studied geographic area; obtaining the final documents, in cartographic, statistic or tabular form. Preliminary activities comprise collecting and inventorying the available cartographic documents and statistic data connected to land cover: topographic, land survey, forestry, and other thematic maps at various scales. To obtain the land cover/land-use map, satellite images of fine spatial resolution and rich multispectral information have to be used. In case of IRS and SPOT data, the preparation stage consisted in merging data obtained from the panchromatic channel, which supply the spatial detail (spatial resolution of 5 m for the IRS, 10 m for the SPOT), with the multispectral data (LISS for IRS, XS for SPOT), which contain the multispectral richness. For this application TERRA/ASTER data have also been used. These data proved to be suitable for detailed maps of land cover/land-use, especially the visible and near infrared bands (1, 2, 3B) with 15 m resolution. Figure 3 shows a flowchart on the generation of land cover/land-use maps using high-resolution satellite data.

Discriminating and identifying various land occupation classes relies on the classical procedures of image processing and leads to a detailed management of the land cover/land-use, followed by a generalizing process. The satellite-based cartography of the land cover/land-use is important because it allows for periodical updating and comparisons, and thus evaluates the human presence and provides elements of vulnerability, at the same time assessing the impact of the flooding.

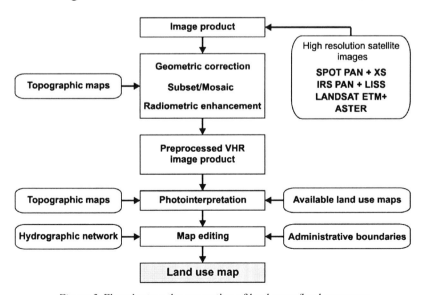

Figure 3. Flowchart on the generation of land cover/land-use maps

3.2. METHOD FOR IDENTIFYING AND MAPPING FLOODED AREAS

The methodology to identify, determine and map areas affected by floods is based on the different classification procedures regarding satellite images. The advantage of using high-resolution satellite images consists in the possibility of selecting precise spatial information on the respective area (through merging images) and to localize and define the flooded or flooding-risk areas (through classifications). Radar images can bring useful information regarding flooded areas, even during periods of abundant rainfall and clouds. The multi-temporal image analysis combined with the land cover/land-use information enables us to identify the water-covered area (including the permanent water bodies) and then the flooded areas. Figure 4 presents the flowchart on the generation of flood-extent maps using satellite radar (SAR) images.

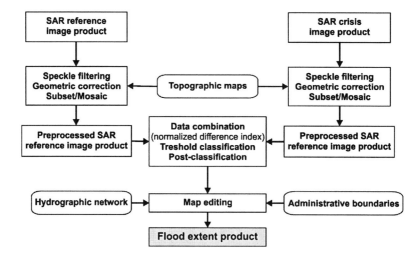

Figure 4. Flowchart on (for) the generation of (the) flood-extension (extends) maps using satellite radar (SAR) images

Using this methodology to identify and map flooded areas allows monitoring and investigating flood evolution during different phases, but especially after the crisis, in order to make a damage inventory and to take recovery actions.

3.3. METHOD TO PREPARE FLOODING-RISK MAPS

The flooding-risk assessment requires a multidisciplinary approach; coupled with the hydrological/hydraulic modelling, the contribution of geomorphology can play an exhaustive and determining role using the GIS tools. The GIS database structure for the study area was designed to be used for the study, evaluation and management of information on flooding occurrence and development, as well as for the assessment of damage inflicted by flooding effects (Lanza and Conti, 1994). In this regard, the spatial geo-referenced information ensemble (satellite images, thematic maps and series of meteorological and hydrological parameters, other exogenous data) was structured as a set of file-distributed quantitative and qualitative data focused on the relational structure between the info-layers. The GIS database is connected with the hydrological database, which will allow synthetic representations of the hydrological risk, using separate or combined parameters (Brakenridge et al., 2004). Figure 5 summarizes the procedure using hydrological/hydraulic model outputs and GIS info-layers for the preparation of flooding-risk maps.

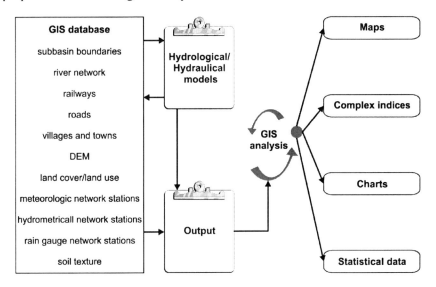

Figure 5. Integration of hydrologic model outputs and GIS info-layers for preparing flooding-risk maps

Building this GIS for the study area was based mainly on maps and topographic plans, updated on the basis of recent satellite images (e.g. the hydrographic network, land cover/land-use) or by field measurements (e.g. dikes and canals network). The GIS database contains the following info-layers: sub-basin and basin limits; land topography (organized in a Digital Elevation Model—DEM); hydrographic network, dikes and canals network; network of communication ways (roads, railways), localities, network of weather stations, rain-gauging network, network of hydrometric stations; land cover/land-use, updated from satellite. Various morphological para-meters have been extracted from the DEM, such as altitudes, slopes, trans-versal profiles, etc. This information is useful in evaluating local or cumulated potential flowing into a zone of the basin, as well as in obtaining a realistic simulation of floods, taking into account terrain topography, the hydrological network and water levels in different transversal profiles on the river, obtained from the hydraulic modelling. Using the GIS database for the study area, several simulation outputs of hydrological or hydraulic models could be superimposed in order to elaborate the flooding-risk maps.

4. Conclusions

Flood-risk analysis needs to make use of and integrate many sources of information. The integrated flood management approach is in harmony with the recommendations of the International Strategy for Disaster Reduction and those of the EU Best Practices on Flood Prevention, Protection and Mitigation.

The optical and microwave satellite data supplied by the new European and American orbital platforms like the EOS-AM "Terra" and EOS–PM "Aqua", DMSP, Quikscat, SPOT, ERS, RADARSAT, Landsat7 provide information and adequate parameters contributing to the improvement of hydrological modelling and warning.

Considering the necessity of improving the means and methods of flood hazard and vulnerability assessment and mapping, this paper presents the capabilities offered by EO data and GIS techniques to manage flooding and related risks.

The specific methods, developed within the NATO SfP project "*Monitoring of extreme flood events in Romania and Hungary using EO data*", to derive satellite-based applications and products for flood risk mapping (maps of land cover/land-use, thematic maps of flooded areas and affected zones, flooding risk maps) have been presented.

The satellite-based products will contribute to a preventive scheme for extreme flood events by aiding the development of plans for flood miti-gation, building infrastructure in flood-prone areas and by optimising the distribution of spatial flood-related information to end-users. At the same time, the project will provide decision-makers with updated maps of land cover/land-use, hydrological networks and with more accurate/comprehensive thematic maps at various spatial scales showing the extension of flooded areas and affected zones.

References

Balint, Z., (2004), Flood protection in the Tisza basin. Proc. of the NATO Advanced Research Workshop – Flood Risk Management Hazards, Vulnerability, Mitigation Measures, Ostrov u Tise, Czech Republik, pp.171–183.

Brakenridge, R.G., Stancalie, G., Ungureanu, V., Diamandi, A., Streng, O., Barbos, A., Lucaciu, J., Kerenyi, M. and Szekeres, J., 2003: Monitoring of extreme flood events in Romania and Hungary using EO data. NATO SfP Progress report, May. Hanover NH, USA.

Hendry, F., 2004: Best Practices for Web Mapping Design, The second MapServer Users Meeting, Ottawa, Canada, June 9–11, 2004. Proceedings of the second MapServer Users Meeting, Otawa, Canada.

Isaaks, E.H. and Srivastava, R.M., 1989: *An Introduction to Applied Geostatistics.* Oxford University Press, New York, 561 pp.

Lanza, L. and Conti, M., 1994: Remote Sensing and GIS: Potential Application for Flood Hazard Forecasting, EGIS. http://www.odyssey.maine.edu/gisweb/spatdb/egis/eg94208.html

Lee, D.T. and Schachter, B.J., 1980: Two algorithms for constructing a Delaunay triangulation. *International Journal of Computer and Information Sciences*, **9(3)**: 219–242.

Nirupama, K. and Simonovic, S.P., 2002: Role of remote sensing in disaster management, *ICLR Research Paper Series* No. 21, 152–160.

Povara, R., 2004: *Climatologie generala.* Bucuresti, Romania, Ed. Fundation "Romania de Maine", 244 pp.

Samuels, P.G., 2004: A European perspective on current challenges in the analysis of inland flood risks. In: Proceedings of the NATO Advanced Research Workshop–Flood Risk Management Hazards, Vulnerability, Mitigation Measures (Ostrov u Tise, Czech Republic), 3–12.

Stancalie, G. and Craciunescu, V., 2005: Contribution of Earth Observation data supplied by the new satellite sensors to flood disaster assessment and hazard reduction. In: P. van Oosterow, S. Zlatanova and E.M. Fendel (eds.), *Geo-information for Disaster Management*, Springer, Berlin/Heidelberg/New York, 1315–1332.

Townsend, P. and Walsh, S.J., 1998: Modeling of floodplain inundation using an integrated GIS with radar and optical remote sensing. *Geomorphology*, **21**: 295–312.

THE USE OF REMOTE SENSING AND GIS TECHNIQUES IN FLOOD MONITORING AND DAMAGE ASSESSMENT: A STUDY CASE IN ROMANIA

A. Irimescu[1]*, **Gh. Stancalie**[1], **V. Craciunescu**[1], **C. Flueraru**[1] **and E. Anderson**[2]

[1] *National Meteorological Administration, 97 Soseaua Bucuresti-Ploiesti, 013686 Bucharest, Romania*
[2] *Dartmouth Flood Observatory, Department of Geography, Dartmouth College, Hanover, NH 03755, USA*

Abstract. The use of remote sensing and GIS techniques for rapid mapping and monitoring is an important tool of information for decision-makers. In 2005 and 2006, the worst floods in the last 40 years affected several regions in Romania. In this paper, we describe the methodology used to map the flood-affected areas. MODIS TERRA and AQUA images were acquired during the flood events and used as the main input to map the affected areas. Considering the fact that MODIS is an optical sensor, one of the main difficulties encountered was the cloud cover. The temporal resolution proved to be one of the strongest points of the MODIS imagery. The water identification is dependent on NDVI threshold-values.

Keywords: Remote sensing, GIS, flood, monitoring, land cover

1. Introduction

Floods are among the most devastating natural hazards in the world, affecting more people and causing more property damage than any other natural phenomenon (DMSG, 2001). In the period of time between 1900 and 2006 a total of 415 major flood events occurred in Europe alone, with an average death toll of 22 and 35,159 affected people (www.em-dat.net). In Romania as well as in many other European countries, prevention and monitoring floods are activities of national interest, taking into account the frequency of occurrence and the degree of the effects. There are several types of floods:

- Floods generated by natural overflows of water courses, caused by the increase of flows or blocking produced by ice packs, floating material, alluvial deposits, snow avalanches and flowing from the slopes
- Floods caused by accidents or damage to hydro-technical structures

---

* To whom correspondence should be addressed. e-mail: anisoara.irimescu@meteo.inmh.ro

J.A.A. Jones et al. (eds.), *Threats to Global Water Security*,
© Springer Science + Business Media B.V. 2009

- Severe meteorological phenomena: downpours, heavy snowfalls, storms and blizzards, ice caps, rime, glaze, early and late frosts, hail and hydrological drought
- Accidental pollution of surface and ground water and the sea in coast areas

For planning any flood management measure, the latest, most reliable, accurate and timely information is required. The Earth observation satellites are able to provide comprehensive and multi-temporal coverage of large areas. The real-time mapping of flood extent, which requires rapid acquisition, processing and analysis of data fulfil the requirement of fast supply of data during floods. Availability within a few hours supports planning of emergency relief and helps the coordination of the response activities of various decision makers. Basically, two different types of products of rapid mapping are claimed by users: overview maps of flooded areas and damage maps combined with additional information, such as flood extent variation or land-use types within the flooded area (Allenbach et al., 2005).

Remote sensing provides information that has proved useful for a wide range of applications in disaster management. Usually, ground-based monitoring networks are inadequate due to a low density of observation points; space-based platforms provide wide spatial coverage without access limitations. Satellite observation facilitates the regular monitoring of the extent of floods and mapping of flood risk zones (Brakenridge et al., 2001).

Remote sensing data can easily help in prevention, with mapping of hazardous areas, drainage networks and land cover mapping, and precise river basin modelling, or in a post-event evaluation of damaged areas. On a case-to-case basis, one may need to provide cartographic products derived from available satellite data with the shortest delay.

The paper describes the methodology used to map the flood extend by using MODIS data and to estimate the affected land cover/land use categories.

2. Background

In the recent years, significant efforts have been devoted to the implementation of a monitoring methodology. The management of flood hazards and risks is a critical component of public safety and quality of life. The cost of natural disasters is increasing; their impact is invariably higher in developing countries and where people are concentrated. Many factors contribute to increasing vulnerability, such as unsustainable development practices, degradation of natural resources, weak disaster management and forecasting techniques or inadequate communications and transport infrastructure. At the same time, there are several technological trends that can serve to decrease the vulnerability to disasters. These include: better understanding of hazardous processes, improved analytical methods and communications.

One important type of flood mapping work is carried out on a world scale/level at Dartmouth Flood Observatory in Hanover, New Hampshire, USA (www.dartmouth.edu/~floods/). At the European level, flood-related research has been undertaken in several important projects like:

- The European flood forecasting system – EEFS
- Operational Solutions For The Management Of Inundation Risks In The Information Society – OSIRIS
- Near real-time flood forecasting, warning and management system based upon satellite radar images, hydrological and hydraulic models and *in situ* data – FLOODMAN
- Real Time Flood Decision Support System Integrating Hydrological, Meteorological and Radar Technologies – FLOODRELIEF
- Integrated flood risk analysis and management methodologies – FLOODsite
- Monitoring of extreme flood events in Romania and Hungary using EO data etc.

Over the last 4 years, Romania has been involved in the NATO SfP project, *Monitoring of extreme flood events in Romania and Hungary using EO data*. One of the project's main objectives was to analyse, adapt and test methods and algorithms for processing and interpretation of the medium and high spectral and spatial satellite resolution data, from the optical and microwave regions. The specific goal was to identify, delineate and map floodwater and excess soil moisture areas (Brakenridge et al., 2001).

3. Study area and weather conditions

The Danube is the second longest European river, flowing for a distance of 2,857 km and 1,075 km (38%) is on Romanian territory. The study area is located along the Romanian sector of the Danube with six major affected areas: Ghidici – Macesu de Jos, Bechet – Corabia, Oltenita – Calarasi, Balta Ialomitei, Insula Mare a Brailei and the Danube Delta (Fig. 1). The maximum discharge on the Danube over the last 100 years, 15,800 m^3/s, was generated by heavy rains and melting snow in Central and Eastern Europe at the end of winter and beginning of spring (Meteorological bulletin, 2006a, b). More than 50 l/m^2 of precipitation in 24 h has been recorded in Romania.

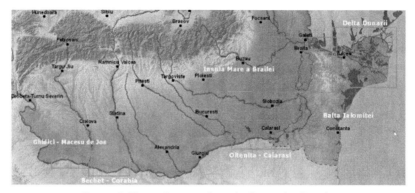

Figure 1. The study areas along the Romanian Danube

4. Data used

The information needed by the user (authorities, NGOs etc.) should be delivered in short time with a good spatial resolution and integrated in a user-friendly cartographic environment. In order to obtain this kind of product different types of data have been used:

- Topographic maps have been used to extract different types of info-layers (Fig. 2):
  - Administrative elements (national limits – border, administrative limits, localities).
  - Natural elements (hydrographical network: rivers, dykes, channels, lakes etc.).
  - Communication network (roads and railways).

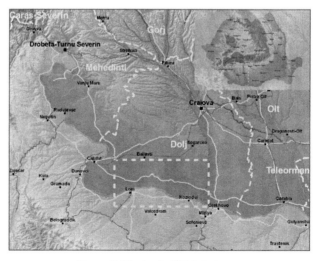

Figure 2. The basic GIS info-layers

- Satellite images: MODIS (Moderate Resolution Imaging Spectroradio-meter) is a sensor carried on the TERRA (EOS-AM) satellite, launched in December 1999 and AQUA (EOS-PM), launched in May 2002. It is an important tool for regional-scale flood mapping. MODIS is a passive imaging spectro-radiometer carrying 490 detectors, arranged in 36 spectral bands that cover the visible and infrared spectrum, with a spatial resolution ranging from 250 m (in visible and near IR) to 500 m and 1 km (at nadir). The data format used for MODIS data is HDF-EOS (Hierarchical Data Format – Earth Observing System), which is not always found among the data format of all image-processing software. This is why it is neces-sary to read the image header or to use specific software that imports the image and exports it in a known format. For our purpose, we used a Level 1B, visible bands 1 and 2 (250 m resolution) calibrated reflectance data (format MOD02QKM). The MODIS images offer daily global cover-age and are provided by NASA free of charge. They can be ordered on the web site of the EOS data server (http://edcimswww.cr.usgs.gov/pub/imswelcome/). The images can be browsed by different criteria: location, date, etc. Because of MODIS characteristics, the data have been used for operational applications in hydrology (Brakenridge and Anderson, 2006): flood detection, flood disaster response and damage assessment and flood disaster prevention or mitigation.
- CORINE LAND COVER 2000 was used to identify the land cover/use classes in the flood-affected areas. The CORINE LAND COVER cate-gories have been grouped into six classes: localities and industrial areas, complex cultivated patterns, orchard and vineyard, pasture, forest and bare soil.
- LANDSAT 7 ETM+ as thematic background.

5. Methodology

5.1. MODIS DATA: ADVANTAGES AND DISADVANTAGES IN FLOOD MAPPING

The advantages of MODIS data for flood mapping are related to the temporal resolution (daily coverage), spectral resolution (36 bands, for visible and infrared spectrum), the radiometric resolution of 12 bits and the cover area (the swath dimensions of MODIS are 2,330 km across track by 10 km along the track at nadir) (http://modis.gsfc.nasa.gov/). To provide the infor-mation to the authorities in real time is possible because the MODIS data are available in real time or near real time and free of charge, which means low product costs.

To monitor the Danube flood during end of March to middle of May, 36 MODIS images were processed. Cloud-free images were selected and the flood extent for six areas extracted. The disadvantages of MODIS data are related to spatial resolution (250 m) and cloud cover.

5.2. MODIS DATA PROCESSING

The methodology used to extract the flood-affected areas along the Romanian sector of the Danube has been developed in the framework of NATO SfP Project (Brakenridge et al., 2003). It has also been used to map the flood-affected areas on the Timis River in south-western Romania and on the Siret River in eastern Romania) in 2005. In order to extract the water bodies, the MODIS data products MOD02QKM (two bands, NIR and R, 250 m resolution) have been used along with the processing capabilities of ENVI^{TM} software. The steps in the processing chain for rapid mapping are:

- *Correction of the "bow-tie" effect*. The "bow-tie" effect distorts images at large viewing angles. The "bow-tie" is a phenomenon associated with sweeping instruments, where the instantaneous field of view increases as the sensor sweeps away from nadir. As the instrument looks further out to the side it "sees" a larger area on the ground. A consequence of this is that subsequent scans will have overlapping observations on the outer edges of a swath (http://modis.gsfc.nasa.gov/) (Fig. 3a). The "bow-tie" correction tool of MODIS has been used to eliminate the effect (Fig. 3b). The "bow-tie" correction has to be done for each band, separately. If the "bow-tie" correction is performed together with geo-reference, the results are not satisfactory and the distortion still exists.
- *Image geo-referencing*. This is the first step in order to integrate the image with other geo-spatial data. MODIS products are distributed using the sinusoidal projection in 10° tiles. The MODIS scenes have geographic coordinate information embedded in the file. To geo-reference them, the Georeference MODIS 1B tool has been used.
- *Compute the NDV*. The NDVI (Normalised Difference Vegetation Index) is a normalized ratio of the NIR and red bands. The reflectance values are the surface bi-directional reflectance factors for MODIS bands 1 (620–670 nm) and 2 (841–876 nm). NDVI was computed using red and near-infrared channels as (band 1 – band 2)/(band 1 + band 2). Usually the NDVI is used to monitor the vegetation, but the water bodies always appear in dark on the NDVI band, which can be a very good indicator of the water.
- *Separate the water from other cover classes*. The next step is to decide on the NDVI threshold used to separate water from other surface classes. The water threshold varies from image to image. So, it was necessary to test the threshold values for each image. In the case presented here, the water threshold values varied between −0.20 to +0.24.

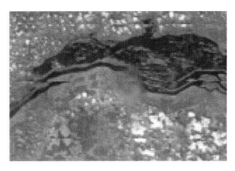

Figure 3a. MODIS "bow-tie" effect

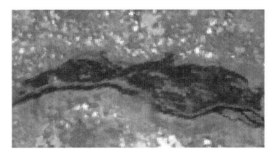

Figure 3b. MODIS "bow-tie" effect eliminated

- *Extract the water-mask.* The NDVI water values have been transformed into a .shp format in order to overlap it on the other geo-data (administrative elements, natural elements, communication network, CORINE LAND COVER etc.) (Fig. 4).

Figure 4. The water-mask derived from MODIS image from 25.04.2006, for the Ghidici-Macesu de Jos Sector

- *Result validation.* The result validation has been done using topographic maps, hi-resolution satellite data (e.g. ASTER, DMC, RADARSAT and other satellite data with better spatial resolution than the 250 m of MODIS data, provided by the International Charter) and GPS field surveys.

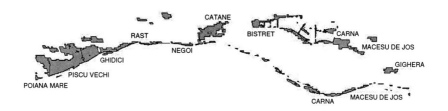

Figure 5. The flooded villages extend derived from MODIS image from 14.04.2006, for the Ghidici-Macesu de Jos Sector

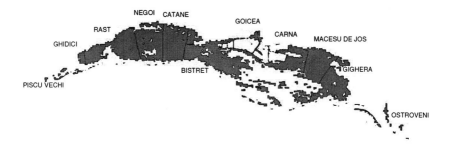

Figure 6. The flooded villages extend derived from MODIS image from 26.04.2006, for the Ghidici-Macesu de Jos Sector

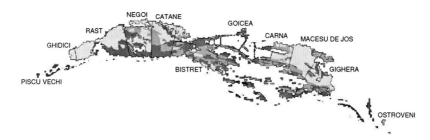

Figure 7. The flooded villages derived from MODIS image from 26.04.2006, for the Ghidici-Macesu de Jos Sector

- **Economic damage estimation.** The water-masks extracted daily between 30.03.2006–31.05.2006 and village limits were imported into the GIS environment to map the water extend for each village. An example of the village flood extend is shown in Figs. 5 and 6. The results were then integrated with CORINE LAND COVER (Fig. 7) to estimate the economic damage and other valuable information – surface area flooded and land cover/land use categories affected (Table 1).
- **Data dissemination.** Distribution of flood-related products to the interested authorities (national, regional and local) was in print form or via Internet (http://nato.inmh.ro).

TABLE 1. Villages flooded in Ghidici-Rast-Bistretu-Macesu de Jos area

Village name	Administrative area (ha)	Affected area (ha)	Affected part of administrative area (%)
Rast	8,452	4,500	53.2
Bistret	12,214	6,632	54.3
Gighera	13,154	6,894	52.4
Ostroveni	8,245	590	7.2
Macesu de Jos	5,639	3,428	60.8
Carna	8,475	5,603	66.1
Goicea	5,841	324	5.6
Catane	4,702	2,481	52.8
Negoi	5,019	3,062	61.0
Ghidici	4,469	1,585	35.5
Piscu Vechi	5,782	1,506	26.1

6. Results

The polygons classified as flooded were combined with different geographical data (administrative or natural elements) to generate a layout that the decision makers are familiar with, and allow them to add further information. During a flood event, more than 100 products have been produced (http://nato.inmh.ro). An example of this kind of product, with the water-mask superimposed on a LANDSAT ETM+ mosaic, is presented in Fig. 8. This procedure enables extraction of information about the type of affected area (e.g., settlements, forest, pasture, Fig. 9).

Figure 8. The Ghidici–Macesu de Jos Sector flooded area, 04.05.2006, 11.26 LT

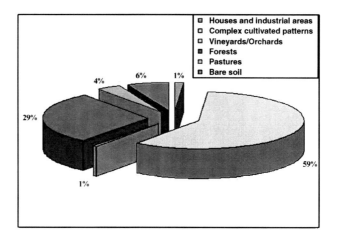

Figure 9. The flood-affected categories

7. Conclusions

Rapid mapping products aim to be quickly available and easy to interpret by the end users. For Romania, as in many countries, floods in several parts of the country cause significant damage over large areas every year, including loss of human lives and economic impacts. The evaluation and management of floods constitute the indispensable first step and the rational basis for mitigation measures against flood damage. The use of new satellite source data adds significantly to ground observations and traditional map-based methods that are not able to offer a quasi-real time updated cartographic products of the flooded area, land cover and land use, as well as the water works state. The products serve as valuable tools for the restoration and rehabilitation of rivers, whose environment has been altered by flooding events, and also for future analyses and decisions concerning the efficiency and environmental impact of the structural works. The distribution of the graphic and cartographic products to the interested authorities, media and public is an important issue. These products contribute to preventive consideration of flooding in land development and special planning in the flood-prone areas, and for optimising the distribution of flood-related spatial information. The products obtained with the methodology described above provided the decision-makers with accurate and comprehensive thematic maps at various spatial scales containing the extent of the flooded areas and the affected zones in near real-time. The method for satellite data processing and interpretation for the analysis of flooding is now used in operational activities by the Romanian National Meteorological Administration.

References

Allenbach, B. et al., 2005: Rapid EO Disaster Mapping Service: Added value, feedback and perspectives after 4 years of Charter actions, SERTIT, France.

Brakenridge, R.G. and Anderson, E., 2006: MODIS-based flood detection, mapping and measurement: The potential for operational hydrological applications. In: Marsalek, J., Stancalie, G. and Balint, G. (Eds.), Transboundary floods: Reducing risks through flood management. *NATO Science Series: Earth and Environmental Sciences*, **72**: 1–12.

Brakenridge, R.G., Stancalie, G., Ungureanu, V., Diamandi, A., Streng, O., Barbos, A., Lucaiu, M., Kerenyi, J. and Szekeres, J., 2001: Monitoring of extreme flood events in Romania and Hungary using EO data (project plan). Bucharest, Romania.

Brakenridge, R.G., Stancalie, G., Ungureanu, V., Diamandi, A., Streng, O., Barbos, A., Lucaiu, M., Kerenyi, J. and Szekeres, J., 2003: Monitoring of extreme flood events in Romania and Hungary using EO data (progress report), Hanover, NH.

DMSG, 2001: The Use of Earth Observing Satellites for Hazard Support: Assessments & Scenarios. Committee on Earth Observation Satellites Disaster Management Support Group, Final Report, NOAA, Dept Commerce, USA.

EM-DAT: The OFDA/CRED International Disaster Database www.em-dat.net Université Catholique de Louvain, Brussels, Belgium.

MODIS World-Wide WEB site: http://edcimswww.cr.usgs.gov/pub/imswelcome/.

Meteorological bulletin (in Romanian), 2006a: year IX, No. **4** March, Bucharest, Romania.

Meteorological bulletin, 2006b: year IX, No. **5** – April, Bucharest, Romania (in Romanian). www.dartmouth.edu/~floods/ http://modis.gsfc.nasa.gov/.

FLOODING IN AFGHANISTAN: A CRISIS

E. Hagen[1,2,*] **and J.F. Teufert**[1]
[1] *NATO Consultation, Command and Control Agency (NC3A), The Hague, The Netherlands*
[2] *Department of Geography, National University Singapore (NUS), Singapore*

Abstract. Afghanistan is a nation prone to natural disasters such as floods and droughts. Yet the severe floods of 2005 and 2006, which displaced thousands, highlighted how little was known about floods in the mountainous nation. At present no flood hazard maps exist pinpointing the location and extent of the inundated areas, nor are there data on occurrence and impact of floods. This paper explores the flooding crisis and its impact on Afghanistan; furthermore it analyses the causes of floods and some aspects of flood mitigation.

Keywords: Afghanistan, NATO, crisis, GIS, flooding, inundation, DEM, hydrology, floodplain, remote sensing

1. Floods in Afghanistan

Afghanistan is a country not only crippled by decades of conflict, but also prone to natural disasters. Earthquakes, droughts, floods and extreme snowfall affect the lives of thousands of Afghans. Severe flooding in 2005 and 2006 made thousands homeless, and destroyed agricultural land, livestock and infrastructure. In the post-Taliban era alone 50 floods have already been reported. However, except for short situational reports many of these floods are not well documented, thus little is known about the Afghan flood issue.

Floods are mostly created by heavy, intense rainfall, and by snow meltwater, or the combination of both. Technical failures due to increased precipitation or meltwater add to flooding as well. Dam and levée breaks are not uncommon, and their destructive power lies not solely in the inundation, but also in the sheer strength and mass of the floods. Afghan flash floods are powerful enough to sweep entire houses and bridges away, making their impact severe and costly. An assessment of past flood events highlights that flash floods are the primary cause of loss of life due to inundation in Afghanistan.

---

[*] To whom correspondence should be addressed. e-mail: emlyn.hagen@nc3a.nato.int

Currently no accurate flood hazard maps are available, yet the local population is generally aware that they live in flood-endangered areas. They know this either through personal experience or from the oral history of their lands. In some regions Afghan farmers will even irrigate their fields with flood water, which requires detailed knowledge of flood behaviour. Yet the population is at times caught off-guard as floods can come quickly and without much warning. Heavy rainfall could be a flood indicator, yet if it rains many kilometres upstream the inhabitants of the flood region might not notice and will not be warned. Flash floods leave very little time between the swelling of the river and floodplain overflow, whereas sluggish or gradually rising floods leave more time for the people living in the danger areas to get to safe zones.

One of the major constraints for flood assistance is that many floods are at times reported only days after the onset of the flood. This is due to limited communication and road infrastructures, which are very often damaged by the floods themselves. This delay makes helping the flood victims efficiently all the more difficult. Organizations such as the International Red Cross (IRC) and the UN will assist the local population in case of flooding. The IRC will provide medical attention and assist in search and rescue operations, whereas the UN will provide funding for reconstruction and flood mitigation measures.

Not all floods are notified after a delay; floods that affect major towns are reported immediately and extreme floods are anticipated by aid agencies. The NOAA National Weather Service has set up the Famine Early Warning System Network (FEWS-NET), which also encompasses flood warning. For major flood events, the International Security Assistance Force (ISAF) will often be the first organization to provide assistance. With aerial reconnaissance capabilities, it is easier for the military to perform search and rescue operations and assess the situation on the ground. ISAF's heavy and amphibious vehicles can bring people into safety and deliver emergency aid where normal trucks and jeeps could not venture. It is not only a matter of having appropriate equipment, but the availability of efficient communications and coordination designed for work in difficult circumstances makes the military favoured to assist in extreme floods. In April 2007, ISAF helped rescue 350 people from rising flood waters and evacuated 1,300 families (OCHA, 2007). Helicopters quickly delivered 30 days worth of food and other humanitarian aid to 2,500 people. But for reconstruction, long-term aid and relief, organizations such as the IRC are better-equipped than ISAF.

2. Impact of floods

Afghanistan's economy is mainly sustained by agricultural and livestock produce, therefore floods hinder the economy to a considerable extent. A small flood can take away the source of income of an entire village and plunge its livelihood into uncertainty. Larger floods can cripple the economy of entire regions or even have influence on a nationwide scale. The March flood of 2007 was especially devastating in this regard as it crippled the larger part of the nation. Such large-scale floods destroy communication lines, the electricity grid, and road links including bridges; these floods also contaminate drinking water in village wells, and disease spreads if the water is left untreated. Owing to limited assessment by aid agencies and the lack of insurance coverage, there are few estimates available of the financial impact of flooding. However, the major flood of 1988 that displaced 3,000 people caused an estimated damage of US$260 million (Brakenridge et al., 2006), which is a considerable amount of money for a country that has an approximate GDP of US$5.5 billion.

Prolonged inundation of agricultural lands will destroy the harvest and with it the source of income for entire families. However, non-damaging, less-intensive short floods provide the arable land with new nutrients, as these fields are irrigated during a flood period. Furthermore, water is stored in reservoirs for later use. Turbulent floods, however, can carry away the top soil or deposit amounts of non-fertile sediment (USDA, 1996), thus depriving the land of crop-carrying capacity for long periods. The Afghan irrigation infrastructure is basic, and is easily destroyed by flooding. It consists mainly of primitive log dams, fascines and sand-pebble-filled bags used for construction of aryk [qanat style] networks (Zonn, 2002). Soviet specialists and other organizations have calculated that every year up to 10% of the crop is lost due to failure of and interruptions in the use of canals, while up to 30% of land in the lower reaches of these canals remains without water because of the absence of regulation structures on them (Zonn, 2002).

Afghanistan is one of the most heavily landmine-affected areas in the world. The Mine Action Programme of Afghanistan indicated that 2,368 Afghan communities are exposed to 716 million square meters of suspected hazardous areas, affecting as many as 4.2 million people (UNMAS, 2005). Severe flooding can cause displacement of these mines, and deposit them in other areas. Lighter landmines, mostly anti-personnel mines, can be swept away over large areas, moving downstream. As these mines exist almost entirely out of plastic some can even float in water (Wareham, 2002).

3. Flood map and mitigation

At this moment it is not known where floods will occur, and there is little data available on past floods. It is therefore important that basic information be gathered on past floods and that a flood hazard map be created that highlights areas potentially threatened by flooding.

Currently there is a time delay before an emergency report reaches the disaster management centres. However, the delay could be reduced considerably if endangered settlements had emergency communication lines. This would also allow an early warning system to be used to warn villagers of approaching danger. But first the settlements that are in danger need to be identified, and a flood map would provide authorities with information about when and where floods are likely to occur. Currently the aid agencies are not aware of which areas are safe from further flooding, and which would be ideally located to help flood victims.

A flood hazard map is the primary input map for developing and identifying locations for mitigation measures. Dams could be built to protect densely populated areas, or to allow the authorities to start reforestation measures to slow down water runoff. The map will not only assist identification of locations to build dams, levées or flood walls, but will, more importantly, directly indicate where not to build hospitals, schools, bridges, etc., thus preventing wasting reconstruction money on doomed infrastructural locations.

4. Past floods

Information about past floods can reveal a wealth of information. It can indicate which provinces and districts have suffered the most, and which have never been affected. Information about which settlements have been affected can further enhance insight into flood-prone areas of Afghanistan. This information is critical to allow decision makers to allocate funds for flood mitigation measures, or to finance a flood-awareness program.

For the work reported in this paper, information about past Afghan floods was collected in a geographical database. The data originated from two main sources: crisis databases, such as the Active Archive of Large Floods managed by the Dartmouth Flood Observatory (DFO), and text sources, such as news articles and status reports from relief agencies. This geo-database of flood events was then spatially linked to provinces, and if possible to districts and villages.

From 1988 to 2006, a total of 77 floods were recorded; these are all the floods that were known to the media or aid agencies. There were probably many more floods in this period, since between 2002 and 2006 alone 50 floods were registered. These recorded floods have caused 5,036 deaths; however the number of actual fatalities is likely to be much higher. Almost 400,000 people have been displaced since 1988, and of these more than 200,000 people have been displaced since 2002. Small floods can occur sporadically all year round, with the least number of floods in the period October to December. From February to July more frequent and prolonged floods can be expected, and these are the floods that cause the most damage and devastation.

The Dartmouth Flood Observatory detects and measures major flood events world-wide using satellite remote sensing, and maps them into their World Atlas of Flooded Lands (Brakenridge et al., 2006). Initial investigation of these data revealed that the inundated areas are populated areas and are the primary agricultural production centres. The inundation map was particularly useful for assessing the number of settlements affected by flooding. Most floods occur in flat areas, mostly in the foothills of a mountainous region, at the outlet of a major watershed. For the work reported in this paper, the DFO's data were particularly helpful in creating a first impression about past floods in Afghanistan. It provided necessary basic information such as the flood frequency, human losses and damage. The analysis of the data shows that floods affect all Afghanistan provinces, and that many areas experience recurrent flooding. The provinces that are most affected by the number of floods do not necessarily suffer the most in terms of fatalities and displaced persons. This is due to adaptability of the local population, who even use cyclical inundation for irrigation purposes.

For effective flood control not only is a flood map needed, but a population-density map and flood-preparedness map are necessary as well. Being able to warn people in advance of coming floods will reduce expensive search and rescue, evacuation and reconstruction operations.

5. Causes of floods in Afghanistan

Flooding, exacerbated by the combined effects of conflict and climate change, will continue to threaten Afghanistan in the future.

Afghanistan is likely to be subject to increased flooding due to more snowfall in winter and warmer summers. In recent years snow basins contained a lot more snow than usual (USGS, 2003), and in 2006 this combined with rising temperatures and heavy spring rains resulted in a great deal of meltwater and consequent flooding. The Afghan climate is typical of an arid or semiarid steppe, with a wet-cold season in winter and early spring, and in summer a dry season often marred by droughts. Precipitation fluctuates greatly over the year, and is characterized by sudden rainstorms that transform the episodically flowing rivers and streams from puddles into torrents. Invading armies have at times been trapped in these flooding waters, and nomadic and semi-nomadic Afghans have sometimes succumbed to the sudden flooding of their camps (Blood, 2001).

The effects of global warming on floods are not fully understood and no studies have researched this specifically for Afghanistan, although research conducted in other parts of the world has found that global warming can contribute to flood events. In Afghanistan and surrounding countries, the area of glaciation in mountains has decreased by nearly 40% over the past

40 years (Zonn, 2002). Glavgidromet mathematical models enabled evaluation of the effect of climate changes on river basins and an evaluation of
the runoff (Chub, 2002). The results showed that a marked reduction of
flow in may be expected during the growing season, and that increased
intensity of rainfall in combination with higher temperatures will lead to a
greater number of floods and mudflows (Chub, 2002).

The following information was deduced from a survey of interpolated
climate data from the NOAA's National Climatic Data Centre: In Afghanistan,
the temperatures have been steadily increasing over the past decades.
Not only the summers have become hotter, but also the winters. The daily
temperature differences have also increased. The precipitation has also
increased over the past decades, yet it must be noted that this increase is
solely in winter and spring. The combination of increased snow precipitation in winter and higher temperatures might be the cause of more intense
spring floods in the valleys of Afghanistan.

War can magnify the proportions of disasters for populations, but
natural disasters can also be partly caused by the impact of conflict. As a
result of conflict, an estimated 30% of Afghan farmlands and pastures have
been lost by either abandonment or degradation (Glantz, 2002). Agriculture
plays an important role in mitigating the effects of floods; the small
irrigation channels divert a lot of runoff, and the vegetation on the fields
absorbs and slows the runoff. If these fields are abandoned they erode
quickly, which increases the chance of flooding. In time of war, dams have
often been a primary target of military strategists. At this moment the
Helmand River's Kajaki hydroelectric dam, which supplies almost all the
energy to southern Afghanistan, is a primary target for Taliban insurgents
(Edelman, 2007). The Kajaki dam's strategic importance is its energy production, and a dam break flood (1.8 km^3) resulting from its destruction
would be immense. Without the mitigating effect of the dam, the Helmand
River's agricultural lands would suffer each year from uncontrolled flooding.

Wars and conflicts have prevented mitigation infrastructure and
planning from being undertaken. No major dam construction projects have
been undertaken in Afghanistan for the past 30 years. The political situation
has led to the neglect in maintenance of many dams, and the regular occurrence of dam bursts in Afghanistan emphasizes the necessity for maintenance.

Acknowledgments This contribution is part of a Afghanistan flood mapping research project, supported by the NATO C3 Agency. Furthermore the
authors would like to thank Robert Brakenridge and Elaine Anderson of the
DFO for sharing past inundation extents with us.

References

Blood, P.R., 2001: Afghanistan: A Country Study. Washington, DC: GPO for the Library of Congress.

Brakenridge, G., Anderson, E., and Caquard, S., 2006: Active Archive of Large Floods. Hanover: Dartmouth Flood Observatory.

Chub, V., 2002: The consequences of climate changes for water resources of the Aral Sea Basin. Aral Sea and Circum Aral Problems – Imperative for International Cooperation. Tashkent.

Edelman, E., 2007: A Comprehensive Approach to Modern Insurgency: Afghanistan and Beyond. PfP Consortium of Defense Academies and Security Studies Institutes, Quarterly Journal Summer 2007. Garmisch-Partenkirchen, Germany. March 2007.

Glantz, M., 2002: Water, Climate and Development Issues in the Amu Darya Basin. Workshop report of the Informal Planning Meeting 18–19 June 2002, Philadelphia, PA.

OCHA, 2007: Afghanistan floods Report No. 1. Office for the Coordination of Humanitarian Affairs, April 2007, New York.

UNMAS, 2005: The United Nations Mine Action Service: Annual Report 2005.

USDA, 1996: Soil Quality Resource Concerns: Sediment Deposition on Cropland. *Natural Resources Conservation Service, Soil Quality Information Sheet.* April 1996.

USGS, 2003: Water Resources Assessment, Monitoring and Capacity Building. USGS Projects in Afghanistan, December 2003.

Wareham, M., 2000: 'Landmines in Mozambique: After the Floods'. Paper prepared for Conference on Mozambique After the Floods, 28 March 2000. Washington, DC.

Zonn, I.S., 2002: Water resources of Northern Afghanistan and their future use. *Water, Climate, and Development Issues in the Amudarya Basin.* The Franklin Institute.

HUMAN FACTORS IN THE FLOODS OF ROMANIA

B. Constantin-Horia[1*], S. Simona[1], P. Gabriela[2] and S. Adrian[3]
[1] *"Lucian Blaga" University of Sibiu*
[2] *Water Management System, Sibiu*
[3] *Regional Weather Center, Sibiu*

Abstract. During recent years, Romania has experienced many floods, causing huge material damage and many losses of life. The paper presents in detail some of the main factors enhancing Nature's "protest against global warming": poor sewage within the towns, illegal deforestation of hills or mountains slopes, dirty river banks, lack of citizens' education and responsibility, slow communication between different authorities, absence of specialized equipment, and in some cases, political or economical interests. Romanian authorities, having in mind the floods of 2005 and 2006, try to prevent these disasters or to mitigate their effects, but up to now there has been a big gap between theory and practice, mainly because of the complexity of the actions to be taken, which require time and money.

Keywords: Floods, flash floods, deforestation, responsibility

1. Introduction

In Romania, the floodplains of the undiked rivers are the most vulnerable to flooding hazard. High flood risks are also present in areas crossed by intermittent, torrential watercourses in hilly and mountainous regions. Floods occur especially in springtime when snow melting is associated with rainy events. Erratically, floods may take place in every other season if heavy rains occur (Dumitru et al., 2001). Most of the flash floods occur in mountain and hill watersheds, with deep valleys, deforested slopes and significant elevation range (Stancalie et al., 2008).

When discussing the relationship between humans and floods, one should consider not only the damage or casualties caused to humans by floods, but also the role human actions (or inaction) have played in causing or enhancing these effects of excess water.

During the last few years, Romania has suffered many floods of different types:

- 2004: massive water flows from/on mountain and hill slopes
- 2005 and 2006: high discharges in many rivers
- Almost every year, within cities, due to poor sewage systems unable to drain the heavy rains

---

* To whom correspondence should be addressed. e-mail: horiab@rdslink.ro

Even if flood control is one of the main objectives of national development, and 54% of the area of flooding hazard is already provided with works for preventing flood (river dykes, reducing the sinuosity of watercourses, afforestation of torrent catchments, building dams for water retention, etc.), this has proven insufficient.

2. The floods of 2004

After several years when floods were reduced in importance, and especially after 2003, an extremely dry year, floods have devastated a large area of Romania. Their effects were terrible: 19 deaths, 18,727 households affected, more than 116,000 ha of agricultural land flooded etc., the value of these damages being more than 1 billion euros.

According to the official conclusions (Environment and Water Management Ministry, 2004), the main causes of these floods were:

- Heavy rainfall in small areas and in very short periods of time, especially in the hills and mountain areas, as well as on slopes and torrents
- Exceedence of the transport capacity of bridge sections, due to their undersizing and clogging with wooden materials, debris or waste, originating from *ad hoc* deposits near rivers or on slopes
- Absence or bad maintenance of ditches for rain water drainage in nearly all rural communities
- Bad maintenance of watercourses
- Large scale deforestation and bad agricultural practices, favoring soil erosion and water runoff on slopes
- Building of houses and domestic annexes near rivers, in the floodable areas
- Building using poor quality materials (adobe), on shallow foundations
- Absence of hydrotechnical works
- Absence of materials and means for intervention in case of emergency, floods included
- Poor warning systems

Based on these official conclusions, a major debate has started in Romania on illegal deforestation, caused by the absence of a coherent law enforcement system, and it is clear that the law needs to be strengthened. One factor that favoured the deforestation was the State handing back the forests to the former owners, from the total of 6,337 million hectares, representing 26.7% of the total land surface (Borlea et al., 2003), 30% now being private property.

The role of forests in preventing floods has been "rediscovered", emphasizing that trees stabilize soils, forest soils act like a "sponge", and leaves hold large amounts of water, delaying infiltration into the soil (coniferous 80%; beech 50%; oak 20%) (Gaspar, 2003). Besides the role played by deforested slopes in floods, it has been ascertained that in many cases tree trunks deposited on slopes have acted like rams; during transportation tree trunks often cause soil compaction and micro-ravines; a major problem was created by trunks stored near the water and by the piles of sawdust.

Another sad conclusion of the subsequent investigations was that many people refused to leave their homes, being afraid of looters, and some of these people paid for this with their lives. One of the major actions following this year's floods was the re-thinking of the national Emergency Situation System, enhancing its response capacity, on a permanent basis, and the allotment of large amounts of money for flood prevention and control (Jelev, 2005).

3. The floods of 2005

2005 was a very rainy year. All Romania suffered floods of greater or smaller magnitude. According to the Report of the Environment and Water Management Ministry (2007), the floods caused by excess water in the rivers, damage to or overtopping of dikes, and drainage on hillslopes affected 1,734 communities, the total value of damages exceeding 1.5 billion euros. The death toll was of 76 persons, and 93,976 houses and domestic annexes were damaged; more than 650,000 ha of agricultural land was severely affected.

One of the most affected areas was Timis County, Western Romania. Here the volume of the April 2005 flood was *three times greater* (about 720–770 million cubic meters) than that of the 2000 flood, though the peaks of these two were very similar. As a result of the high peak, the embankment on the right-side bank of Timis River was overtopped and multiple ruptures of the dikes were reported. Due to the huge volume of the high waters on the inundated floodplain, the flooding covered an area of about 30,000 ha and the volume of inundation was about 250–300 million cubic meters. People called it the "Sea of Banat", where the water stagnated for more than 1½ months, because there were not many routes for gravitational drainage. As a result of this, the pumps installed in the area and the drainage ways made downstream in Serbia-Vojvodina were needed to evacuate the water from the inundated localities and the surrounding areas. Four villages were completely flooded, 3,644 homesteads and 2,344 distressed people were reported (Stanescu and Drobot, 2006a). It is worth mentioning that in 2000 water breached the left-side dikes, flooding a less populated areas, without great damage. Reluctance to create a controlled breaking of these dikes in 2004 allowed water to breach the right-hand dike, with severe consequences (Jelev, 2005).

Besides the floods generated by excessive rain (218 l/m^2 in 24 h), a special situation was created on the Siret River in Eastern Romania, by the sudden opening of the Calimanesti dam (4,300 m^3/s for many hours), the released water flooding large areas of land and destroying many house-holds (Stanescu and Drobot, 2006b). Only the quick response of the authorities and population downstream reduced the huge potential damage.

The Ministry Report identified the main causes of these floods as the excessive rainfall combining with the rapid melting of the deep snow layer (more than 1 m):

- Dike breaking, due to the great volume of water, exceeding the designed retention capacity and resistance in time.
- Accidental holes in fisheries dams, due to the faulty maintenance.
- Large scale deforestation and bad agricultural practices, favoring soil erosion and water drainage on slopes.
- Absence or bad maintenance of ditches for rain water drainage in nearly all rural communities.
- Bad situation of the municipal sewage system (faulty maintenance, under-dimensioning in case of torrential rain).
- Building of houses and domestic annexes near rivers, in the floodable areas.
- Building using bad quality materials (adobe), on shallow foundations.
- Exceeding of the transport capacity of bridges sections, due to their under-dimensioning and clogging with wooden materials, debris or waste, originating from *ad hoc* deposits near rivers or on slopes.
- Absence of materials and means for intervention in case of emergency, floods included (large volume pumps).
- Bad warning systems.
- In some cases population and authorities did not know the actions to be taken in case of emergency, floods included.

It can be seen that many causes are common between the years 2004 and 2005, proving the difficulty of changing mentalities and mending things that have occurred for decades.

4. The floods of 2006

The causes of this year's floods can be divided in three categories:

- Intense rainfall in the upper part of the natural watershed in the mountain regions
- Intense rainfall in catchments altered by humans
- Excessive rainfall in the natural watershed areas in the plain regions (Stancalie and Drobot, 2006a)

After the previous year's floods, it seemed that this rainy year could not bring something new, but, besides floods on Somes (11 deaths) and Siret (6 deaths, 6 disappeared) and huge material losses, this year Romania was confronted with a "one in a hundred years" flood on the Danube (Report of 2006). The meteorological cause was simple: heavy rains in the upper part of the watershed, plus rains in the Romanian tributary catchments, all these leading to flows of 15,775 m^3/s, close to the historic maximum of 15,900 m^3/s, in 1895. The great difference between these 2 years was the absence of wetlands on the Romanian bank. After 1950, a huge programme of "giving back" (!) these lands to agriculture resulted in the loss of 400,000 ha of wetlands that played a critical role in buffering high amounts of water. The presence of more than 1,100 km of dikes alongside the river was intended to protect communities, but the amplitude of the flood exceeded these dikes in certain places, in several cases even breaking the dikes. The deliberate breaking of certain dikes temporarily released the water pressure downstream and prevented the loss of human lives, but the material damages were huge. As a result of these floods, a new strategy was conceived for the Danube banks, including the creation of new wetlands, in order to absorb large amounts of water and to reduce the danger posed to human communities.

5. Conclusions

This brief description of the main floods of Romania during the last years and the human influence may reveal the following aspects:

1. After many years of neglecting the hydrotechical works (not building new ones or letting the old ones to be damaged), heavy rains exceeded the defense capacity of these with disastrous consequences.
2. The absence of clear laws and of means to enforce the existing ones has led to massive deforestation within certain watersheds, thus enhancing the effects of heavy rains.
3. It is good that the causes were discovered, but it takes effort, time and money to control floods in Romania.

References

Borlea, Gh.F., Brad, R., Merce, O. and Turcu, D. 2003: Wood energy in Romania, Annals of I.C.A.S., **46**: 321–326.
Dumitru, M. et al., 2001: Country Report – Romania http://www.icpa.ro/fao_glwi/.
Gaspar, R., 2003: The role and contribution of the Forest Research and Management Institute in the establishment of the Romanian school for torrential watershed management. *Annals of I.C.A.S.*, **46**: 377–383.
Jelev, I., 2005: Can floods be controlled? (in Romanian) **16**: 12–13.

Ministry of Environment and Water Management, 2005: Report on the effects of floods and dangerous meteorological phenomena occurred in 2004: actions taken for rebuilding the damaged objectives and for prevention or avoiding future losses (in Romanian).

Ministry of Environment and Water Management, 2006: Report on the effects of floods and dangerous meteorological phenomena occurred in 2005: actions taken for rebuilding the damaged objectives and for prevention or avoiding future losses (in Romanian).

Ministry of Environment and Water Management, 2007: Report on the effects of floods and dangerous meteorological phenomena occurred in 2006: actions taken for rebuilding the damaged objectives and for prevention or avoiding future losses (in Romanian).

Stancalie, Gh., Irimescu, A. et al., 2008: *Geophysical Research Abstracts*, **10**: 811.

Stanescu, V.A. and Drobot, R., 2006a: Hydro-meteorological characterization of the flood from the period 14–30 April 2005 in the Timis-Bega River Basin, Proceedings of the BALWOIS Conference, Ohrid, FYR Macedonia, May 23–26, 769–773. www.balwois.mpl.ird.fr/balwois/administration/full_paper/ffp-769.

Stanescu, V.A. and Drobot, R., 2006b: High floods in Romania in 2005 – Siret River Basin case study. Lessons on preparedness and prevention role in flood control. Proceedings of the 23rd Conference of the Danube Countries on the hydrological forecasting and hydrological bases of water management, Belgrade, Serbia, 28–31 August.

DEVELOPMENT OF DANGEROUS GEODYNAMIC PROCESSES IN THE SOUTH CAUCASUS AND THE PROBLEM OF MITIGATING THEIR CONSEQUENCES

I.V. Bondyrev[1]* and E.D. Tsereteli[2]
[1] *Vakhushti Bagrationi Institute of Geography, Ministry of Education and Sciences of Georgia, Tbilisi, Georgia*
[2] *Ministry of Environment Protection and Natural Resources of Georgia, Centre of Monitoring and Forecasting, Tbilisi, Georgia*

Abstract. The specifications of distribution and character of dynamic hazards due to water processes is reviewed in the paper: floods, mudflows, landslides, and swamping. The assessment of the damage caused by these processes is also described.

Keywords: Floods, landslides, mudflows, South Caucasus

1. The problems

The most pressing problem in most countries is to protect people and infrastructure from natural disasters and ensuring their sustainable development. This is particularly true in the mountainous region of South Caucasus, where complex natural conditions and excessive human interference cause a whole gamut of destructive processes.

In this respect, the South Caucasus deserves special attention, given its geopolitical importance on the Eurasian Transport Corridor and the "Great Silk Road". Sustainable development here is totally dependent on reliable protection from catastrophes, of which the most dangerous are earthquakes, floods, landslides, mudflows, snow avalanches and water safety. More than 50% of the territory and more than 5,000 inhabited locations are subject to these processes. Just in the last 30 years, more than 35,000 people perished as a result of dangerous geodynamic processes and over 150,000 were resettled to safer regions (eco-migrants). Also, as a result of armed conflicts, more than 630,000 refugees and internally displaced people have been officially registered (331,000 in Karabakh, 280,000 in Abkhazia, 26,000 in South Ossetia).

Erosion of riverbanks and the accumulation of alluvium play a significant role in the rivers of the South Caucasus. Such rivers as the Rioni, Alazani, Araks and Mtkvari, flow in channels packed with alluvium and rising 1–1.5 m above the floodplain, with occasional disastrous floods.

---

* To whom correspondence should be addressed. e-mail: ybondyrev@yahoo.com

The annual damage caused by mudflow processes in Georgia alone is estimated at US$0–20 billion. Rare, high magnitude events occurring every 3–5 years or every 6–11 years cause damage worth hundreds of billions of dollars. One short mudflow near Telavi in the upstream part of the River Tskhenistskali caused US$130 billion of damage in 1977. Mudflows in 1982–1984 and 1989–1991 cost nearly US$10 billion. Less than 3% of mudflows are caused by snowmelt, but 12–13% are due to glacier meltwaters (Bondyrev and Kiknadze, 2001).

Over the last 100 years, more than 200 destructive mudflows have occurred, with damage to the economy of Azerbaijan of US$1.5 billion. More than 100 settlements, with a total population of nearly 700,000 people, are under constant threat from mudflows. Nearly 1,200 of the 3,000 major highways in the country pass through areas with high risk of mudflow, and are constantly subjected to damage. Disastrous mudflows occurred due to rainstorms in 1993 in the basin of the Tskhenistskali, Rioni and Alazani rivers, causing significant damage in the cities of Telavi, Lentekhi and others. In Georgia and Armenia, the maximum intensity of precipitation is 3–6 mm/min, in Azerbaijan it is 10 mm/min, but in Rize (Turkey) 18 mm/min.

Mudflows around Mleta in the Aragvi basin have existed for a long time, but they entered a period of disastrous activity after the demise of the USSR. The mountainous slopes of the upstream reaches of the River Aragvi have become unable to feed the huge flocks of sheep, which in the past spread onto the Kizlar steppe in Russia. The closing of the borders has brought unprecedented new pastures for cattle. The mountain meadows have not had time to regenerate the layer of grassy sod, and this has caused extremely active erosion, and, as a result, destruction of whole ecosystems that took ages to develop and with this the loss of practically the whole livestock of sheep and what is more important, irreversible erosion. The village of Mleta (Table 1) will have to move to a safer place (Bondyrev et al., 2007).

TABLE 1. Feature of mudflow basins in South Caucasus

Mudflow basin	Area of basin (km^2)	Area affected (%)	Peak volume (1,000s km^3)	Recurrence (mudflows per year)
Georgia				
R.Durudji (Kvareli)	84.0	9	640	2–10
R.Aragvi (Mleta)	3.2	85	5–18	1–3
Azerbaydjan				
R.Kish-tchai	165	45.5	1,000–2,175	3–10
R.Orubad-tchai	40	27.5	500–1,000	15
"Selav" to Talish	223.7	–	200	1

In South Caucasus, disastrous floods are caused either by rapid warming after a snowy winter and intensive snow melting or by bursting of dammed lakes in the nivo-glacial area (Figure 1). Flooding on the Rioni River in 1982, inundated 130 km^2 and cost US$12 million. The 1987 Rioni flood cost US$300 million. A similar situation is found in Azerbaijan and Armenia. Lack of flood defences and absence of monitoring and forecasting systems are the main reasons for increased flood risk.

Before irrigation measures were carried out in the middle of last century, floods on the Kolkheti lowland occurred a number of times a year. As a result of a disastrous flood on the River Rioni in 1811–1812, the population in Imereti was reduced 30–35%. Recent data show that raising of the water level in the River Rioni (Figure 2) caused principally by the above-mentioned processes, in 1839 reached 9.6 m, in 1911 – 2–3 m, in 1922 – 2.8 m (discharge of 2,420 m^3/s), but in 1968 – 7.7 m (discharge of 5,220 m^3/s). In 1982 130 km^2 were flooded causing US$12 billion of damage.

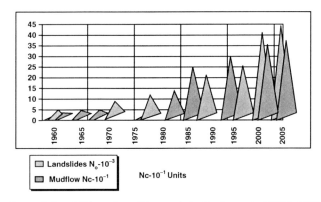

Figure 1. Landslide and mudflow activity for the period 1960–2005

Figure 2. Eroded banks of the R. Mudkhuri near v.Jamushi (Svaneti)

Floods in Azerbaijan are mainly observed on the rivers Kura and Araks. Since construction of the Mingechaur reservoir on the R. Kura in 1953 and the Araks on the R. Araks in the 1970s, the scale of flooding has vastly decreased. However, intensive regulation of the flow does not provide total protection. Increased frequency of floods and accelerated surface erosion in the Kura and Araks basins, speed up silting of these reservoirs and reduce their flood storage effect. Thus, the maximum depth of the Mingechaur reservoir today has been reduced from 83 to 63 m (Bondyrev et al., 2004).

Due to lack of necessary hydrometeorological information on the River Kura from other countries upstream, the large reservoirs are used ineffectively and the floods in the downstream reaches of the river in 2001 and in 2003 are evidence of this. It has been estimated that 56% of Georgia, about 45% of Armenia and 15% of Azerbaijan are situated in avalanche prone areas (Table 2).

TABLE 2. Precipitation (mm) in basins of main rivers of Caucasus-Pontius region (Vladimirov, 1991)

Basin	Height zones (m)							
	0–500	500–1,000	1,000–1,500	1,500–2,000	2,000–2,500	2,500–3,000	3,000–3,500	3,500–4,000
Mzimta	1,500	1,700	2,050	2,440	2,780	3,290	3,380	–
Psou	1,500	1,780	2,050	2,300	2,460	2,640	2,760	–
Bzibi	1,860	1,680	2,060	2,580	2,980	3,390	3,780	–
Codori	1,580	1,660	1,860	2,250	3,100	3,840	4,020	4,160
Inguri	–	–	1,080	1,310	1,640	1,980	2,120	2,200
Rioni	–	1,180	1,500	1,800	2,020	2,100	2,180	2,180
Iori	–	680	860	1,100	1,460	1,920	2,140	–
Alazani		900	1,100	1,550	2,000	2,180	2,200	–
Chelti	–	1,200	1,640	1,860	2,060	2,260	2,300	–
Supsa	1,840	1,920	2,060	2,220	2,280	2,300	–	–
Natanebi	2,000	2,180	2,420	2,660	2,840	2,900	–	–
Chakvist-skali	2,720	3,060	3,400	3,660	–	–	–	–
Machakhelist-skali	–	2,200	2,340	2,620	2,820	2,940	–	–
Adjaristskali	–	1,450	1,400	1,700	1,980	2,300	–	–
Nakhichev-antchai	–	–	320	420	530	920	1,220	–
Lencoran	1,440	1,400	1,360	760	540	–	–	–

Clean water reserves per capita in Georgia and Armenia are amongst the highest in Europe, but by consumption level and the supply of clean water they are bottom of the league. About 50% of drinking water sources in the region have no proper sanitation facilities and are breaking the sanitary rules. The reason for this is complete breakdown of water-supply systems.

A similar situation obtains in Azerbaijan. A system of so-called "red areas" should be introduced to identify areas of ecological disaster, which are the most urbanized and ecologically most sensitive areas (Figure 3).

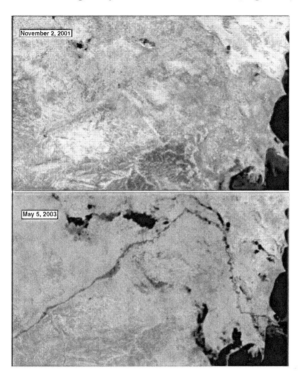

Figure 3. Flood on R. Kura in Azerbaijan, 3 May 2003 (*bottom* photo) and the normal condition of the river network (*top* photo). (NASA images)

2. Conclusions

It is necessary to create a united centre in South Caucasus for monitoring and managing these risks, since they are transboundary in nature. The priorities are: (1) development of a conceptual model for managing the environment, based on principles acceptable to all countries; (2) collection, analysis and generalization of information on all types of geodynamic hazards in South Caucasus; (3) determining the mechanisms, energy sources and trends of development of these processes; (4) analysis of the efficiency of all previous defensive actions in the region and a quest of new, more effective methods of protection, considering their economic value and engineering possibilities; (5) creation of a general regional databank for South Caucasus, developed on the basis of GIS technologies and constantly updated on a data base of GPS.90.

References

Bondyrev, I.V. and Kiknadze, A.G., 2001: Problems in monitoring, analysis and dissemination of environmental information in Georgia, *NATO Adv. Res. Workshop Integrated technologies for environmental monitoring and information production* (est. ARW 977883), Marmaris, Turkey, Dokuz Eylul University.

Bondyrev, I.V., Tavartkiladze, A.M., Tcereteli, E.D., et al., 2007: *Geography of catastrophes and risk in area of humid subtropics the Caucasian-Pontides region.* Tbilisi, Poligraf, 376pp.

Bondyrev, I.V., Tatashidze, Z. K., Singh, V.P., Tsereteli, E.D. and Yilmaz, A., 2004: Impediments to the sustainable development of the Caucasus-Pontdes region. *Journal of International and Comparative Social Welfare*, **20**(1): 33–48.

Vladimirov, L.A., 1991: *Water balance of Caucasus.* Tbilisi, Metsniereba, 106pp.

THREATS IN THE TIENSHAN-PAMIR REGION OF KYRGYZSTAN

I. Hadjamberdiev[*], V. Shablovsky and V. Ponomarev
Central Asia NGOs Network "Water and Pollution", P.B. 1451, Bishkek 720040, Kyrgyzstan

Abstract. There are over 5,000 mountain lakes in the region, and 724 of them can potentially destroy villages and farmland. In the case of Alpine lakes, failure can produce discharges containing many stones, which have a large destructive potential. Several cases of such disasters have occurred over the last 10 years – a tenth of the dwellings in several villages and two towns had been destroyed. Many water reservoirs were built up to 20–30 years ago, now partly ruined and situated in a zone of high seismic activity (up to 9 on the Richter scale). Thus, these water reservoirs may be the starting point of disasters and floods in the event of earthquakes or terrorist acts. Other disasters can happen on the riversides. There are many warehouses and tailings situated on the banks of rivers. A major fish kill caused by poison from a pesticides warehouse in Son-Kul, which is a high-altitude lake (storage – 2.64 km^3) in Kyrgyzstani Tien-Shan in 1977.

There are two extremely dangerous warehouse or tailings areas in TienShan. The first one is due to cyanide contained in waste from gold mining in the Kum-Tor area, 4,100 above sea level. The Petrov-Davidov Glacier is situated above the dump and can destroy it, and the waste flow down to the River Ara-bel and thence down to the Narin River. Such an event would be dangerous for 80,000 riverside residents. The other area, with the uranium storehouses, is situated in the Mailuu-Suu River region. There are tailings, number 3 and 7, which are especially unstable, and in danger of flowing down to the town and the River Mailuu-Suu. All the dangerous processes mentioned are sensitive to global warming: glaciers and snow melting, increases in river discharges, etc.

Keywords: Debris flows, mine tailings, fish kills, earthquakes, reservoirs

1. Introduction

There are over 5,000 mountain lakes, and 724 of them can flatten villages in TienShan-Pamir region. Bursting of the Alpine lakes can carry many stones, and have great destructive potential. Several cases of such disasters

---

[*] To whom correspondence should be addressed. e-mail: igorhodj@mail.ru

J.A.A. Jones et al. (eds.), *Threats to Global Water Security*,
© Springer Science + Business Media B.V. 2009

have occurred during the last 10 years – dwellings in several villages and two towns had been destroyed. There are 13,954 water-landslide danger points in the TienShan-Pamir mountain system. Many water reservoirs in the region were constructed many years ago. There are many warehouses with toxic wastage situated on the banks of rivers. In the event of failure, it would be like a "water-Chernobyl". The cause of these catastrophes can be natural (groundwater rise, soil saturation, landslides, avalanches or earthquakes), or man-made (unstable waste dumps or damage to drainage channels).

2. Alpine water reservoirs and lakes

Many water storages were constructed 20–30 years ago, are now in a ruinous state, and situated in highly seismic zone. Examples include Toktogul on the Naryn River (storage 19.5 km^3, dam height 215 m) and Andijan on the Kara-Darja River (storage 1.94 km^3, dump height 115 m). These water reservoirs may be starting point for disasters and floods in the event of an earthquake.

There is often formation of a saturated soil-stone mass on the slope above river bank, and after this slides down the hillside it moves very quickly along the river valley and destroys everything. There are at least 5,500 such water-landslide danger points in Kazkhstan (called "sely"), with about 4,000 in North TienShan (Vice Minister, 2006). As a result of these processes, 246 villages in Kyrgyzstan and Tadjikistan were endangered by landslides in winter 2004, and about 3,000 people were evacuated in that unfortunate season.

But the most dangerous point in the region is Sarez Lake (in the Pamir system, Republic of Tadjikistan). It was created in 1911 by a large rock fall on the Murgab River. Water volume rises year by year and storage has reached 17 million cubic meters. The level of the lake rises up to 2 m annually. There is a real danger of more rock falls. This would lead to a big wave of water running through the territory of Tadjikistan, Uzbekistan, North Afghanistan, endangering 6 million people.

3. Warehouses with toxic wastage

There are many warehouses (16 storing uranium, 54 obsolete pesticides, 2 storing cyanide in our region) situated on the banks of rivers. There was a release from a pesticides warehouse that killed fish in Son-Kul, which is a high-altitude lake (3,016 m above sea level, storage – 2.64 km^3) in Kyrgyzstan TienShan in 1977.

Some examples follow. There is a large waste warehouse containing cyanide from gold mining (Kum-Tor area, 4,100 m above sea level, warehouse stores a total volume 100 million cubic meters, in an ice-gravel dump), see Fig. 1 (Torgoev et al., 2003). The Petrov-Davidov Glacier is situated above the dump and can destroy it, in which case all the waste will flow down to the River Ara-bel and on to the Narin River (length 807 km, average discharge 432 m^3/s) (Torgoev et al., 2003). Such an event would endanger 80,000 riverside residents.

The second group are uranium warehouses in the Mailuu-Suu region (Uranium tailings in Table 1, especially numbers 3 and 7), which can be washed down to the nearby town and into the River Mailuu-Suu, endangering 48,000 riverside residents.

Third example is the largest sewage lake in former USSR territory, holding wastewater from both industrial and domestic sources (Izvestia Kazakhstan, 2006). It is called Sorbulak, in the district of Almaty. Its length is 15 km, width 82 m, and volume about one billion cubic meters. It contains heavy metals and other toxins. The security of lake is supported only by a sand-dump, which is degrading year by year. There is a great danger of it bursting, as one did in 1988 in Jamankum, when 30 billion cubic meters were released. If it occurs in Sorbulak, several settlements would be washed away in 10–20 min, beginning with Kurty and Akchy, with a total population of about 50,000.

4. Water pollution study at the former uranium factory near Issyk-Kul Lake

The formation of a pollution leakage mostly by underground waters in the Kadzi-Sai tailings repository area is caused by the complex interrelation of cumulative stratigraphic, hydrological, and climatic factors (KAJI-SAY uranium tailings demonstration, 2006). The most probable primary source of environmental pollution is a very permeable water-bearing horizon of proluvial-talus sediments laid down in former channel beds. The pollution is possibly a result of washout and removal from places where it is buried on the right bank of the Western Sai by rain and freshet waters, and also a result of washout of a protective dam, heap number 2, which is in an unstable condition, by freshet waters from the Dzilubulak-Sai. A secondary influence on the pollution could be water-bearing horizons of lake sediments and the Issyk-Kul Lake, but the main source of pollution would be surface freshet waters.

The hydrological network of the area is represented by temporarily active or ephemeral channels. From the eastern part of the toxic tailings burial site, the Eastern Sai channel flows to join Western Sai channel. The

Dzilubulak-Sai channel, being formed at the merging of the two river branches, runs into the Issyk-Kul Lake. The basic hydrological characteristics of the Dzilubulak-Sai area are given below.

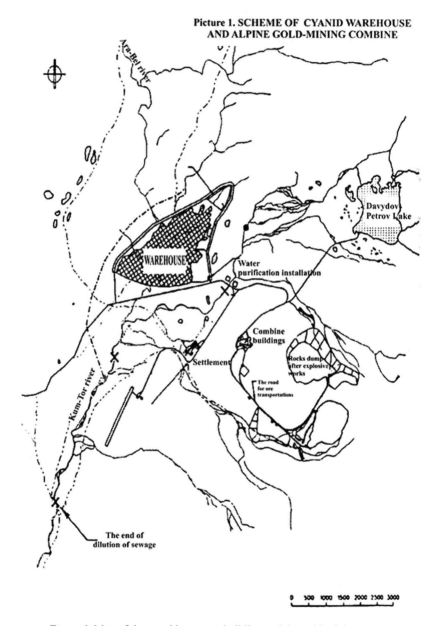

Figure 1. Map of the cyanide storage building and the gold mining works

TABLE 1. Uranium tailings on the banks of the Mailuu-Suu River

Name and place	Date of construction	Volume (1,000 m^3)	Exposition dose min/max (micro-roentgen per hour)
Number 2a, Ailampa	1967	85	20/40
Number 2b, Ailampa	1967	65	20/40
Number 3, "Izolit"	1954–1958	110–150	20/800
Number 4, Ailampa	?	115	25/330
Number 5, Mailu-Suu River, right bank	?	111	20/400
Number 6, in the same	1970	35	15/30
Number 7, same	1958	600	15/55
Number 8, same	?	90	15/30
Number 9a, Mailuu-Suu River, left bank	?	115	30/60
Number 9b, in the same	?	50	40/70
Number 10, Mailuu-Suu River, right bank	?	70	20/30
Number 12, Ailampa-Say	?	62	20/30
Number 13, the same	?	40	30/360
Number 14, the same	?	99	15/30
Number 15, Suget-Say	?	47	15/90
Number 16, Ashvaz-say	1968	303	16/20
Number 17, Mailuu-Suu River, left bank	destroyed in 1994 by landslides		
Number 18, near no. 3	?	3	25/800
Number 19, Mailuu-Suu River, left bank	?	1	15/25
Number 20a, Mailuu-Suu River, right bank	?	1	15/25
Number 20b, Mailuu-Suu River, right bank	?	2	18/85
Number 22, Mailuu-Suu River, left bank	?	2.2	25/18

Figure 2. Dwelling destroyed by landslide on banks of Mailuu-Suu River

The beginning of the pollution event is located approximately 150 m below the bridge across the Dzilubulak-Sai channel. At the start of the zone, the channel is entrenched into red sandstone and gritstones of the Kyrgyz ravine to a depth of about 10 m. As a result of the road being constructed on ash heaps, the channel easily breaches the barrier. The holding capacity created by the barrier is estimated at approximately 1.6–2,000 m^3. As inflow increases above the holding capacity, the water would pour across the ash heap along a length of about 220 m, concentrating erosion at the foot of the slopes. A breach across the barrier is quite probable and the debris flow wave from higher up the drainage system would add to this. In this case, the sediment-laden water flow is transformed into a mud-flow, full of sand and ash heap products. Recharge with sediment occurs as a result intensive destruction of the banks by debris flows, both by direct abrasion and washout, and as a result of cave formation and shock or corrosive action involving crack formation and subsequent collapse of whole blocks of the bank.

At present, understanding of the characteristics of debris flows is limited to natural conditions, i.e. without taking into consideration the human-induced factors. The conditions of debris flow movement in the zones undergoing human economic activity are very complex, and urgently require careful scientific research, which at present is practically absent.

5. Conclusion

All the dangerous processes mentioned above will be aggravated by global warming: glaciers and snow melting, increasing discharges and velocities in rivers, etc. The preventive measures needed include: prevention of over-irrigation (Shablovsky and Polyak, 2003); making riversides and waste dumps stronger with cement, although this can only withstand earthquakes of up to 6 on the Richter scale; monitoring of melting rates; creating new lake; and simulation of water and landslide events before they can create a dangerous incident.

References

Hadjamberdiev, B. and Hadjamberdiev, I., 1986: *Eco-geography of Kyrgyzstan.* Bishkek, Geogr. Society, Acad. Sci., 284pp. (In Russian.)

Izvestia Kazakhstan, 2006: 31 July. (In Russian.)

Kaji-Say Uranium Tailings Demonstration, 2006: Techl Rep ISTC, Bishkek, Scientific Research Center "Georisk".

Shablovsky, V. and Polyak, E., 2003: Automatization System of Irrigation Water Accounting. In: Agro Press, Bishkek, Ministry of Rural Water Economy of Kyrgyzstan, No. 6, pp. 33–34. (In Russian.)

Torgoev, I., Aleshin, U. and Moldobaeva, B., 2003: *Geological security and risk.* Bishkek, JEKA, 288pp. (In Russian.)

Vice-Minister Emergency Situation of Republic of Kazakhstan, 2006: V. Petrov Interview. In: *Kazakhstan Today*, 9 February. (In Russian.)

LONG-TERM PROGNOSIS OF HIGH FLOWS IN THE MOUNTAIN RIVERS OF GEORGIA

T. Basilashvili[*]
Institute of Hydrometeorology, Tbilisi, Georgia

Abstract. Climate change over the last century combined with increased human interference has increased the frequency of droughts and floods. These extreme events should not necessarily constitute a threat to the human population, provided water is carefully managed. The lack of hydrological data, especially for mountain rivers, is a major hindrance to management. The research reported here aims to provide a predictive framework to assist water management in the various mountainous regions of Georgia. Timely warning about an oncoming danger enables us to avoid material loss incurred by the element. Early forecasts (a few months in advance) save us from the above as it enables to take certain safety measures in time and to minimize the loss.

Keywords: River runoff, multifactor statistical model, long-range forecasts

1. Introduction

Landscape and climatic conditions in the mountain rivers result in high and prolonged floods from April to August in the highlands and from March to June lower down. In turn, these floods rather frequently lead to ecological damage, do immeasurable harm to the environment and bring about enormous economic loss for the country. Therefore, forecasting these floods is of vital importance from both the scientific and economic point of view. The basins of these rivers differ greatly from each other in size, varying from 280 to 2,800 km^2, and also they differ in physical-geographic and climatic conditions. Unfortunately, the information available on these mountain rivers is inadequate for proper development of forecasting methods, which are so important for national economy. Therefore, management of much of the water resources of Georgia has to proceed with limited data.

The specific, local conditions for the formation of river runoff here makes using the information from neighboring basins impossible. Absence of definite regularities of space-time distribution of atmospheric precipitation, the main runoff-forming factor, makes it impossible to extrapolate or interpolate, and also the application of any mathematical model for the estimation of their values in the ungauged basins. Solution of water balance equation

---

[*] To whom correspondence should be addressed. e-mail: jarjitabatadze@yahoo.com

for the critical prognostic period, which comprises numerous factors, such as runoff, precipitation, evaporation, condensation, infiltration, etc., is the physical basis of runoff forecasting. Modern mathematical prognostic models have been developed taking into account these factors. They may be used in areas with well-developed hydro-meteorological networks and in those regions where the information is regularly recorded. In some countries, this process is automated (WMO, 1986).

2. Development of statistical models

Because of the complicated mountainous relief of Georgia and the multitude of riverflow factors, modern forecasting models cannot be used. Therefore, for water flow forecast we have to use simple, multifactor statistical models based on limited data of standard observations on water discharge Q, m^3/s, atmosphere precipitation P, mm, water content of snow W, mm and air temperature θ, °C, during a certain time:

$$Q_{t+T} = f(P_{to}, \theta_{to}, Q_{to}, P_{to+1}, \theta_{to+1}, Q_{to+1}, \ldots, P_{t-1}, \theta_{t-1}, W_{t-1}, Q_{t-1}, P_t, \theta_t, W_t, Q_t, \theta_T, P_T)$$

$$(1)$$

where Q_{t+T} is a forecast of water flow at time t with T being the duration of the forecasting period, and t_o – starting time of data calculation (Basilashvili, 2000).

As a result, the statistical model considers all available runoff-forming factors with allowance for the time dynamics of the hydrometeorological elements in the formation of the runoff. In the model, breaking the year up into different periods is necessary, because the responses change. For instance, the flood flow is differently affected by precipitation in autumn, winter and spring. Therefore, it would not be right to treat them the same. For forecasting there can be used only early or delayed information. Sometimes, some of the factors can be excluded from the forecasting model, e.g., P_T and θ_T. Thus, forecasts containing different factors should be worked out in advance for different periods. The model is suitable for making all kinds of forecasts, both short-term and long-term. While working out a certain kind of forecast, we modify the model taking into consideration the forecasting flow formation and factors affecting it.

We developed the program, which is still used, with this aim (Basilashvili, 1977). In order to select forecast predictors, a correlation matrix and the weighting of individual factors have been identified. Applying certain mathematical criteria, no effective factors are excluded (Dreiper and Smith, 1973). For choosing an optimal system of predictors out of all the factors, we use the multi-step sifting method along with Fisher's and Student's criteria (Basilashvili and Plotkina, 1985).

For example, for the flood period from April till August on the River Enguri near the village of Khaishi, which is calculated to assist management of the Enguri power station, a multifactor model has been established with 12 potential factors. The model has been established with the help of physical and correlative analyses (Equation 2), where the indices are used to indicate the month and weeks under consideration. Having selected the predictors, four factors remain, the multiple correlation coefficients (R) of which range from 0.71 to 0.90 (Equation 3).

$$Q_{IV-VIII} = f(Q_{IX.2}, \theta_{X-XI}, P_{XI-I}, Q_{XII.3}, \theta_{II.2}, W_{II}, P_{II}, Q_{II.2}, \theta_{III}, P_{III}, W_{III}, Q_{III}) \quad (2)$$

$$Q_{IV-VIII} = f(P_{II}, Q_{IX.2}, P_{XI-I}, Q_{II.2}) \quad (3)$$
$$R = 0.71, 0.84, 0.88, 0.90.$$

Optimal models for runoff forecasting are worked out for each forecast period. These models consist of mainly three to four factors, which are quite acceptable, as the available series of observations covers 45 years. The digital value of the prognoses relations and estimation are defined using a corresponding program (Alekseev, 1971), where direct and reverse expansion of multifactorial equations are used to investigate all the possible alternatives, while calculating two systems of equations through adding up earlier and delayed information. This enables us to simultaneously investigate the effects of decreasing or increasing the length of all the predictors or excluding some factors acting during the forecast period. As a result, we can get prognosis formulae with different information, precision and length of advance warning.

We should note that for obtaining the prognostic equations we consider standard values rather than the direct ones in Dreiper and Smith (1973). In this case, statistically stable, empirical coefficients are obtained. The coefficients fully characterize the precision of the relationships, including curved line correlation as well. We can also explain some errors related to the dimension. Solving Equation 3 allows us to obtain the following prognostic equations:

	S/σ	p%	r
$Q_{IV-VIII} = 175 + 0.41\, P_{II}$	0.72	66	0.71
$Q_{IV-VIII} = 136 + 0.36\, P_{II} + 0.39\, Q_{IX.2}$	0.61	81	0.82
$Q_{IV-VIII} = 111 + 0.36\, P_{II} + 0.35\, Q_{IX.2} + 0.12\, P_{XI-I}$	0.52	86	0.87
$Q_{IV-VIII} = 101 + 0.23\, P_{II} + 0.44\, Q_{IX.2} + 1.27\, Q_{II.2}$	0.48	94	0.89

According to Gidrometeoizdat (1962), the criteria for estimation of the prognosis are: S/σ – average square error of the prognosis in ratio to the average square deviation of the forecasting element, p – correctness of forecasting, and r – correlation coefficient between real and forecasting values.

While forecasting riverflows, the multilateral character of the forecasting equations enables us to select them according to the available information. The results obtained can control each other and the expected interval of the riverflows can be defined. It should be noted that according to this program (Gidrometeoizdat, 1962), there can be obtained data, delivering the forecasts both ordinary and probabilistic with the provision of 5–95%.

From these investigations, a methodology has been worked out for 12 rivers in Georgia to predict flows for different periods in advance. The resulting economic benefit exceeds the results obtained from normal forecasting methods by 10–35%. No consideration has been given to the improvement in efficiency produced by increasing the forecasting period, which in our case is: maximum of high-floods 12–24 h, daily discharge 1–2 days, weeks for 2–3 weeks during 2–3 months, quarterly 3–4 months, average flooding 3–5 months, maximum discharge over 2–4 months, discharge for the growing season 6–7 months.

Forecasts of high water flows for the most important public units have been achieved through application of the above method. For the yet unstudied rivers, common regional forecasts have been worked out. Various, prognostic equations have been worked out for each separate range and region, depending on a variety of information with different advance periods (4–6 months) and accuracy. In turn, this enables us to select them according to the information available, considering the required earliness and accuracy.

Maximum high water discharge has also been carefully studied. Discharges increase with intense snow melting and rainfall downpours that is unpredictable beforehand, i.e., 2–4 months earlier (in March) before they reach their peak. Thus, no forecasts have been made with good confidence levels for many of the 17 rivers investigated.

A few examples are given here of the forecasts obtained with their criteria of estimation for the River Mtkvari at the city of Tbilisi (Equation 5):

	S/σ	p%	r
$Q_m = 463 + 1.41W_{III}$	0.77	59	0.65
$Q_m = 760 + 1.23W_{III} - 69\theta_{III}$	0.73	63	0.71
$Q_m = 636 + 0.93W_{III} - 110\theta_{III} + 2.39Q_{III}$	0.69	66	0.75
$Q_m = 752 + 0.63W_{III} - 117\theta_{III} + 2.78Q_{III} + 2.6P_{XII-II}$	0.64	70	0.80

Reservoirs, of which there are 37 in Georgia, are the most efficient means of managing water resources. Hydroelectric stations, irrigation and drainage canals and aqueducts for water supply have been constructed. Their management is very dependent upon the quality and timeliness of hydrological forecasts. Efficient methods of predicting inflows to the reservoirs is of paramount importance, as only they can ensure there is no stoppage in operation of the hydropower stations and other systems.

Computer software has been designed to develop optimal prognostic equations. The investigations have led to working out long and short term forecasts of riverflows for different periods of the year. For the main flows, the software predicts the average and maximum discharges of high waters, and the growing season, quarterly, monthly, weekly and daily discharges of rain-generated high waters. General flood forecasts for the country have been drawn up for river basins that have not yet been investigated, as they rather frequently create danger for reservoirs and the environment.

The existing environmental conditions and the gradual process of deforestation increase floodflows in Georgia. The rivers Rioni and Enguri have been chosen for developing the methodology of prognostication, as these rivers are the most hazardous and the most important from the economic point of view. The mathematical methodology has already been worked out for the Rioni, because 12-hourly data are available. However, currently the necessary network of investigation has been reduced and we face the need of devising more realistic methods to get operational forecasts from limited data.

Analyses of rain and rain-snow flows into the river networks for various seasons of the year show that the conditions of their formation vary greatly and appear to be individual for every single flow depending on the time and area. Therefore, the possibility of getting absolutely identical flows from a given input is practically ruled out. In the upper reaches of the Rioni, the flows can be observed only in the warm season, while in the lower reaches high flows are common throughout the year when daily precipitation reaches 20 mm. The past maximum discharge with 1% probability (100-year flood) exceeds the average peak discharge by 2–6 times.

Due to lack of data, forecasting the riverflows based on statistical methods appears to be of paramount importance. For the aim of forecasting the peak of high waters, a practical forecast model has been drawn up taking into consideration physical analysis and practical opportunities. Special attention was paid to the influence of hydro meteorological factors on the amount of maximum water discharge (Q, m^3/s) according to the existing daily data: precipitation (P, mm), average ($\theta°C$) and maximum air temperature ($M°C$) and the atmospheric moisture deficit (D, mm):

$$Q_{max} = f(Q_b, P_b, \theta_{n-1}, P_{n-1}, M_{n-1}, D_{n-1}, \theta_n, P_n, M_n, D_n) \qquad (6)$$

The subscripts here indicate the following: b – initial (before the flood), $(n-1)$ – a day before the maximum and n – on the very day of the flood peak.

However, the large number of variables encourages a greater spread of results. Therefore, by application of various mathematical criteria and of the method of step by step sifting of each parameter, optimal forecasting models have been established, through which the basic prognostic equations for different seasons have been developed.

The results are generally satisfactory: $S/\sigma = 0.80 \div 0.30$, correctness $p = 60\% \div 100\%$, coefficient of correlation between the actual and the forecast values $r = 0.60 \div 0.97$. For instance, for the River Enguri, for different seasons of the year, the following prognostic equations have been worked out:

	S/σ	$p\%$	r
$Q_{max(IV-VI)} = 33.9 + 0.32P_b + 1.25Q_{n-1}$	0.64	77	0.76
$Q_{max(VII-IX)} = 77.2 + 1.04Q_b + 8.57P_{n-1}$	0.56	80	0.84
$Q_{max(X-III)} = 1.78 + 0.57Q_b + 1.0Q_{n-1}$	0.36	85	0.94

It should be pointed out that for predicting the peak for each season, a few versions of the equations have been devised based on various initial data. This enables us to pick up a version for forecasting according to the initial information available and cross-control the results. For instance, the following prognostic equations with the corresponding values have been calculated for the River Rioni for the spring–summer period (III–VIII):

	S/σ	$p\%$	r
$Q_{max} = 269 + 0.71Q_b$	0.65	75	0.76
$Q_{max} = 233 + 0.73Q_b + 0.78\theta_{n-1}$	0.63	77	0.77
$Q_{max} = 249 + 0.72Q_b - 9.14\theta_{n-1} - 5.72M_{n-1}$	0.62	79	0.78
$Q_{max} = 296 + 0.73Q_b + 14.0\theta_{n-1} - 13.8M_{n-1} + 12.3P_b$	0.61	80	0.79
$Q_{max} = 283 + 0.71Q_b + 7.49\theta_{n-1} - 9.2M_{n-1} + 13.9P_b +$	0.60	82	0.80
$\qquad 1.7P_{n-1}$			

The fact that the satisfactory equations have been worked out for the spring–summer period appears to be of paramount importance as the maximum discharges occur in these seasons. Moreover, these discharges are rather dangerous and harmful for the environment. Application of these forecasting methods in operational work enables us to prevent expected dangers in good time and to carry out safety measures that will significantly reduce material losses.

If the prognosis substantially exceeds the average of maximum discharge, certain measures should be taken to make timely warning about the oncoming disaster. The timely prognosis allows taking certain measures in an attempt to save population as well as property. More importantly, the reservoirs should gradually be emptied so as to let them receive and withstand another high flood. This is particularly true and important in the case of Jvari reservoir weir, as an emergency here is very likely to result in calamity in Georgia in particular and the Black Sea coastline in general.

References

Alekseev, G., 1971: *Obiektivnye metody viravnivania normalizatsii korelatsionykh sviazei*, 365 p., Leningrad: Gidrometeoizdat (in Russian).
Basilashvili, Ts., 1977: Anotirovani ukazatel algoritmov i program. Obninsk, VNIGMI-MCD, 42–43 (in Russian).

Basilashvili, Ts., 2000: The method of working-out hydrological prognosis in the conditions of limited information availability. *Bulletin of Georgian Academy of Sciences* **62**(1), 110–112.

Basilashvili, Ts. and Plotkina, I., 1985: *Anotirovani perechen novikh postupleni v OFAP Goskomgidrometa, vipusk* **4**: 21–22, Obninsk: VNIGMI-MCD (in Russian).

Dreiper, N. and Smith, G., 1973: *Prikladnoi regresioni analiz*. Moscow: Statistica, 392pp. (in Russian).

Gidrometeoizdat, 1962: Nastavlenie po slujbe prognozov. *Razdel 3, ch. I*, 193pp., Leningrad (in Russian).

WMO, 1986: Inter comparison of models of snowmelt run-off operational hydrology. *Report No. 646, 36 pp.*

WATER DISASTERS IN THE TERRITORY OF ARMENIA

T. Vardanian[*]

Department of Physical Geography, Yerevan State University,
1 Alek Manoukian Street, Yerevan, 375025, Armenia

Abstract. This work is dedicated to the study of flood and mudflow formation on the territory of Armenia. It studies and evaluates the peculiarities of flood and mudflow formation in the recent 70–80 years, the regularities of their extension as well as the ways of struggling against them. The integral curves of changes of annual absolute maximum runoffs and their module coefficients have been calculated for several rivers in provinces of Armenia having different bio-climatic conditions. According to the curves, for four of the five rivers studied the absolute maximum runoff has a tendency to decrease. However, their runoff recurrence frequency has increased. The same pattern is observed in the case of mudflows.

According to the data of the State Committee on Emergencies (STE), during the period of 2004–2006, the damage caused by floods and mudflows was estimated at US$5–6 billion.

Keywords: Water disasters, floods, mudflows, dynamics of annual absolute maximum runoff, integral curves of mudflows and floods, global warming of climate, growth of frequency of mudflow and flood occurrences, mitigation of water disaster hazards

1. Introduction

In the age of present-day scientific and technical achievements, man's safety is still scarcely ensured. This is specially connected with the process of forecasting and managing natural phenomena. Today, man is unable to stand up to some natural disasters, such as earthquakes and floods, of which he can become a victim anytime. Science has developed thousands of mathematical models for forecasting natural disasters, which, nevertheless, cannot entirely explain and predict the cause-and-effect relationship of the emergence of these phenomena. The reason is that natural phenomena are affected by many factors and cannot be managed by science; they have complex mechanisms, like a living organism.

The territory of the Republic of Armenia is characterized by intense and widespread natural hazards that are a result of diverse natural climatic and geological conditions, as well as the influence of the anthropogenic factor.

---

[*] To whom correspondence should be addressed. e-mail: tvardanian@ysu.am

Of natural disasters affecting water security, floods, mudflows and land-slides are typical of Armenia.

According to the data of the State Committee on Emergencies (STE), the vulnerability of the Republic of Armenia to natural disasters shows the following picture: landslides 2.2%, high hazard mudflow territories 0.17%, low hazard mudflow and flood-prone territories 20–30%, inundated terri-tories about 11% (http://www.ema.am).

This work studies and evaluates the dynamics of changes in the annual absolute maximum runoff of some rivers in Armenian provinces with dif-ferent natural climatic conditions (Fig. 1, Table 1).

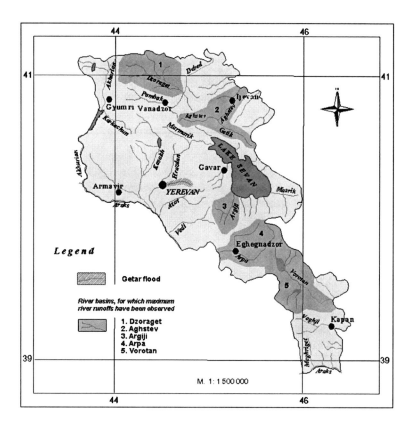

Figure 1. The hydrological network of the Republic of Armenia

This study covers the period of last 70–80 years and includes integral curves of their module coefficients. The location of mudflows in Armenia, the peculiarities of their formation as well as the damage caused by them were also studied. The work suggests some methods of mudflow prevention.

TABLE 1. Several hydrometric, hydrological and meteorological characteristics of the studied rivers

No. on map (Fig. 1)	River observation post	Area of drainage basin, km²	Mean altitude of drainage basin, m	Observation period	Maximum discharge Q, m³/s	Runoff, Q, m³/s	Annual average	
							Precipitation, P mm (station)	Temperature, T, °C (station)
1	Dzoraget below the Gargar	1450	1,860	1932–2006	395 (1959)	16.4	673 Stepanavan	7.9 Stepanavan
2	Aghstev – Dilijan	303	2,000	1939–2006	61 (1989)	3.2	620 Dilijan	8.2 Dilijan
3	Argiji – Getashen	366	2,470	1927–2006	265 (1938)	5.5	480 Martuni	5.7 Martuni
4	Arpa – Areni	2040	2,110	1931–2006	365 (1994)	21.9	349 Areni	12.9 Areni
5	Vorotan – Angegha-kot	2020	2,280	1927–1994	130 (1960)	22.6	382 Sisian	7.5 Sisian

2. Methods and materials

The results of hydro-meteorological observations by the Armenian Hydro-Meteorological Department, other departments and organisations in the country, climatic and hydrological atlases, as well as materials of the author's field observations on different rivers of Armenia have been used as basic material in this work.

The main principles and methods of spatial-temporal analysis and the synthesis of respective data and generalisations have been used for processing the basic data. To assess the consequences of natural disasters, such as floods, mudflows and landslides in Armenia, documents of state institutions and local government, namely, emergency reports, statements on losses and damage caused by floods and mudflows, have been used.

3. Results

Water disasters in Armenia are mainly floods (Fig. 2), and mudflows (Fig. 3).

In Armenia, floods are mainly caused by snowmelt in spring and downpours in summer. The extent of floods and the maximum discharge, duration, the start and end, and nature of formation, as well as other distinctive features

Figure 2. After the flood   *Figure 3.* Recently formed mudflow

are dependent on the amount of precipitation, the depth of the active snow layer and its water content, the air temperature, relief specifics, hydro-geological structure and other factors. Snow cover begins to build up in October, continues throughout winter and lasts till May. The depth of the active snow layer and water content also increase with altitude. In medium and high mountainous areas, the maximum depth of snow can reach 80–100 cm, while in the Aragats mountains it may reach 140–150 cm. The maximum value of water resources in the snow of Mount Aragats is 470 mm (Hydrography of Armenian SSR, 1981).

Snowmelt periods vary from basin to basin, depending on the altitude, slope position, and air temperature regime. In high mountainous areas, snow cover may persist for about 140–160 days. Rains also play an important role in flood formation. They accelerate snowmelt, increasing the maximum discharge, extend the duration of floods and often increase the maximum volume of snowmelt discharge.

The volume of flood discharge in individual Armenian rivers depends on the physical-geographical and geological features. Floods account for 25–80% of the total annual runoff in the rivers. Relatively high values were observed in the Gegharot, Dzknaget, Arpa, Vardenis and Marmarik rivers, where the total volume of the flow during the flooding period amounts to 70–80% of the total annual flow.

In the country, flooding is over in late June and early July. On average, the flooding period can last 80–120 days, maximum 150 days (Hydrography of Armenian SSR, 1981).

Besides the spring freshets, which occur as a result of snowmelt, floods can occur due to downpours in summer and autumn. There are many examples, when the river discharge caused by downpours, exceeded the maximum discharge of the spring freshets and in annual flood series they could have the highest value of all.

Such cases have been observed in the Meghri, Voghji, Geghi, Getar, Jrvezh, Shahverd, Hakhom, Tavush, Alaverdi and other rivers. During floods, every square kilometer of the river basins of these rivers yielded 3–4,000 l/s of water. For instance, the Getar River, which has a lower probability of maximum discharge formation, formed about 200 m^3/s of discharge and the flow module comprised 5,700 l/s.km^2 (Hydrography of Armenian SSR, 1981).

Observations on several rivers in Armenia show that over the last 70–80 years annual maximum discharges have tended to decrease (Fig. 4). Only the Aghstev River has shown a tendency to increase, which appears to be due to some local factors.

The analysis of the module coefficients' integral curve (Fig. 5) for the maximum discharges of these rivers shows that from 1952 to 1968, there was a general increase in maximum discharges. These wet years continued for about 20 years, after which, starting from 1987 till 2006, maximum discharges started decreasing and a low-water phase set in for the rivers.

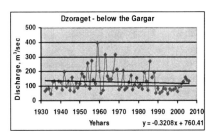
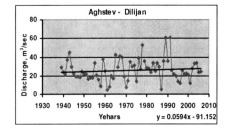
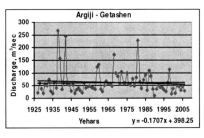
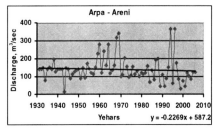
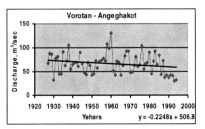

Figure 4. The changes in the maximum discharges of some rivers in Armenia

Despite the decrease in the maximum discharge, the frequency of their re-occurrence has increased, which has resulted in greater damage to the economy (Table 2). Table 2 shows that in 2004–2006 the damage caused by spring floods, river flooding, and mudflows in Armenia amounted to US$5,684,793.

Mudflows are formed under conditions of mainly dry continental climate and complex relief and lithology. A considerable amount of material is accumulated as a result of weathering under such conditions, which cause mudflows during heavy rain. In Armenia, brief rainstorms are typical during the months of May to August, and they may initiate mudflows. Mudflow-carrying basins occupy about 60% of the territory of Armenia. However, the mudflows mainly occur in the smaller basins, up to 100 km^2 in area.

TABLE 2. Damage caused by spring floods, river flooding, and mudflows in Armenia (2004–2006) (www.ema.am)

Years	Locations damaged	Degree of damage (in US dollars)
2004	Aragatsotn, Gegharkunik, Ararat, Lori, Kotayk, Shirak, Syunik, Vayots Dzor, Tavush	4,735,540
2005	Yerevan, Aragacotn, Gegharkunik, Ararat, Lori, Kotayq, Shirak, Vayots Dzor	830,293
2006	Yerevan, Aragatsotn, Ararat, Lori, Tavush	118,960
Total		5,684,793

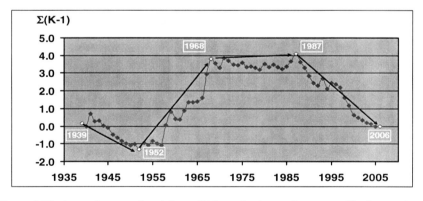

Figure 5. The integral curve of module coefficients for the maximum runoffs of some rivers in Armenia (Dzoraget, Aghstev, Argiji, Arpa, Vorotan)

Over 180 mudflows have been recorded in Armenia. In terms of mudflow activity, one should single out the south-western slopes of Mount Aragats (the Mastara and Talish mudflows), the rivers in the eastern part of Lake Sevan basin (the Artanish, Pambak, Babajan, Jil and others), the active mudflows of the Urts-Yeranos region (Shaghap, Chorselav, Kotuts and others), the mudflows of the Arpa river basin (Yelpin, Malishka and others), the mudflows of the Voghji and Meghri basins (Geghii, Musalum, Meghri and others). There are mudflows of average intensity in the northeastern areas of Armenia as well (Pambak, Gargar, Alaverdi and others) and in the Hrazdan river basin (the Getar, Jrvezh, Voghjaberd).

In terms of intensity and degree of destruction, one should mention the mudflow of the Getar River in 1946, which caused serious damage and destruction to Yerevan. 800 houses were ruined and another 630 seriously damaged. There were about 200 casualties. The main flow of the mudflow was 415,000 m^3, and the maximum discharge reached up to 200 m^3/s (Hydrography of Armenian SSR, 1981). Rock fragments with a diameter of 2–3 m in the center of Yerevan helped to shape public opinion about the intensity of the mudflow. A series of hydro-technical and afforestation activities were carried out to combat the mudflows of the Getar River; they helped to stop the destructive episodes in the riverbed.

In recent years around Yerevan, the riverbed has been mainly covered (Figs. 6 and 7). We believe one has to take into consideration the environmental and technical safety norms.

Figure 6. The riverbed of the Getar River in an open section

Figure 7. Covered section of the riverbed of the Getar River

Climate change is likely to increase the damage caused by both floods and mudflows in Armenia. The wide use of floodplain areas for crop cultivation and settlements, for construction of transportation communications, irrigation systems, etc., is going to increase the mean annual damage from flooding. Damage may persist if highly productive agricultural lands are not used because of a high probability of flooding (Anon., 2000). Climatic aridification, an increase in the intensity of downpours and growing anthropogenic load on the landscape, create favourable conditions for increasing hazard from mudflow events. This was proved by the studies on mudflow hazardous areas in Armenia by Nazarian (2003). The analysis of maps of mudflow processes and mudflow hazard on the Republic's territory for the 1950s up to the 1990s showed that for the period of 40–50 years, within the borders of the Amassia depression, in the basin of the Gavaraget River, on the outskirts of the villages of Yeranos, Vardadzor, Zolakar in Martuni province, on the left bank of the upper-stream section of the Masrik River, in the upper part of the Arpa river basin, and many formerly potential mudflow-prone zones have acquired the status of active areas. At the same time, new potential mudflow-prone areas of river basins have been formed (Nazarian, 2003).

In a temporal sense, these occurrences coincided with the period of increased water runoff in the areas of the Akhuryan, Alvar, Gavaraget Martuni, Argiji, Masrik and Arpa rivers. Observations show reduction of mudflows, but an increase in mudflow power and the number of mudflow basins.

From the analysis of the integral curves of module coefficients of mud-flow activity in Armenia (Fig. 8), it follows that big mudflows were more frequent before 1974. Their formation became much rarer in 1974–2001; this is explained by growing climatic aridification. Due to precipitation decrease and climatic aridification, the number of mudflows with weak alluvium saturation will reduce and the number of mudflows with high mineral particle will grow. This means intensification of erosion activity during mudflow occurrences.

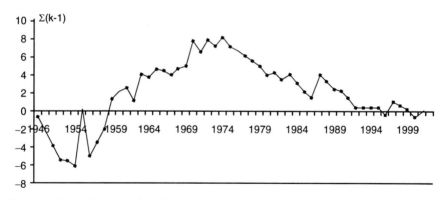

Figure 8. Integral curve of module coefficients of mudflow activity on the territory of Armenia (Ter-Minasyan, 2003)

For hazard mitigation and prevention, for mudflow formation and flood-ing, it is necessary to carry out a complex combination of anti-mud-flow, anti-flood and anti-erosion activities. Among them are construction of reservoirs for temporary accumulation of runoff in the period of downpours and mudflow masses, creation of transversal terraces in riverbeds of mud-flow rivers, flattening of sections of riverbeds, consolidation of river banks, restoring vegetation on mountain slopes, especially planting trees, optimization of pastureland use, introduction of respective agro-technical rules into agricultural practice (Nazarian, 2003). If these activities are not carried out, there will be a much higher cost for the state budget in correcting the damage after these hazardous processes; in its turn, this will have a negative impact on the implementation of the urgent socio-economic programmes initiated by the Armenian Government.

The engineering activities, directed at damage warning for natural disasters, require significant investments. In some case, the methods for prevention of such disasters may be less expensive, but they still require funding (Transboundary Floods, 2005; Vardanian, 2005).

As Professor Ter-Minasyan (2003) justly notes, tax revenues can serve as an important and primary source of financing for this kind of engineering and technical activities. Unfortunately, this is not practiced in Armenia, though in many countries of the world it is the main source of financing projects to protect the population and economy from the negative effects of hazardous natural processes.

4. Conclusions

As a result of investigation of water disasters in the territory of Armenia, we arrived at the following conclusions:

- Water disasters are observed in river basins and they are mainly due to floods and mudflows.
- In recent years, floods have become more frequent in Armenia, and this is certainly conditioned by global hydro-climatic changes.
- The studies carried out for several rivers in Armenia showed that in the last 70–80 years, the annual maximum runoff volumes for four of the five rivers studied have displayed a tendency to decrease.
- Despite the decrease in the maximum runoff volumes, the frequency of their recurrence has increased, which has caused greater economic damage.
- According to the data of the State Committee on Emergencies (STE) for the period of 2004–2006, the damage caused by floods and mudflows was estimated at US$5–6 billion.
- From the analysis of the integral curves of module coefficients of mudflows as well as maximum runoff activity in the territory of Armenia, it follows that large mudflows were more frequent before 1974, while the maximum runoffs were greater before 1987. After these respective years, their formation became much rarer; this is explained by growing climatic aridification.

As we see, water disasters in Armenia are widespread, and the struggle against them is of economic importance.

References

Anon., 2000: *The Global and Regional Changes of Climate and Their Natural and Socio-Economic Consequences.* GEOS, Moscow, 263 pp. (in Russian).

Hydrography of Armenian SSR, 1981: The Hydrography of Armenian SSR. Yerevan: 178 pp. (in Armenian).

Nazarian, Kh., 2003: Activation of mudflow phenomena in the territory of the Republic of Armenia as an indicator of climate change. *Armenia: Climate Change Problems/Collected Articles, II Issue*, Yerevan, 314–318 (in Armenian).

Ter-Minasyan, R., 2003: Activity of hydrological disasters in the territory of RA under the climate change conditions. *Armenia: Climate Change Problems/ Collected Articles, II Issue*, Yerevan, 277–283 (in Armenian).

Transboundary Floods, 2005: *Reducing Risks and Enhancing Security through Improved Flood Management Planning. Proceedings of the NATO Advanced Research Workshop,* Oradea (Baile Felix), Romania, 492 pp.

Vardanian, T., 2005: The Application of Mathematical Methods for Flood Forecasting. *Proceedings of the NATO Advanced Research Workshop "Transboundary Floods: Reducing Risks and Enhancing Security through Improved Flood Management Planning"*, Oradea (Baile Felix), Romania, 250–257.

THE FLOODS ON THE RIVER DANUBE IN 2006: SOCIAL IMPACT AND REMEDIAL PROCESSES

M. Lazarescu[*]

National Research and Development Institute for Environmental Protection –ICIM Bucharest, Bucharest 060031, Romania

Abstract. The 2006 floods in the Danube basin affected Romania badly. The increased severity of flooding in recent years is partly explained by climatic change. This paper describes the events and their causes, before analyzing the measures since put in place to prevent a repetition of the disaster. These measures include improving dike security, providing more coordinated early warning and evacuation procedures, allowing localized controlled flooding, restoring wetlands and forests and installing dams for temporary flood storage.

Keywords: Flood protection, flood control, climatic change, Danube, Romania

1. Introduction

In winter 2006, very high amounts of snow fell in Germany, Czech Republic, Slovakia and the subsequent warming overlapped with significant rainfall amounts in the whole Danube basin. Such extreme historic flows appear to be the result of sudden climatic changes that have occurred in the last 20 years.

The Romanian Environmental Ministry published the situation and the hydrological prognosis for the national rivers and the Danube. The rivers flows were increasingly high due to the snowmelt and water accumulation in the upper and middle basins of the Lot, Rage and Alumina rivers, as well as the Moldavian rivers. The existing ice formations in the upper and middle Siret basin were also melting generating artificially high water levels on some river sections. The flows would be increasingly high due to the rainfall expected mainly on the rivers in Maramuresh, Crişana, Banat and Oltenia, where frequent exceeding of the warning levels is possible.

2. Increasing discharges

A number of developments within the basin have resulted in "denaturing" by draining the ponds and marshes that used to be flooded by the Danube every year and provided a vast storage for the water overflow. In order to

---

[*] To whom correspondence should be addressed. e-mail: mihaela.lazarescu@yahoo.com

reduce flood severity, the experts will design new controlled floods in the Danube Holm on the Borcea branch.

Flood inundation is the result of river embankment works and land draining made in order to extend the land area during the communist period. Last year, areas where embankment works did not exist were flooded. This year, works like the Danube embankment are the main reason for floods, since the Danube has nowhere to overflow if the water flow is too high. The water elevation in the Danube has exceeded the flood level.

After performing the necessary topographic surveys, a breach will be made in the Danube Holm compartment III, to flood a surface area of 8,000–10,000 ha, between the Borcea branch and the Danube. A dike located in Potelu village has also been broken. Consequently, about 15,000 ha of land within the Potelu, Ianca and Grojdibadu localities were flooded.

In Romania as well, the increased discharge on the Danube threatening the riparian localities. In the first week in April 2006 the forecast showed that the Danube riverflow entering the country would increase and reach 12,200 m³/s. The hydrologists declared that the warning levels would be exceeded in the following localities: Gruia, Fetești, Isaccea and Tulcea. The final hydrological measurements indicated that the Danube level reached 595 cm, i.e. 3 cm more than previous day.

The ice formations is the upper Siret basin were melting and disappearing, generating artificial level increases in some river sections. The Danube flow where it enters the country (in the Bazaar section) was stationary (at 10,500 m³/s), above the multi-annual average for March (6,800 m³/s). Along the whole sector downstream from Iron Gates One, the Danube flow was increasing, i.e. in the Gruia – Calafat, Hârșova – Tulcea and Bechet – Cernavodă sectors. The Danube flow was 12,600 m³/s and increased to 16,000 m³/s over the next 2 weeks. The Danube rose by 1–2 cm in just 24 h at Galați. The Prefecture representatives declared that the defense dikes were safe for the moment.

3. The 17-18th April 2006 peak flows

The peak flows recorded in Romania are the result of significant amounts of water coming from the Danube tributaries before the river enters the country. These tributaries grew particularly due to the abundant rainfall recorded in the countries crossed by the Danube.

The external flow increases together with the increases within Romania resulted in a discharge of 15,400 m³/s at Baziaș on the Thursday morning. The increasingly high flow values reached 15,800 m³/s and during 17–18 April maybe 16,000 m³/s. After 18 April the values remained constant. After that, discharge in the upper Danube decreased, e.g. at Novisad, so that flood levels would not rise even if there were more rainfall.

The National Land Reclamation Authority monitored the situation at the critical points on the Danube in order to intervene in case of floods if the dikes were damaged. The National Fish Fund Management Company monitored the water level in the fish planning units and prepared systemized development works to take floods flows. The water level can be decreased by half with systemized development works to take the flood flows.

The year 2006 started with a succession of floods, the most affected areas being in the southern and western regions of the country. The experts forecasted that abundant rains would fall in April and almost the whole territory of the country would be threatened by floods.

The prefectures of 19 counties received hydrological warnings on water level increases during 22–25th March.

At Bechet, the city dike was going to break and the authorities at Bistreţ evacuated several hundred people. Due to the rainfall recorded in Romania, the river flows began to increase, in some places exceeding the defense levels. Dolj County was the most damaged. Thousands of people were evacuated from Bistreţ, Plosca, Săpata and Măceşu de Jos. 1,400 people rallied in the middle of a field near Rast village. The village of Rast was totally relocated after the flooding. On 28th March 2006, the Danube water elevation rose 50 cm above flood level at Turnu Măgurele. A large part of Mare a Brăilei Island was flooded as the water overtopped the protection dikes and Mică a Brăilei Island was completely covered by water. For this reason, the Stânca ferry crossing point from Brăila city has been moved near to Gropeni.

4. The results of the floods

One hundred and thirty seven localities were affected by floods, 693 houses flooded, 198 destroyed and 120 threatened with collapse, 3,368 outbuildings, 30 socio-economic installations, 378 km of county roads, and 107 km of village roads, as well as 19 bridges and 130 footbridges damaged. 21,000 ha of land were flooded deliberately under controlled flooding. At Turnu Măgurele, the flood elevation was exceeded by 68 cm and at Zimnicea by 55 cm. On the south-western side of Constanţa city, the Danube flooded several households and hundreds of hectares of farming land. At Cernavodă and Hârşova, the flood elevation indicated in April a level exceeding by a few tens of centimetres. Almost half of the surface area of the Galaţi shipyard was flooded by the Danube. At Tulcea, the river water rose up the city cliff. In Tulcea county, over 150 households were damaged in I.C. Brătianu, Măcin, Smârdan, Pătlăgeanca and Tudor Vladimirescu. The Pod Rout – Pătlăgeanca road and the Channel Bridge at 36 Danube miles, both of them located on the Chilia Branch were covered by water for a distance of 1.5 km.

5. Evacuating and returning the people

Almost one thousand people were moved from Bistreț, Dolj County, due to the floods. The Minister of Administration and Internal Affairs informed Mediafax. According to the Ministry of Administration and Internal Affairs, 5,750 persons were moved and housed at their relatives, schools, health units, tents or floating pontoons. At a press conference, the Ministry of Public Works declared that if the embankment were to break at Bistreț, the upper areas of this locality would be isolated (if not flooded) and this situation would make the work of the authorities more difficult. During the next few days, one of the most vulnerable points on the Danube was Bistreț. If the dike were to break, four settlements would be destroyed by water and 10,000 persons would need to be evacuated. As a precaution, the authorities began to move people, but the population refused to comply with the recommendations. Some inhabitants refused to leave the villages and remained to guard their properties. Only 1,800 of the approximately 4,800 inhabitants were moved. They were sure that, being located in the valley, their houses would be flooded.

One of the most difficult situations was at Bechet, where a few hundred inhabitants spent the night at relatives or friends after the local authorities' warning to evacuate the lowlying area of the city. The rain and flood on the Danube caused the water to overflow beyond Ostroveni-Part Bechet dike. The dike was strengthened; the materials were carried in only by helicopters. If the dike broke, over 1,000 people should be evacuated and 400 households would be flooded. In Brăila County, the Danube exceeded the warning level by 27 cm, according to AMOS News. Dobrogea Water Authority announced that on the 24th March the Danube level would reach 594 cm and the discharge would be 11,947 m^3/s. Some of the inhabitants living nearby the Danube began to reinforce the banks and dikes, others ran away.

6. Some causes which produced the large flood impact

6.1. A BIG MISTAKE – GREEN CORRIDOR DESTRUCTION

The Danube "green corridor" destruction and development for farming purposes was a mistake. Thus, along the course of the Danube, all former ponds and commons have been embanked and used as farming lands. It cannot be said if this was a good or bad action, but the remaking of the Danube green corridor has to be taken into consideration.

There are two types of floods: upstream and downstream floods. In 2005, Romania faced severe cases of both these types of floods. The prevention and mitigation strategy of such phenomena should be considered on several grounds, such as: wetlands reconstruction, floodplain identification and dam building to partly store the water flow. Protection against floods cannot be provided only by dikes.

The people are responsible for the disaster generated by water on the upper streams. Large and uncontrolled forest clearings were made; roads were built for carrying the logs. These roads now encourage the water accumulation like torrents. It is difficult to raise a dam capable of stopping a torrent.

Romania has special mutual agreements with its neighbors concerning the international waters. A mutual agreement was signed with Serbia, for example. This stipulates that we should use the Iron Gates reservoir as floodwater storage, so that in case of high flows at the reservoir outflow Belgrade city would be not flooded. This was one of the conditions requested when the dam was first built. But the prime purpose of the reservoir is to generate hydroelectricity, and when the water levels become too high, the excess water has to be released in order to avoid damage to the dam. Thus, the hydroelectric power plant does not profit from extreme historical flows. It is therefore necessary to create polders and afforested areas in the upper basin, to retain water. On the lower course, where the river beds are large and there are broad, flat floodplains, the best solution is to create side reservoirs able to receive additional water for short time periods. The water will be discharged gradually later after the flood has passed. In this manner, a time redistribution of the flows is made without jeopardizing the riparian area.

6.2. CLIMATIC CHANGES – ONE OF THE REASONS

In Romania since 1998, floods have occurred in the west and drought in the south. For the past 20 years, the World Meteorological Organization has been performing studies which indicate that the global mean temperature is increasing permanently. During the last 20 years the temperature increased above the levels for the past 2,000–3,000 years. The climatic change currently being experienced is now generally acknowledged to be due to increasing concentrations of greenhouse gasses, especially carbon dioxide in the atmosphere. Such climatic change can generate rainfall amounts of over 100 l/m^2 within 24 h.

This situation is unique; hence it follows that a long term prognosis cannot be made. It seems inevitable that Romania will experience heavy rainfall during the coming summers. Substantial funds are being allotted from the State budget and an external loan has been given by the World Bank to develop strategic measures, and to repair, maintain and build dikes. 7,000 billion RON were allotted for dike repairs and maintenance. This strategy is being developed to reduce further flood impacts that will be frequent in Romania from now on. The Romanian Government borrowed over 1 billion euros from the World Bank. Action departments and an action regulation system have been set up. An urgency list has been drawn up.

7. Warning system for floods on the Danube River

Tests for setting up a new alert system for floods, which became operational by the end of 2006 were performed on the Danube hydrographic basin. The International Commission for the Danube Protection (ICPDR), an international agency located in Vienna, which coordinates this project, indicated that this system now replaces the national alert systems, aiming to coordinate all defense devices against floods existing on the whole Danube basin.

Up to the present, the alert systems for floods were developed only at national level. An ICPDR flood expert has stated that the new system is designed to send flood alerts 10 days before a flood occurrence. Initially, a pilot program was developed on the upper and middle reaches of the Danube and this is then being extended to the whole river. The recent snowmelt and abundant rainfall have generated flood levels on the Danube higher than any experienced during the last 100 years. Hundreds of families have been evacuated in Romania and many thousands more were at risk from increasing water pressure against the dikes.

8. Conclusion

At the time of the large flood, the Swiss Red Cross gave over 63,000 euros to support Romania and Bulgaria for mitigating the damage generated by floods, as the web site *Movinite.com* reports. The money was used to provide food, beds and shelters for the victims of the calamity. Additional measures were taken having in view the increasingly high flows exceeding the prognosis. Special analyses were made for critical locations having in view the rather high expected discharge of nearly 16,000 m^3/s at the entry point into the country at Bazias. Special measures were taken to avoid the dikes breakings in critical areas or for trying to make controlled overbank flows, if necessary.

The dikes are not sized uniformly. Generally, the large dikes built along the Danube should resist discharges no higher than 16,000 m^3/s. The dikes should be maintained, but due to the lack of money, many of them are in bad condition. All such dikes in Romania have been catalogued and the critical points have been located. Investments should be made in order to ensure the protection, but, with the one exception of people's lives, the protection investment should be lower than the potential damage costs. New Standards have been promoted and special governmental decisions adopted related to the process of rebuilding some works which could not be delayed until the necessary investment regime was available.

A simulation was drawn up for a scenario of a 16,000 m³/s discharge on the Danube. The possible areas of flooding in each locality have been identified according to the protection plan specifying which lands could be flooded and the businesses already flooded. All the details have been set up. People should be evacuated from those areas where overflowing is predicted to occur.

Uncontrolled clearfelling of forests are one of the reasons generating floods. The forest was able to retain a part of the rainwater for a while in its crown. Moreover, the leaf bed may also be a water-retaining layer as well. Scientifically, the hillslope hydrograph has a certain duration. If this duration is short, a high flow occurs in a short time. If this high flow is widespread the transmission capacity of the river may be exceeded. Due to the lack of funds, the river beds were left untouched and riverbed deposits have increased. Man should make the duration of this hydrograph as long as possible. If the hydrograph is distributed during a longer time period at the base of the slope, floods will not be produced. Afforestation can be undertaken.

Gravel pits also represent a difficult problem. In some foreign countries, it is forbidden to excavate the gravel even along the side of a major riverbed, but in Romania this is allowed just 100 m from the base of a bridge. Such activities should be stopped. Landowners should also return floodplains to nature, as the crop yields are poor but the damage to houses, villages, and cities caused by floods can be huge. All measures, plans and strategies will be difficult to apply and this needs a long time to implement.

Urban and rural settlements, businesses, roads, routes, railways developed after the floods of the 1970s should be carefully analyzed. Unfortunately, not all the works started at the time were finished. It should be taken into consideration that any activity is under evolution and should be up-dated. There were cases when the people refused to be evacuated, or to help in cleaning the river after high floods. This was due to a lack of trust, ill-will or simply ignorance.

THE ANALYSIS OF DANGEROUS HYDROLOGICAL PROCESSES FOR THE TEREK RIVER BASIN

N.I. Alexeevskiy and N.L. Frolova[*]
Department of Hydrology, Moscow State University, Russia

Abstract. Analysis of dangerous hydrological processes is one of the essential components of future regional water resource systems planning. One of them is developed now for the Terek river basin. Contrasts of relief and climate create the prerequisites for the development of various dangerous natural processes, constantly endangering the safety of people. Climate changes during recent decades have been reflected in increasing frequencies of dangerous glaciological, erosive and hydrological processes in various parts of this river basin. Evaluation of the variability in river runoff, sediment load and water level, and development of forecasting methods for these phenomena are the most important scientific and practical tasks.

Keywords: Floods, forecasting river discharge, glacial melting

1. Introduction

Dangerous hydrological processes are events which cause social, economic, and/or ecological damage. Among them are changes in extreme water runoff and level and in the capacity of river channels. Dangerous rises in water levels are often caused by ice-jams and wind-induced surges, or backwater phenomena. Activation of slope, ravine and channel erosion, silting of channels and water storages, negative changes of hydraulic and thermal conditions of water bodies and their trophic status are also dangerous processes, which limit security for nearby social and industrial units. Negative changes in water quality should also be considered as dangerous processes.

Dangerous hydrological processes result from combinations of both natural factors and human impact. The risk of damage is equal to the possible damage multiplied by its probability. The size of the damage depends on the population of the area, the density of social and economic activities nearby, their status, etc. According to information provided by the Russian Ministry of Natural Resources, annual damage from floods alone in the Kuban, Don, Terek and Volga basins average 2.1, 2.6, 3.6 and 9.4 billion rubles, respectively.

One of the main tasks for hydrological science is to provide hydrological security for the country, to eliminate or minimize the conditions for

---

[*] To whom correspondence should be addressed. e-mail: frolova_nl@mail.ru

hazardous hydrological processes and water deterioration. This requires determining the conditions that lead to these events, estimation of their repeatability and scope, trends or cyclicity. This is the basis for complex analyses of the probability and magnitude of floods, extreme low flows, and other dangerous processes, for the development of a dedicated monitoring system and of proposals for minimization of social, economic and ecological risks due to actual or possible impacts.

This sort of work was conducted for the Terek river basin.

2. Peculiarities of dangerous hydrological processes in the Terek river basin

The Terek river basin is among the most difficult regions in Russia for water supply. Analyses undertaken in the Scheme of Complex Usage and Protection of Water Resources revealed that in the basin's mountain section hazardous processes are determined by glacial events. Decrease in the glaciers' area creates the necessary prerequisites for violent mudflows, a high probability of disastrous avalanches and glacier surges. There is no lack of water resources, but all social and economic activity takes place at lower altitudes. In the foothills, water management is subjected to the negative effects of gully erosion and mudflows. During a very wet season, there is a high probability of floods, channel changes and landslips. In low water periods, a lack of water resources is possible. A sharp increase in population and the density of industrial buildings here results in increased water consumption and large amounts of water abstraction have negative affects on river water quality. In the lowlands, there is an increased probability of catastrophic floods partly related to excessively high sediment loads. A sharp expansion of the irrigated land area here means a greater dependence on water resources during critical seasons of the year. Intensive channel deformations in the lowlands threaten social and industrial activities on the riverbanks and result in silting up of irrigation and supply channels, as well as radical reformation of the channel network in the delta.

Events in 2002 and 2005 testify to the high probability of a new phase in the reorganization of the channel network in the delta. Under a semi-desert climate, when agriculture depends exclusively on the inflow of freshwater from the Terek, shifting of the delta waterways causes severe economic and ecological stress by creating localized water deficiencies. The problem of water quality can also become extremely grave because of the intense economic activity here. In this connection, it would be pertinent to remember the recent problem of oil pollution in the Terek from the Chechen republic. This is particularly notable because of the great importance of the Terek for fisheries: valuable breeds of fish spawn in the lower reaches of the river.

3. Climate change and dangerous hydrological processes

Reduction of glaciation began at the end of 19th century, after the peak of the Little Ice Age (Voitkovskiy and Volodicheva, 2004). Now, glaciation is in a regressive phase and the general reduction removed about 20% from the glaciated area during the first 70 years of the 20th century. Thus, there has been a reduction in the area, volume and length of glaciers, and an increase of their number and the altitudes of their snouts. The size of glaciers has continued to change in recent decades (1970–2000), despite an increase in precipitation in the Large Caucasus (of 10–15%) and air temperatures of +1°C. From 1970 to 2000, the area of glaciation was reduced by 12.6%, the volume of ice by 14.9%, and the number of glaciers has increased for 2.4%. Glacier contraction continued over the 30 years and they have on the average reduced in length by 100 m. The increase in precipitation, especially during winter, has resulted in an increase in snowfall and the frequency of snow avalanches during the last 20 years.

Over the last 50 years, a general increase has been seen in the naturally renewed water resources of the region. A consequence of glacial area reduction has been a reduction of the discharges of the rivers that depend on glacial meltwaters. In basins where glaciation is actively degrading, the runoff is increasing. For the rivers with mixed snow-rain and glacial feed, the trend of runoff is mostly significantly positive. The increase of sediment yield has been particularly marked in recent years, reflecting the regressive changes in snowlines and glaciation, increasing the area of high-altitude zones producing huge volumes of sediments and mudflows.

Floods on the rivers of the Terek basin occur in the spring–summer period during intensive snow and ice melting and heavy rainfalls. Extreme catastrophic floods are always connected with the maximum height of the spring flood being accompanied by significant or outstanding rainfall floods. Such floods were observed in the Terek basin in 1931, 1932, 1958, 1960, 1963, 1966, 1967, 1970, 1974, 1998, 2000, 2001 (Taratunin, 2000; Dobrovolskiy and Istomina, 2006) and in 2002. In the mouth of the Terek, high spring and summer floods result in the breaching of the levées and flooding. In the Terek basin, over 200,000 ha of agricultural land, including 84,000 ha of irrigated land, and a population of 140,000 lie in the zone of possible inundation.

As a result of the catastrophic water level rise in 2002, the territories of Kabardino-Balkaria, North Ossetia, Chechnya, Ingushetiya and Dagestan were flooded, 114 people were killed, 300,000 people were injured and over 100,000 people were evacuated. Dozens of thousands of homes, hundreds of municipal buildings, educational establishments and other units were destroyed and damaged. Flood protecting constructions and dams were seriously damaged as well. The financial damage amounted to about $530 million (Lourie, 2002). In the mountain regions of Kabardino-Balkaria, North Ossetia-Alania, Ingushetiya, Chechnya and Dagestan numerous mud flows occurred. Breaching of protection dikes took place at the Kargalinskaya station when the water discharge was 1,530 m^3/s. Vast areas were flooded in the delta of the Terek, homes, bridges and other constructions were destroyed, kilometres of protection levées and dams were scoured, great ecological and economic damage was caused in the affected areas (Gorelits et al., 2005). The duration of peak level in the rivers varied from 3 to 288 h. The probability of the maximum discharge during the flood at the lower gauging section of the Terek amounted to 3%. In spite of the fact that greater discharges were observed at the head of the Terek mouth in 1931 and 1967, the flood of 2002 was unique in its duration as well as its discharge and sediment load.

The effects of climate change on flood characteristics has been estimated by Christoforov et al. (2007). Probable climate changes will not affect the magnitude and shape of the flood peaks, so much as their average number and distribution during the year. For the rivers of the Terek basin, the average number of peaks increases during the winter season and decreases in the summer with the rise in air temperature, and may even be reduced to zero with extreme warming. An increase in annual precipitation will lead to a general increase in the number of floods in all seasons for all the rivers studied. Calculations show that the risk of catastrophic floods on the rivers of Northern Caucasus will increase in coming decades with the rise in air temperature.

Owing to the change of climatic conditions in this region, there has also been a change in the level of the Caspian Sea. Over almost 500 years, it fell sometimes quickly, sometimes more slowly, until in 1977 it reached the lowest position at −29.01 m BS (Fig. 1). Then it promptly began to rise (on average by 13 cm/year), reaching a maximum in 1995. As a result, part of coastal territories has been flooded, and conditions of hydraulic interaction between river and sea waters and sediment yield have changed significantly. During the last decade, the level of the Caspian Sea has stabilized. The amplitude of annual fluctuations was 0.4 m. On average, for the period of 1996–2006 the level of Caspian Sea was −27.03 m BS.

As a result of floods, sedimentation and reorganization of the channel network in the delta occurs periodically. Every 60–80 years, the main channel in the delta changes direction considerably. All known cycles of channel orientation change in the delta have coincided with periods of increases in flood numbers (Alekseevskiy et al., 1987).

For the last 500 years in a lower reaches of Terek, seven such catastrophic breaks have taken place. Each of them has led to the occurrence of systems of channels named the Kuru-Terek (16th century), Sullu-Chubutli (17th century), Old Terek (the beginning of 18th century), New Terek (the end of 18th century), Borozdinskaya Prorva (the beginning of 19th century), Talovka (the end of 19th century) and lastly the Kargalinskiy breach. The most recent cycle in the Terek delta formation occurred during the catastrophic flood of July–August, 1914, at discharges of more than 2,000 m³/s. For some years, the river has developed a channel which carries 80–100% of the discharge. The worst water shortage has been in the cultivated left-bank part of Terek delta. Attempts to block the Kargalinskiy breach continued until 1927. In addition, it was decided to secure the water supplies in this part of delta with an expensive dam and irrigation canals.

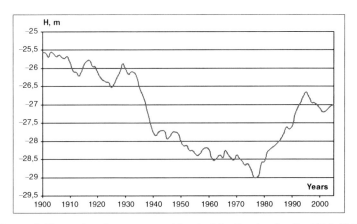

Figure 1. Fluctuations in mean annual level of the Caspian Sea for 1900–2005

The history of the evolution of the Terek delta shows that the most complicated problems in a lower reaches of the river are caused by the high sediment yield and extremely high turbidity of river waters. In conditions of gradual reduction of a bed slope and a water surface the stream loses ability to transport all of its sediments. They deposit in the channel, reducing its carrying capacity. As a result even rather small discharges of water can lead to the flooding of districts adjacent to a river channel. For the observation period 1924–1988, the average suspended sediment discharge in the Terek delta was 588 kg/s, which corresponds to an annual sediment load 18.5 million tonnes. In individual years, the suspended sediment yield was as high as 38 million tonnes and as low as 6 million tonnes.

The increase in damage from dangerous hydrological processes in recent decades in the Terek basin is connected with lack of data about these processes, neglecting the risks, development of areas for the potential occurrence of these processes, mistakes in scientific support of projects, changes in natural and anthropogenic factors, and shortage of funds for the monitoring schemes.

In order to decrease the risks, it is necessary to estimate the size and variability of the components of runoff – the main reason of changes in the safety of the population and economy; to prepare a modern information basis for the characteristic of probability and scales of dangerous hydrological processes; to create methods of estimation, calculation and forecasting of characteristics of the hydrological hazards at local, regional and basin level.

4. Maximum water level variability in the rivers of the Terek basin

The specific features of the régime of the Terek are caused by the combination of groundwater, snow, rain and glacier feed. Maximum levels of the alpine rivers are mainly provided by the spring flood caused by glacier and snow patch melting in May and June. The origin and the time of flood peaks change lower down the basin. In the plains and foothills, the highest levels are the result of summer rains, but floods occur throughout the year as a result of heavy rainfalls.

To study the peculiarities of flood formation, series of daily maximum H_{max} and minimum H_{min} values of annual water level were studied from data up to and including 2005. Using these data, water levels with 1% probability $H_{1\%}$, the maximum amplitude of the level change $\Delta H_{max} = H_{max} - H_{min}$, the amplitude of the change of the level of 1% probability $\Delta H_{1\%} = H_{1\%} - H_{min}$, the edge of flood-lands H_{fl}, the maximum flood layer of 1% probability $H_{jam} = H_{1\%} - H_{fl}$ and the average data of establishment of maximum water levels D_{Hmax} were calculated for each observation station. On average, for the Belka, Fortanga and Assa rivers the maximum levels are observed at the beginning of June, for the Sundza River in the middle of June, for the Argun, Ardon rivers at the end of June, for the Malka River in the middle of July, for the Baksan, Tseya, Cherek rivers at the end of July. The maximum water levels change in accordance with the altitude of the catchment area (Fig. 2). The change of ΔH_{max} depends on the peculiarities of water régime, river bed and the valley at the observation station, as well as hydrological and morphological characteristics of the river (Fig. 3).

All the territory of the Terek basin can be conditionally divided into three parts:

- The upper reaches of the Malka River, the upper and the middle stream of the Baksan River, the basins of the Cherek, Chegem, Ardon Fiagdon rivers, the upper reaches of the Terek River (up to Vladicavcaz): the maximum recorded amplitude of the level oscillations ΔH_{max} is less than 2 m.
- The middle (below the River Gedmysh) and the lower part of the Malka River, the lower section of the Baksan River, the middle and the lower reaches of the Terek River (from Vladicavcaz to Kargalinskaya station): within this territory the maximum amplitude of water level oscillations is 2–3 m.

– Sounzhy river basin: the amplitude of water level oscillations in this region is not high and amounts to 3–5 m. In the lower reaches of the Sounzhy River (village of Braguny) ΔH_{max} = 510 cm, of the Argun River (village of Duba Yurt) ΔH_{max} = 487 cm, for the Belka River (town of Goudermes) ΔH_{max} = 789 cm.

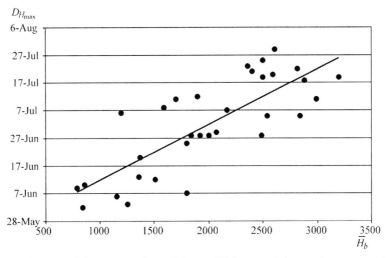

Figure 2. Change of the average date of the establishment of the maximum water levels D_{Hmax} on the rivers of the Terek basin related to average elevation of the watershed H_b

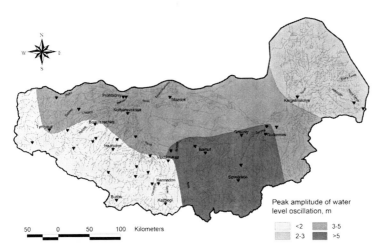

Figure 3. Regionalizing of the territory according to the variability of the maximum water level ΔH_{max} (m)

In the Terek delta the amplitude of water level oscillations is 15–350 cm. The regularity in many respects corresponds to the distribution scheme of water régime type (Fig. 4a) and to the distribution of the mean turbidity of the rivers and sediment flow (Fig. 4b).

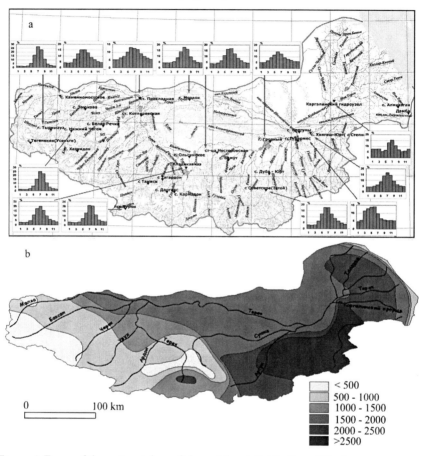

Figure 4. Types of the water régime of rivers (**a**) and distribution of the long-term average turbidity (g/m³) of the rivers of the Terek basin (**b**) Map by L. Anisimova

The average long-term values of water turbidity in the Terek basin are subject to substantial changes from 0.2 kg/m³ on the Chegem River at the village of V.Chagem up to 3.19 kg/m³ on the Sundza River at the village of Braguny. Malka waters are characterized by higher turbidity, which increases rapidly downstream. The average turbidity at the village of Khabaz is 0.52 kg/m³, towards the station Prokhladnaya it increases to 1.73 kg/m³. The mean annual turbidity in the upper reaches of the Baksan River (village of Zayukovo) is 0.60 kg/m³. It increases downstream and reaches 0.87 kg/m³ at Prokhladnaya station. The turbidity of the Sounzha River in its upper reaches at Karabulak is 0.55 kg/m³. It rapidly increases downstream and reaches 3.19 kg/m³ at the gauging station in the village of Braguny. Upstream on the Terek River, the annual averaged turbidity is 1.3 kg/m³. It changes rapidly downstream: at Vladikavkaz it increases to 2.6 kg/m³. By the village of Eilchotovo, turbidity decreases up to 0.6 kg/m, apparently as a result of the less saturated suspended particles of the streams. At the station of Kotlyarevskaya, the turbidity of water reaches 1.14 kg/m³. Considerable increase of turbidity takes place downstream of the Sounzha River outfall.

The largest amplitude of water level oscillations is observed in the basin of the Sounzha River. It is connected with the formation of catastrophic floods in the summer. Estimation of the dependence of the maximum water level amplitude from a number of hydrological and morphological characteristics is no less interesting and important. The first dependence reflects the influence of watershed area on water level changes (Fig. 5). Irregular high value ΔH_{max} = 789 cm is recorded for the Belka River at the town of Goudermes.

The altitude of the catchment area determines the degree of its glaciation and also influences the range of the maximum water level changes in rivers (Fig. 6). The amplitude of level oscillations decreases up to 100–150 cm with the increase of glaciation area. At the median height of the basin of more than 2,500 m, the amplitude of water level oscillations is about 250 cm. In the alpine regions, the regulatory control of glaciation is felt. For the basins with mean heights lower than 2,500 m, the level amplitudes are about 300–400 cm. On the plain, large oscillations are connected with great temporal and spatial heterogeneity of the rain and snow feeds. In this situation, local factors, like meso-relief and microclimate, play the major roles.

The dependence of ΔH_{max} on the mean slope of the basin is clearly apparent, reflecting the general tendency for decreasing water level oscillations as the mean basin altitude increases.

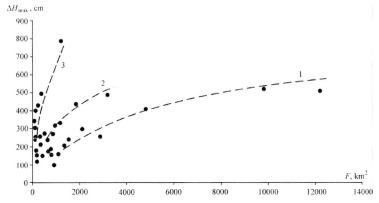

Figure 5. Dependence of ΔH_{max} on the catchment area (1) for the large Rivers Sounzha, Malka, Bakhsan at plains observation stations, (2) medium size mountain rivers (F_b from 700 to 3,200 km²) and (3) small mountain rivers (F_b < 1,100 km²)

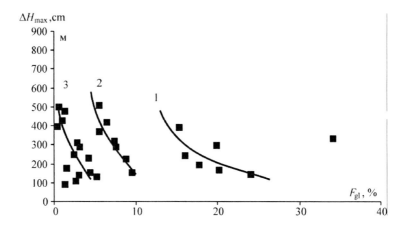

Figure 6. Dependence of ΔH_{max} on the glaciation coefficient of basins F_{gl} (%) ($1 - F_{gl} = 15-25\%$; $2 - F_{gl} = 5-10\%$; $3 - F_{gl} < 5\%$)

The analysis of temporal variability of the peak flows showed that in the 20th century almost at all observation stations in the Terek river basin peak levels increased (Fig. 7). The reasons for the increase may be both natural and anthropogenic. According to Lourie (2002), more frequent recurrence of extreme hydrometeorological phenomena is one of the negative consequences of the climatic fluctuations. During the last 50 years, the increased precipitation in the cold and warm periods has averaged 5–10% in the Large Caucasus. The steady increase in maximum water levels is connected with fluctuations in precipitation and air temperature. For the Terek basin, the tendency in peak water levels depends on the peculiarities of the sediment load and accumulation processes, resulting in the gradual elevation of water level marks for a given discharge, as observed at the Kotlyarevskaya gauging station on the Terek River.

According to the analysis of long-term variability of the sediment load, the average annual values have decreased in the Terek River, near the village of Kazbegi, the Argon River (village of Taminsk), the Malka River (village of Kamennomostskoye), the Soundza river (village of Braguny) and others. In contrast, at some gauges on the Terek River at Vladicaucas city and Kotlyarevskaya station, a considerable increase in the sediment load reflects the specific river bed erosion upstream the gauging stations. The maximum increase of suspended sediment load is observed in the Terek River at Vladicavcaz city, 396%. Under the inconsiderable average annual water discharge oscillations, the increase in maximum water levels and sediment load is observed here. Sediment load has increased here by more than twice over the last 25 years.

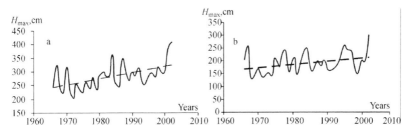

Figure 7. The maximum water levels in 1966–2002 for the Terek River, station Kotlyarevskaya (**a**), and the Baksan River (village of Zayukovo) (**b**)

The diking system was built on the Terek River for flood protection. Nowadays, continued diking increases the eventual financial losses when rare and very rare floods do overtop them. Breaching of the dikes takes place every year and catastrophic ones occur every 11–14 years. Such events were observed in 1958, 1963, 1967, 1970 and 1977.

Floods in lower reaches of the Terek over recent years are mainly related to sedimentation in the river channels. Decreases in channel capacity raise the risk of flooding during the period of higher runoff even when water inflow is relatively stable. Conversely, erosion of floodplains and riverbeds during floods lowers water level. The influence of alternating processes of erosion and accumulation on changes of water level ($Q = const$) is most evident in the delta.

The history of development of this area shows periods of reduction in flood risk (1928–1961, 1978–1990) and periods of increase, even up to flooding every year (1962–1978, 1991–2004). Changes in flood risk correspond with stages in the delta's evolution and sediment loads.

5. Means of maximum level and discharge forecasting

In an effort to improvement protection from both floods and lack water resources, methods of short-term forecasting of water level and discharge were developed for the Terek basin. For mountainous rivers, where runoff depends on snowcover and glacial thawing, a physical-statistical model of runoff during a summer high water period has been produced, based on air temperature. The model realized for the Cherek River at Sovetsky allows summer discharges to be forecast adequately with 2–4 day lead time. For the rivers of the foothills and lowlands, due to lack of hydrometeorological data, the method of corresponding levels is used. This makes it possible to describe the movement and transformation of flood waves in the channel network, based on hydrometric observations data. The method is developed for six sections on the Terek, Malka and Sunzha, allowing forecasts with lead time equal 1–3 days (Christoforov et al., 2007).

6. Trends in dangerous hydrological processes

According to forecasts from regional centers, such as the North-Caucasian Hydrometeorological Service, as well as models of global climate change, the mean annual air temperature in the Northern Caucasus will rise by 3–5°C. The increase in annual precipitation will be c.10–15%, or c.50 mm in the Terek basin. These changes, will increase hazard activity.

It is predicted that river runoff in the Caucasus will increase by 44% up to 2050, including 34% in the Large Caucasus and almost three times in Ciscaucasia. On the northern slope of the Large Caucasus, the runoff will increase 37%, and on the southern 32%, because of the more significant area of current glaciers on the northern slope. The area of glaciation will decrease by 30% between 2000 and 2050, leading to a 37% deduction in glacial runoff.

The increase in discharge will lead to changes in erosion and increases in turbidity and sediment load. The increases in river runoff and precipitation and the reduction in glacier areas will lead to intensification of mudflow activity. The duration of the mudflow period, and the number and volume of mudflows will increase, especially at heights above 1,500 m. Similarly, the frequency of catastrophic flooding will increase. The greatest hazards will be mudflows in the mountain and foothill areas, and shortage of water resources and flooding in the lowlands.

7. Conclusion

Contrasts of relief and climate in the Terek basin create the prerequisites for the development of various dangerous natural processes, constantly endangering safety of the population. High tectonic activity, landsliding, avalanching and mudflows, intensive slope, gully and riverbed erosion, intensive sediment accumulation in the lowland part of the streams, and floods are the main sources of social, economic and ecological damage. Evaluation of runoff and sediment load variability and water level oscillations, estimating discharges and developing forecasting methods for these hydrological characteristics are the most important scientific and practical tasks.

Acknowledgements The present research was financially supported by the Russian Federation President grant for supporting leading scientific schools (project NS-4964.2008.5) and was provided by RFBR (project No. 09-05-00339-a).

References

Alekseevskiy, N.I., Mikhailov, V.N. and Sidorchuk, A.Y., 1987: Delta forming processes in Terek mouth region. *Water Resources*, **5**: 123–128 (in Russian).

Christoforov, A.V., Yumina, N.M., Kirillov, A.V. and Rec, E.P., 2007: Runoff forecasting for Terek's basin. *Water Industry of Russia*, **4**: 25–37.

Dobrovolskiy, S. and Istomina, M., 2006: *Floodings of the World*. GEOS, Moscow (in Russian).

Gorelits, O.V., Zemlyanov, I.V., Pavlovskiy, A.E., Artemov, A.K. and Yagodintsev, V.N., 2005: Disastrous flood in the Terek Delta in June–July, 2002. *Meteorology and Hydrology*, **5**: 62–71 (in Russian).

Lourie, P.M., 2002: *Water Resources and Water Balance of the Caucasus*. Gidro-meteoizdat, St. Petersburg (in Russian).

Taratunin, A.A., 2000: *Floodings in Russian Federation*. Rosniivh, Ekaterinburg (in Russian).

Voitkovskiy, K.F. and Volodicheva, N.A., 2004: Evolution of Elbrus glacial system. *Geography, Society and Environment*, **1**: 377–395 (in Russian).

FLOODS AND THEIR RISK ASSESSMENT IN EAST SIBERIA

L.M. Korytny, N.V. Kichigina* and V.A. Cherkashin
V.B.Sochava Institute of Geography SB RAS, Ulan-Batorskaya St., 1, 664033, Russia

Abstract. We examine the main kinds of floods in East Siberia, and their recurrence rate and distribution across the territory. Floods resulting from dam burst waves on the Angara-Yenisei cascade of hydropower stations are the most hazardous; flash floods are also highly hazardous. Analysis of damage across the territory of East Siberia showed that flash floods and jam floods are responsible for the greatest damage. Areas with a high flood risk in the southern part of the region also show a considerable industrial risk level, which results in a synergy effect when determining total damage.

Keywords: East Siberia, floods, risk, flood damage, synergy effect

1. Introduction

The regions of East Siberia, along with North Caucasian and Far Eastern regions, experience the greatest flood risk in Russia. Floods in East Siberia occupy a highly important place among the natural calamities as regards their distribution area and recurrence as well as actual damage; they are second only to strong earthquakes. 90% of all residential centers, and as much as 75% of arable lands and hayfields are along the river valleys which have long undergone floods.

2. Recurrence rate of floods, their distribution and consequences

The first pioneer settlers of Siberia, who settled along the banks of large rivers, had to combat floods. Using archival and published materials, we carried out an analysis of the floods for the East-Siberian cities and towns in the 19th and 20th centuries. Nearly all large cities of the region under-went floods at some time: Irkutsk on the Angara (32 events), Yeniseisk (33), Krasnoyarsk (10), and Kyzyl (5) on the Yenisei, Kirensk (19) and Yakutsk (15) on the Lena, Abakan on the Abakan (6), Tulun on the Iya (9), Nizhneudinsk on the Uda (14), and Ulan-Ude on the Selenga (5).

In the second half of the 19th century, flood damage increased as a consequence of an increase of their recurrence rate as well as due to housing development on floodplains. It is not an uncommon occurrence

---

* To whom correspondence should be addressed. e-mail: nkichigina@mail.ru

that floods spread over significant areas. For instance, disastrous floods occurred in August 1960 in the basins of the Kan, Mana and Biryusa rivers covering an areas of about 250,000 km^2. In July 1971, floods occurred on most of the rivers in the Baikal region and the Transbaikalia; in the Irkutsk region alone inundations affected 33 residential centers, 82 industrial enterprises, and about 700 km of motor roads. In July 1984, in the Iya River basin, 12 residential centers were flooded, including one third of the area of Tulun, 5,500 ha of pasture lands, and 80 ha of the areas under crops. In 1993, six districts in the Republic of Buryatia were affected by floods, including its capital city of Ulan-Ude: three persons died, 8,250 houses were flooded, and 3,000 houses destroyed, more than 36,000 ha of agricultural lands were damaged, as well as 60 farms, 250 km of roads, 58 railway bridges, 1,800 km of communication lines, and more than 2,800 power transmission lines. In 2001, in the Republic of Buryatia, the Tunkinsky and Okinsky districts were affected and seven persons killed; in the Irkutsk region, more than 150 residential centers with the population totaling 460,000 were flooded, and 12,000 people were evacuated.

The last decade saw severe floods in the Lena basin – they recur almost every year. A disastrous flood occurred on the Lena in 1998, during which 49 persons perished, 95% of the city of Yakutsk was inundated, the city of Lensk and the villages on the riverside were flooded, and 41 persons were evacuated. In 1999, the water level at Yakutsk rose as a result of the jamming, to marks not observed since the previous century. The city of Yakutsk was almost entirely inundated, with 169 residential centers finding themselves in the zone of flooding. In 2001, as a result of a flood, the city of Kirensk was almost entirely inundated.

3. The genesis of floods

East Siberia is characterized the most by floods caused by springtime (spring-summer) freshets and summertime rain-induced high water; by the raising of water levels in the case of ice-jams and blockages as well as by anthropogenic activity and by discharges of superfluous volumes of water over the dams of the integrated hydroschemes. The water level rises on small rivers in East Siberia are often caused by the thawing of icings produced as a result of complete freezing of these streams. Floods are not infrequently caused by several factors at a time, e.g., freshets and jamming events on the plain, thawing of snow, and torrential rains in the mountains (Korytny and Kichigina, 2006).

Floods in East Siberia occur most frequently during high water and freshet periods. Analysis of the relationship of maximum water flow rates for these critical hydrological phases indicates the following salient features: the mountain areas in southern East Siberia (the ranges of the Western and

Eastern Sayan mountains and Khamar-Daban, the North Baikal and Patoma upland, etc.) are dominated by rain-induced high-water maxima; on the Middle Siberian tableland the largest flow rates and levels relate to the phase of spring and spring–summer high water; combinations of factors are possible for the piedmont plains. Freshet floods occur in May and June, and high water periods occur in July and, more rarely, August (Korytny and Kichigina, 1998).

Freshet floods every so often combine with blockage-induced floods, but blockages often give rise to floods on their own. In East Siberia, the number of blockage-induced floods and damage caused by them is quite considerable, because the main East-Siberian rivers flow in a sub-meridional direction. Blockage-affected areas in East Siberia total about 447, with the total length of 744 km, floods occur on 62 of them or 14% of the total number (Gidrometeoizdat, 1976). The largest number of blockage-affected areas (26% of the total number) occur in the basins of the Yana, Indigirka and Kolyma rivers.

In addition, there is a threat of floods resulting from water development projects. The region under consideration has several large hydropower stations that are included on the list of potentially hazardous economic enterprises. Thus, in the last few years the upper reaches of the Angara cascade experienced repeated events forcing the water level above a normal hydrostatic head, which was responsible specifically for a nearly constant inundation of the low-lying sections of the Baikal shore area. On the other hand, discharges of additional volumes of water over the dam in order to prevent inundation in the upper reach leads to inundation and flooding in the lower reach, as was the case in Irkutsk in 1995.

Particularly hazardous are the floods caused by damage to integrated hydroschemes, and by dam burst waves involving huge water volumes and characterized by high speed. Analysis of such situations in Russia (Malik, 1995) reveal that such a threat is also looming for the hydropower stations of the Angara-Yenisei cascade: in 1988, a similar situation occurred in the lower reach of the Krasnoyarsk Hydro. In the event of a dam burst at the Irkutsk Hydro, there would emerge a zone of catastrophic inundation with an area of 62.4 km^2, and 123,600 people would find themselves in the zone of potential flooding; similarly, for the Bratsk Hydro – 104 km^2 (33,300 persons); and Ust-Ilimsk Hydro – 117 km (135,020 persons) (Pavlov, 1998). This hazard should be taken into consideration in view of the fact that some of the hydropower stations are in a seismically hazardous zone.

4. Flood damage

The greatest damage was reported in 1993, which for Buryatia amounted to about 40 billion rubles as of 1993 prices, or several hundred million rubles as of today (according to mass media). Losses are aggravated by the fact that this territory is relatively densely populated.

A total of 49 floods occurred during 1991–2006, inflicting considerable material damage. Victims of these natural disasters were recorded in five of them, and people had to be evacuated in five cases. Those floods were largely occasioned by rain-induced high water, and by long-lasting heavy rains (Fig. 1).

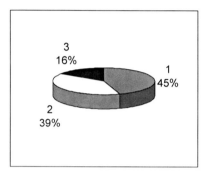

Figure 1. Share of the kinds of floods on the rivers of East Siberia during 1991–2006, %.
1 – freshets, 2 – jamming-induced, 3 – high water

The amount of material losses caused by large floods is very considerable. Thus, for the floods considered above they were: 1993, Buryatia – about $50 million; 1998, Yakutia – $300 million; 2001, Irkutsk region – $9 million. As regards economic damage in the East Siberian economic district (save Yakutia), floods resulting from rain-induced high water (about 50% of all floods) come first; they are followed by jamming-induced floods (30%), and the list ends with floods of a mixed origin (20%) (Taratunin, 2000). Furthermore, losses for each sector of the economy in East Siberia are distributed as follows. The greatest damage is done to agriculture and public utilities – 28% and 30%, respectively, of total damage; then come the losses suffered by transport, 19%, and industry, 16%, and damage to private property of citizens makes up about 7%. These figures vary from region to region of East Siberia and depend on the development level of their particular sectors (Table 1).

TABLE 1. Distribution of losses in the sectors of the economy (as a percentage of the total yearly average)

Sectors of the economy	Krasnoyarsk territory	Irkutsk region	Buryatia Republic	Chita region
Industry	32.0	12.4	17.0	2.3
Agriculture	6.0	18.2	53.8	35.1
Transport	15.0	12.1	11.2	39.6
Public utilities	39.0	47.3	10.0	23.0
Private property	8.0	10.0	8.0	-

5. Flood risk assessment

Flood risk assessment involves calculating the occurrence probability of an event as well as determining the boundary conditions for its buildup, development and manifestation. We have carried out an analysis of the risk for the territory of the Irkutsk region, based on factual statistical data spanning the period from 1999 to 2004. The amount of damage, and the frequency (recurrence rate) of floods was used as the input.

The territorial cell was represented by an intra-regional administrative district. In this case, the risk R_i is used in reference to the product of the number of events n by the ratio of specific damage D (equal to the ratio of the sum of all damage of a given kind of hazard of the region in the monetary form to the number of districts) to the sum of all damage of a district from a given kind of hazard U_i (Akimov and Radayev, 2004):

$$R_i = n*(D \ \Sigma \ U_i)*100\%$$

The calculated values of the risk are divided into three classes, with the rank determined on an expert basis corresponding to each of them. In this case, class one corresponds to an increased level of hazard (rank 1), class two refers to a threatening level (rank 2), and class three corresponds to an alarming level (rank 3). An exponential distribution of risks is used in the ranking. The risk indicators are arranged along the exponential axis, and the abscissa axis is divided into three equal intervals within the specified data.

6. On the synergy effect

When the ranks of hazards of a different kind, natural and technogenic (or human-induced), are considered jointly, a synergy combination is possible when the hazards come into play to acquire new qualities and mechanisms of development, thus considerably increasing the magnitude of total risk. For the Irkutsk region, we determined stable combinations of hazards for its separate districts. We identified two types of intrinsic integration of floods with the other hazardous phenomena which form cascade-like integral emergency situations.

1. Combination of the flood hazard with high emergency rate in industries. This type is characteristic of the Angarsk, Irkutsk and Usolye districts. In this case, a natural emergency situation triggers incidents in the production sphere.
2. Combination of floods with high emergency rate for transport. This is characteristic for the Usolye and Ust-Kut districts. Such a combination is hazardous primarily for hydrological enterprises, because at the emergence of such situations the greatest damage is inflicted to them.

References

Akimov, A.A. and Radayev, A.N., 2004: *Natural and Technogenic Hazards: Risk, Monitoring, Forecasting*. Moscow.

Gidrometeoizdat, 1976: *Catalogue of Plug and Ice-Jam Areas in the USSR*. Leningrad: Gidrometeoizdat, vol. II.

Korytny, L.M. and Kichigina, N.V., 1998: Floods along the river valleys of East Siberia. *Vodnye resursy*, 2: 161–165.

Korytny, L.M. and Kichigina, N.V., 2006: Geographical analysis of river floods and their causes in southern East Siberia. *Hydrological Sciences-Journal*, 51(3): 450–464.

Malik, L.K., 1995: The natural and anthropogenic factors of damage to water-development works. *Izvestiya RAN.Seriya Geog.*, 1: 76–81.

Pavlov, A.V., 1998: The state of industrial safety on the territory of the Irkutsk region. *Ecological Insurance: Regional Features and International Experience*, 72–79.

Taratunin, A.A., 2000: *Floods on the Territory of the Russian Federation*. Yekaterinburg: Russian research Institute of Integral Utilization and Protection of Natural Resources.

REMOVING CHLORINE-CONTAINING ORGANIC COMPOUNDS IN THE ENVIRONMENT AFTER FLOODS

G. Torosyan[1*] and S. Harutyunyan[2]

[1] *Chemical Technology & Environmental Engineering department of State Engineering University of Armenia, 105 Teryan, 375 009 Yerevan, Republic of Armenia*
[2] *Yerevan State Economic University, Yerevan Armenia, 128 Nalbandyan, 0062 Yerevan, Republic of Armenia*

Abstract. This paper presents results of research on the use of natural materials such as zeolites (mordenite and clinoptilolite) as sorbents and as promising materials for protecting flood-prone areas after a flood. Floods transfer large amounts of inorganic and organic water pollutants, creating huge ecological problems in the lower river regions, where human settlements and agriculture are vulnerable to floods. Floods carry bacteria, viruses and parasites, as well as chemicals, along with organic substances that are especially dangerous, e.g., chlorine-containing compounds such as pesticides, solvents, and petroleum-based products. Diseases such as dysentery, hepatitis and giardiasis can be transferred through water sources used by the population that have been contaminated by floodwaters. The paper offers a method of sorption treatment, using natural Armenian zeolites. The advantages of natural zeolites in comparison with other sorbents, include their stability, low cost, availability and filtering properties.

Keywords: Flood, organic pollutants, organochlorine compounds, pesticides, natural natural Armenian zeolites, mordenite, clinoptilolite, adsorption, adsorbent, watsewater treatment

1. Introduction

Over many centuries, floods have caused famine and epidemics in Armenia. Summer downpours cause flooding, especially in the northeastern part of the country. Nowadays, the removal of hydrocarbons, particularly chlorine-containing compounds, from wastewater is becoming more of a problem. The concentrations of chlorine compounds like POPs and pesticides in wastewater have increased rapidly due to agricultural development, and some are human carcinogens.

---

* To whom correspondence should be addressed. e-mail: gtorosyan@seua.am

One of the best methods of wastewater treatment for organic pollutants is to use inorganic and organic adsorbents such as zeolites and other alumino-silicates.

2. Ecological problems

Pollution of the environment by chlorine-containing organic compounds occurs during the burning of fossil fuels (coal, petroleum, peat, combustible shales), from the application of fertilizers, which contain natural and synthetic organic impurities, as example, POPs and pesticides. These substances are particularly problematic in flood-prone areas, because of the large amounts transported during floods and the presence of human settlement and agriculture.

POPs and pesticides cannot be removed by conventional gravity separation technology. The application of treatment technologies to remove pesticides from the water would present a higher degree of complexity, risk and cost to the operations and potentially reduce the viability of many established operations.

3. Using zeolites for river water treatment

A method of sorption treatment is offered here, using natural zeolite. The increasing demand for ion-exchange materials to solve ecological problems has stimulated intensive study of natural and modified zeolites, because they are considered to be cheap (Breck, 1974). The technological stability of zeolites as sorbents is determined first of all by such characteristics as mineral and chemical composition, sorption ability and their mechanical, physical and filtering properties (Breck, 1974).

The advantages of zeolites in comparison with other sorbents (Collela, 1999) are: the large reserves in Armenia, a unique complex of technological properties – sorptional and molecular-sieving – as well as their natural origin, possibilities for their modification in various directions, regeneration and utilization (Torosyan et al., 2002). Natural zeolites and their combination with other systems have been proved to be effective in removing metal ions and organic impurities from water. The efficiency and mechanism of sorption – filtering parameters, length of contact liquid and solid phases ratio and other factors – have been studied. The chemical stability and mechanical strength of the natural zeolites (mordenite and clinopilolite) meet the requirements for filtering materials.

4. Technological decisions

The service provider treating floodwaters must take into account the following:

1. If the onsite wastewater treatment system has electrical components, the ability to restart the system will depend on the flood elevation.
2. If the floodwater only covers the tanks and the components in the tanks, it is possible to restart the system without further evaluation.
3. If the floodwater covers components located on the ground surface (air pumps, panels), the system should be inspected to determine whether it is safe to restart the electrical service and use the system.

Additionally, sorption materials can be placed on the floodplain, e.g., in ponds with mobile walls filled by adsorbents, and removed after the flood. There would be a primary treatment system and a secondary treatment where the containers filled with zeolite are established.

5. Experiments

5.1. SORBENTS

Sedimentary deposits of zeolite are widespread in the Idjevan (clinopilolite) and Shirak (mordenite) regions of Armenia. The natural zeolites clinoptilolite and mordenite were dehydrated at 110°C, and prepared by the heated pile method, using 0.5 and 1 N solutions of $CaCl_2$. Clinoptilolite modified by catamine-AB was prepared according to Torosyan et al. (2000).

5.2. ADSORPTION

Adsorptive isotherms were determined for the equilibrium status of water containing chlorine compounds or pesticides in concentrations of 0.0025–0.070 mol/l on the zeolite sorbents. Adsorption was carried out at a temperature 27.0°C, the average temperature in the Ararat valley in spring and autumn.

To remove the pollutants, sorbents were placed in water containing varying concentrations of the chlorine compounds, and stirred for 1 h and shaken for 8 h. The quantity of the chlorine compounds – chlorobenzene and dichlorobenzene – on the zeolites was determined by the precipitated organic fraction in the filtered solution using UV Spectroscopy, Highly Effective Liquid Chromatography (HELCh) and Refractometry, and the quantity of pesticides removed was calculated. Table 1 and Fig. 1 show the results for dichlorobenzene.

TABLE 1. The adsorption of dichlorobenzene (mg) on the zeolites (10 g)

N/N	Zeolite	Adsorption mg/g	Adsorption %
1.	Mordenite-tuff	0.12	22.0
2.	Mordenite – treated by 0.5N CaCl$_2$	0.19	25.0
3.	Mordenite-treated by 1N CaCl$_2$	0.28	40.0
4.	Clinoptilolite-tuff	0.22	30.0
5.	Clinoptilolite-treated by 1N CaCl$_2$	0.32	45.0
6.	Clinoptilolite-modified by Catamine-AB	0.50	70.0

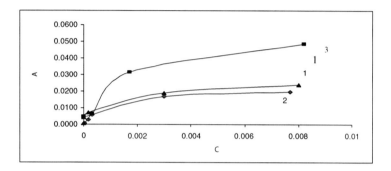

Figure 1. Isotherms of dichlorobenzene adsorption. (1) Clinoptilolite-treated by 1N CaCl$_2$, (2) Clinoptilolite-tuff, (3) Clinoptilolite- modified by Catamine-AB

6. Conclusion

An effective method of removing organochlorine compounds has been demonstrated using natural Armenian zeolites. The method can be applied at relatively low initial concentrations for other chlorine-containing aromatics, such as chlorobenzene and trichlorobenzene.

References

Breck, D.W., 1974: *Zeolite Molecular Sieves: Structure, Chemistry and Use.* New York/London/Sydney/Toronto: Wiley, 782 pp.

Collela, C., 1999: Porous materials in environmentally friendly processes. *Studies in Surface Science and Catalysis*, **125**: 641–655.

Torosyan, G., Sargsyan, S., Grigoryan, A. and Harutjunyan, S., 2000: The phenol sorption on the zeolites. *Bulletin of Armenian Constructors*, **2(18)**: 30–32.

Torosyan, G., Hovhannisyan, D., Shahinyan, S. and Sargsyan, H., 2002: Armenian zeolites and its possibilities in industrial and municipal waste water cleaning. *Ecological Journal of Armenia*, **1(1)**: 93–96.

CALIBRATION OF AN ATMOSPHERIC/HYDROLOGICAL MODEL SYSTEM FOR FLOOD FORECASTING IN THE ODRA WATERSHED

H.-T. Mengelkamp[*]
Institute for Coastal Research, Geesthacht, Germany

Abstract. Hydrologic models commonly contain parameters that can only be inferred by a trial-and-error process that adjusts the parameter values to closely match the input–output behaviour of the model to the real system it represents. Traditional calibration procedures, which involve manual adjustment of the parameter values are labour-intensive, and their success is strongly dependent on the experience of the modeller. This paper describes the application of the objective calibration methodology SCE-UA (Shuffled Complex Evolution – University of Arizona) to the hydrological model SEROS for the Odra watershed. SEROS combines the rainfall-runoff model SEWAB (Surface Energy and Water Balance) and a routing scheme. Simulated and observed time periods of the discharge show reasonable agreement during the calibration period as well as for the verification period. The latter includes the 1997 major flooding event.

Keywords: Objective calibration, rainfall-runoff model, routing scheme

1. Introduction

Hydrologic models commonly contain parameters that can only be inferred by a trial-and-error process that adjusts the parameter values to closely match the input–output behaviour of the model to the real system it represents. Traditional calibration procedures, which involve manual adjustment of the parameter values are labour-intensive, and their success is strongly dependent on the experience of the modeller. Automatic methods for model calibration are objective. However, many studies have shown that such methods have difficulties in finding unique parameter estimates. The consideration of these problems resulted in the development of a robust and efficient global optimisation algorithm called 'shuffled complex evolution' (SCE-UA) global optimisation algorithm developed at the University of Arizona (Duan et al., 1992, 1993, 1994; Sorooshian and Gupta, 1995). The SCE-UA approach is applied to calibrate the hydrological model SEROS for the Odra watershed. Basically, an objective function is optimized (maximized or minimized) by "shuffling" parameter settings in a prescribed feasible parameter space. The objective function is some measure of the correlation between an observed

---

[*] To whom correspondence should be addressed. e-mail: mengelkamp@gkss.de

and simulated time series. In hydrology, streamflow measured at some gauging station is commonly the target variable to be compared to the output of a hydrological model.

2. The hydrological model SEROS

The hydrological model SEROS combines the one-dimensional vertical land surface scheme SEWAB (Surface Energy and Water Balance, Mengelkamp et al., 1999) and a horizontal routing scheme (Lohmann et al., 1996).

The one-dimensional (vertical) land surface model SEWAB calculates the vertical water and energy fluxes between the land surface and the atmosphere and within the soil column for a land surface grid cell. The surface energy balance equation describes the equilibrium of net irradiance, latent heat flux, sensible heat flux and soil heat flux (and in case of snow, the energy available for melting). Precipitation is partitioned into runoff, evapotranspiration and change of snow pack and soil moisture storage. The evapotranspiration is calculated separately for bare soil and vegetated parts of the land surface grid cell. Warrach et al. (2001) incorporated a single-layer snow model to allow for a partially snow covered land surface grid cell. The soil column (Fig. 1) is divided into a variable number of model layers. Within the soil column, temperature diffusion (with a term for soil freezing) and the vertical water transport are calculated. Leaf drip, precipitation on bare soil, evaporation from bare soil and the soil moisture are accounted for. As a one-dimensional model, SEWAB represents a land-surface grid cell of an atmospheric circulation model with dimensions ranging from 1 to more than 10,000 km^2. Runoff from the grid cell soil column is subject to transformation and translation processes before the water reaches the river as streamflow. Runoff may occur from saturated patches inside the grid cell before saturation of the whole soil column or even may be delayed through ponding at the surface. These processes are described by the variable infiltration capacity approach for surface runoff and the concept of linear reservoirs for subsurface runoff and groundwater flow (Mengelkamp et al., 2001).

Linear reservoirs are added to the soil column to describe subsurface runoff and baseflow. Subsurface runoff generation follows the ARNO model conceptualization (Dümenil and Todini, 1992).

The outflow from two linear groundwater storages for the slow and fast component represents the runoff baseflow component. The storages are filled by Darcian flow from the lowest soil layer. Individual time constants for the fast and slow component are determined empirically. This concept of storages allows a subtle adjustment of surface runoff, subsurface runoff and baseflow. However, the large number of calibration parameters (time constants and storage heights) makes the calibration procedure a tedious task.

The routing scheme (Fig. 2) describes both, the time runoff takes to reach the outlet of a grid box and the transport in the river channel system (Lohmann et al., 1996). Inside each grid box runoff is generated by the land surface model through the concepts outlined above or as saturation excess runoff. This runoff is transformed through the impulse response function of the unit hydrograph into box outflow. There is also upstream inflow into the grid cell through the river channel. This part of the flow through the grid cell is transformed by the river impulse response function to box outflow. The sum of streamflow generated inside the grid box and throughflow is the total outflow from the grid box and represents the river inflow to the next downstream box.

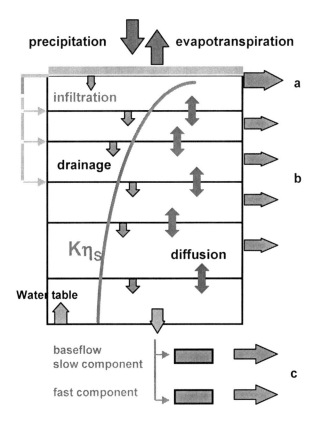

Figure 1. Sketch of the hydrological processes in SEWAB; (**a**), (**b**) and (**c**) represent surface runoff, subsurface runoff and baseflow, respectively.

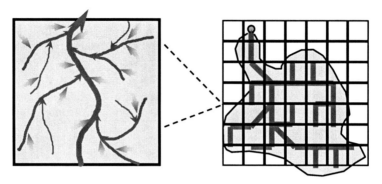

Figure 2. Sketch of the horizontal routing scheme

3. Data and SEROS setup

Data required to set up SEROS for the Odra watershed include geographic information (orography, landuse and soil type), meteorological forcing data from synoptic stations and streamflow at gauging stations. The geographic and meteorological data have to be interpolated onto the model grid. The locations of the gauging stations determine the sub-catchments for which representative parameter sets have to be assigned. We consider a period from 1992 to 1999 with the three years calibration period from 1992 to 1994 and a five year verification period from 1995 to 1999.

The SEROS area covers the Odra catchment with a total number of 5,685 grid cells of 4.5 x 4.5 km^2 size (Fig. 3). Each grid cell is assigned a mean height above sea level, a dominant vegetation and a soil type. The area of the sub-catchments is determined, and the river channel network is established. The natural river length is estimated from the elevation data for each grid cell as well.

The atmospheric forcing for SEWAB are downward shortwave radiation, cloud cover, wind speed, air temperature and specific humidity, surface pressure and precipitation. These data are available from 50 synoptic stations on a three hourly basis. Daily precipitation is also available from 666 additional stations. The forcing data are spatially interpolated onto the model grid. Precipitation is evenly distributed over the whole day to account for SEWAB's three-hourly time step. This should not make up a major uncertainty because input to the routing scheme is daily runoff.

Daily streamflow data are available for 28 primary gauging stations and for 11 reservoirs in the mountainous region.

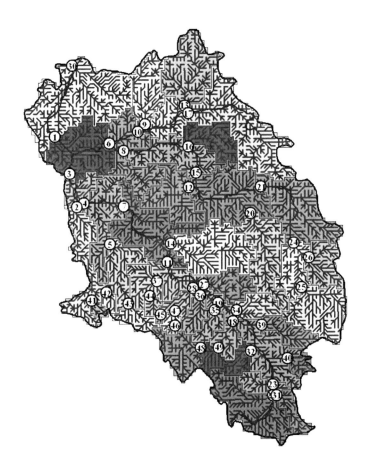

Figure 3. Sub-catchments of the Odra watershed and locations of gauging stations

4. Calibration

The rainfall-runoff (SEWAB) and the horizontal routing scheme are based
on conceptual representations of the physical processes. Conceptual representations are controlled by physical parameters that describe measurable
properties of the watershed and non measurable process parameters.
Despite the detailed information for vegetation cover and soil type the
respective parameters cannot be exactly defined for a single grid nor for a
sub-catchment. This includes parameters of the evapotranspiration, runoff
generation, initial soil water content and the water transport in the channel

system. Some parameters can be deduced from watershed properties (i.e. the length of the river inside a grid box, the partition of major vegetation types). The interception reservoir or the stomata resistance of the vegetation, the retention period of the water inside a grid box or the partition in surface and subsurface runoff can not be known a priori. These parameters are among the ones which are subject to calibration. The choice of parameters conforms to the necessity to include the significant processes but to minimize the number of parameters.

During the calibration period 1992–1994 more than 80% of the sub-catchments show an efficiency over 65%, more than 40% of the sub-catchments reach efficiencies over 90% (Fig. 4). The efficiencies for the validation period 1995–1999 are lower (as expected) in particular in the eastern part of the Odra watershed. An explanation might be that the spatial density of precipitation stations is lowest in the eastern part and that sub-surface water transports in these flat areas and some smaller reservoirs are not adequately accounted for in the model. Additionally, the validation period includes the extreme flooding event of 1997 which in some smaller sub-catchments might not be represented properly.

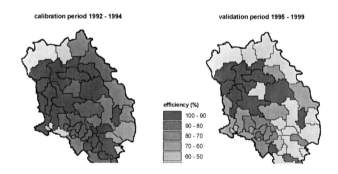

Figure 4. Efficiency of the model system for the calibration period (*left*) and the validation period (*right*)

5. Streamflow at the Otmuchow gauging station

Exemplary the discharge from the sub-catchment Otmuchow is discussed here which contributed much to the 1997 extreme flooding event. The Otmuchow watershed covers an area of 2,361 km^2 in the Sudete mountains (sub-catchment 48 in Fig. 3). Figure 5 shows the simulated and measured discharge for 1993, a year out of the calibration period. The maximum discharge in March mainly caused by snow melt is slightly overestimated (172 instead of measured 134 m^3/s) but exact in time. The total volume of discharge from March 12–31 is underestimated by 10%. Some snow melt events in January and April are underestimated as well. Probably precipitation fallen as snow was not corrected for adequately. Some discrepancies also occur during two heavier precipitation events end of October and early November. The efficiency is estimated as 76%.

A similar picture is shown for the year 1997 out of the validation period (Fig. 6). Streamflow due to snow melt is underestimated (February), while

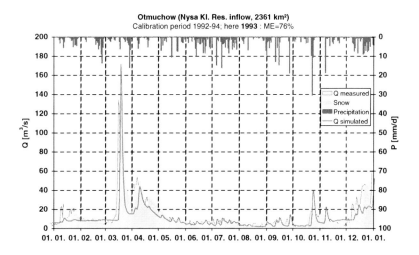

Figure 5. Simulated and observed streamflow at the Otmuchow gauging station for the year 1993 (calibration period)

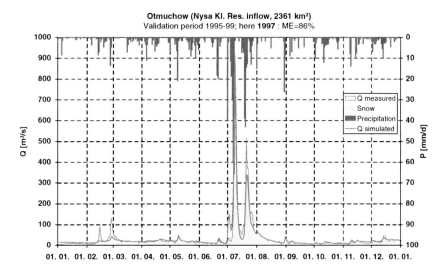

Figure 6. Simulated and observed streamflow at the Otmuchow gauging station for the year 1997 (validation period)

during the rest of the year the total amount of discharge is adequately simulated. With 86% the efficiency is even higher than during the calibration phase. The extreme flooding event in July 1997, caused by two heavy precipitation events, with a maximum discharge of 800 m³/s on July 9 and a second maximum with 520 m³/s on July 21 are simulated on time but with slightly over- and under-estimated peaks. A small maximum on July 3 due to heavy precipitation is simulated but not seen in the observations. The efficiency for July exclusively is 79%.

6. Conclusions

The hydrological model SEROS is described which consists of a grid based rainfall-runoff scheme with advanced features for runoff generation and a horizontal routing scheme. The model is forced by observed synoptic data interpolated onto the spatial grid. The free parameters are found by objective calibration. This results are in reasonable agreement between observed and simulated streamflow during the validation and the calibration period, the latter including the 1997 extreme flooding event.

References

Duan, Q., Sorooshian, S., and Gupta, V.K., 1992: Effective and efficient global optimization for conceptual rainfall-runoff models, *Water Res. Res.*, **24(7)**: 1163–1173.

Duan, Q., Gupta V.K., and Sorooshian S., 1993: Shuffled complex evolution approach for effective and efficient global minimization, *J. Optimiz. Theory Appl.*, **76(3)**: 501–521.

Duan, Q.S., Sorooshian, S., and Gupta, V.K., 1994: Optimal use of the SCE-UA global optimization method for calibrating watershed models, *J. Hydrol.*, **158**: 265–284.

Dümenil, L., and Todini, E., 1992: A rainfall-runoff scheme for use in the Hamburg climate model. In: J.P. O'Kane (Ed.), *Advances in theoretical hydrology, A tribute to James Dooge.* European Geophys. Soc. Series on Hydrological Sciences 1, Elsevier, Amsterdam, 129–157.

Lohmann, D., Nolte-Holube, R. and Raschke, E., 1996: A large scale horizontal routing model to be coupled to land surface parameterization schemes. *Tellus*, **48A(5)**: 708–721.

Mengelkamp, H.-T., Warrach, K., and Raschke, E., 1999: SEWAB: a parameterization of the surface energy and water balance for atmospheric and hydrologic models. *Adv. Water Res.*, **23(2)**: 165–175.

Mengelkamp, H.-T., Warrach, K., Ruhe, C. and Raschke, E., 2001: Simulation of runoff and streamflow on local and regional scales. *Meteor. Atmos. Phys.*, **76**: 107–117.

Sorooshian, S. and Gutpa, V.K., 1995: Model calibration. In: V.P. Singh (Ed.), *Computer models in watershed hydrology*, 23–63.

Warrach, K., Mengelkamp, H.-T. and Raschke, E., 2001: Treatment of frozen soil and snow cover in the land surface model SEWAB. *Theor. Appl. Climatol.*, **69**: 23–37.

THE ROLE OF METEOROLOGICAL MODELS IN THE PREDICTION OF WEATHER HAZARDS – THE EUROPEAN APPROACH

C. Cassardo[*]

*Department of General Physics "Amedeo Avogadro",
University of Torino, Italy*

Abstract. It is a matter of fact that the economic damage and the casualties provoked by extreme meteorological events have dramatically increased in the last few decades, and that, in the last 30 years, a remarkable part of this increment is caused by the increased frequency of extreme meteorological events. The reasons for such an increase are not simply related to the climatic changes, but also to the intensive exploitation of the land areas, like unauthorized building, diffuse urbanization, river canalization, intensive agriculture etc., which has made them much more vulnerable today than in the past. In this context, the use of advanced meteorological instruments could surely give help in the forecasting and in the prevention of extreme phenomena and of their consequences. However, the greatest improvement in extreme events prediction can be achieved by the use of numerical models, which constitute an essential means in the aim of improving both the forecast of the extreme events and the correlated hydrogeological risk assessment. The paper, after a short introduction on the numerical models, contains some considerations on the importance of land surface conditions and of surface layer parameterizations in order to produce a reasonably good prediction of some extreme dry (droughts or heat waves) and wet (floods) events. Among the surface layer parameterizations, the representation of the orographic influences on the atmospheric flow is quite important for estimating the quantity and the intensity of the precipitation. It is explained that for extreme wet events the final objective would be to couple high-resolution meteorological and hydrological models, creating a "meteo-hydrological chain". A detailed discussion will examine the role of the seasonal meteorological predictions, with a part dedicated to the Ensemble Prediction System, and to their utility for forecasting the hydrological variables. Finally, a reflection in the last section is dedicated to the problem of the communication of meteorological and hydrological risks to the public.

Keywords: Flood, drought, heat wave, GCM, LAM, LSPM

---

[*] To whom correspondence should be addressed. e-mail: cassardo@ph.unito.it

J.A.A. Jones et al. (eds.), *Threats to Global Water Security*,
© Springer Science + Business Media B.V. 2009

1. Introduction

Numerical weather prediction can be considered as an application of the experimental Galilean method (Pasini, 2005) in which, due to the complexity and the uniqueness of the terrestrial atmosphere, and thus the impossibility of conducting real experiments, the computer can be seen as a *virtual laboratory*. This idea allows us to develop numerical models in which single differential equations, each one representing a portion of theory describing a portion of reality, are composed in a mathematical model constituted by one or more coupled system of equations. The variables of these equations have a correspondence in the real system where they can be measured. Thus, giving some realistic initial values to each model variable, indicating the boundary conditions, and numerically solving these equations, it is possible to get a prediction of the most relevant thermodynamic and dynamic parameters in the atmosphere.

This approach was used for the first time by Lewis Richardson in 1927, but at that time computers did not exist and also the theory to ensure the numerical stability of the solutions was not yet discovered, thus his attempt was not successful. At the beginning of 1950, John von Neuman and Jule Charney pioneered the use of the electronic computer for weather forecasting at the Institute for Advanced Study at Princeton University. They used a primitive computer, much less powerful that today's personal computers, but they succeeded in forecasting the horizontal air geopotential height pattern at 500 hPa over Northern America.

A meteorological model consists of a series of prognostic differential equations for the main meteorological variables (wind velocity, temperature, humidity and pressure): conservations of momentum, energy and water, state and continuity equations. Some numerical computer models of the atmosphere are designed to operate over different spatial scales depending on the forecast range. For example, a model for short-range forecasts (up to 3 days) can cover an area extending to a continent and needs as initial data the observations carried out only in that restricted area. On the other hand, a model designed for medium-range forecasts (up to 10 days) is generally a global model and thus needs observational data from all over the globe, since within such a forecast range, a weather system can travel over distances comparable to the Earth's radius.

The equations included in the first generation models were simplified and the resolution was very coarse, thus these models were able to diagnose only some large-scale phenomena affecting continental areas. With the continuous increase in computer power, the equations used became more complete, the resolution was increased and the dataset used for the model initialization was also more dense both in time and space. Furthermore, some numerical techniques were developed to assess more stable and detailed solutions. In 1979, the European Centre for Medium-range Weather Forecast

THE ROLE OF METEOROLOGICAL MODELS

(ECMWF) was created in Reading, UK, with the aim of providing medium-range forecast data for the northern hemisphere. Another very important component of the model forecast chain which has been considerably improved is the data initialization. Actually, the ECMWF model can provide the meteorological data on a grid whose size at the latitude of 45° is approximately 25 km, while in 1979 the grid size was only about 120 km. This means that the ECMWF GCM (General Circulation Model) can actually be used to initialize a mesoscale model able to run over a grid size of few kilometres. The correctness of the predicted fields also increased almost regularly over the years (see Figs. 1 and 2). The accuracy of the weather forecasts provided by the mesoscale models is even greater, but is normally limited to the next 2–3 days, because the domain is too small to allow longer runs.

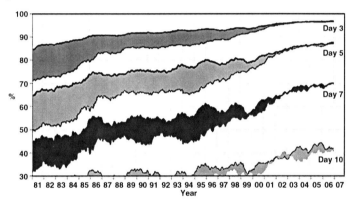

Figure 1. Improvement of the anomaly correlation of the 500 hPa geopotential height depending on the forecast day for the two hemispheres in the period 1981–2007 for the ECMWF GCM (adapted from CSAEOS, 2008)

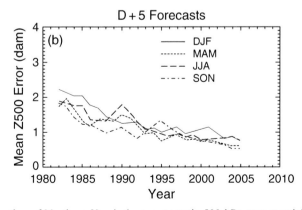

Figure 2. Reduction of Northern Hemisphere systematic 500 hPa geopotential height error for the fifth day of forecast in the period 1981–2006 for the ECMWF GCM (courtesy of ECMWF)

With the availability of these kinds of products, it is actually possible to have highly accurate weather predictions at the large-scale and also accurate predictions at the mesoscale and at the local scale. Thus, in principle, it can be possible to provide some alerts in the case in which some extreme or hazardous events are predicted in the medium-term. In the category of the hazardous or extreme events are included floods or severe rainfalls, typhoons, hurricanes or tropical cyclones, extreme winds, but also prolonged drought episodes or heat waves.

The normal period of time in which a deterministic weather forecast has an adequate validity is normally 3–5 days, depending on the weather situation and on the local characteristics, but using some methodologies, such as the Ensemble Predictions Systems, it is possible to have an idea about the probability of occurrence of a determinate forecast even for the next 5 days. Forecasts over longer periods are currently experimental in some Meteorological Services.

In this paper, the reader will find some information on the most recent developments regarding some characteristics of the numerical models used in meteorology, and a conclusive note showing their importance in the predictions of natural hazards.

2. The prediction of heat waves and drought conditions

During June, July and August 2003, an exceptional heat wave affected western and central Europe. The extreme drought and heat had heavy social, economic and environmental impacts, among them the death of thousands of the elderly. Other serious consequences were the destruction of large forest areas by fire, the depletion of aquatic ecosystems, the retreat of glaciers, power cuts and transport restrictions, and a decrease in agriculture production. In recent years, the number of such extremes was increasing, according to the Intergovernmental Panel on Climate Change (IPCC, 2001), and Meehl and Tebaldi (2004) also said that is possible that the number of heat waves similar to that of 2003 will increase in the future. As the long-range forecasts (greater than 1 month) for this phenomenon were not successful, it is thus important to try to understand which mechanisms can exacerbate or guide this kind of event.

In the specific case of the European heat wave in the summer of 2003, it has been shown that the synoptic circulation induced by the persistence of an anomalously strong anticyclone extended over Western Europe for most of the period May–August 2003 drove hot air from North Africa, producing strong subsidence motions on the right-hand side of the anticyclone (i.e. over central and Western Europe) during most of the time (see for instance Cassardo et al., 2007). The enhancement of the adiabatic compression related to these synoptic motions contributed to the observed temperature increments. A second enhancing factor that exacerbated the heat wave effects was the precipitation deficit recorded during the preceding spring, which caused low soil moisture values to be observed already at the beginning of the heat wave. This fact is not surprising as, according to Black and Sutton (2007), the soil water content in Europe's Mediterranean regions plays a critical role in the climate regulation across Europe and in the development of climate anomalies over Europe.

A thorough understanding of the physical mechanisms of climate is a baseline for their subsequent incorporation in the meteorological and climatological models. For instance, the ECMWF short-, medium- and long-range (up to 30 days) forecasts were successful in inferring the main features of the large-scale flow as well as the temperature anomalies, while the seasonal predictions were a failure, even considering the EPS (see Section 5). Grazzini et al. (2003) suggested that this problem could be due to a combination of (i) failure to predict the observed large Sea Surface Temperature (SST) anomalies in the Indian Ocean, and (ii) the above mentioned dependence on the soil moisture. The continuous refinement of the physical processes in the numerical atmospheric models could allow a more accurate prediction of such extreme phenomena.

3. The importance of land surface conditions and of surface layer parameterizations

In the last 2 decades, it has been recognized that a necessary boundary condition for all Numerical Weather Prediction (NWP) modeling is the representation of the land surface processes. They are crucial in short-term weather forecasts as well as for climatic predictions, as the Earth surface can influence the partition between sensible and latent heat fluxes and consequently the atmospheric stability is related to the surface properties. In this respect, leaving aside the surface and land characteristics, which also are a key factor for the surface processes, soil temperature and especially soil moisture constitute very important parameters. In fact, the Earth surface is the main source of energy and moisture, which regulate the atmospheric motions. Wrong estimates of those parameters lead to wrongly simulated variables in the surface layer, especially for the phenomena related to convection, and thus the forecasts of atmospheric stability and precipitation amount could have a low reliability. Nevertheless, the soil temperature and moisture, as well as sensible and latent heat fluxes, are not measured extensively or for long periods, and also satellite estimates are difficult to generalize over areas with different vegetation cover or types. Thus, a key problem in weather forecasting is the initialization of soil temperature and moisture for the model simulation. Therefore, in the last few years, there have been many attempts to evaluate these parameters in the absence of direct observations, both for weather forecasting and for climatic simulations. One methodology, presented in Cassardo et al. (1997), consists in running for a sufficiently long time a SVAT model over a multitude of synoptic meteorological stations included in a mesoscale area, and in averaging the model predictions of soil moisture and temperatures. This method proved successful in improving the rainfall maxima during an episode of strong convection that occurred near the Alps (Cassardo et al., 2002a), showing that the initial soil moisture field can affect precipitation predictability in short-term weather forecasts. Another method, which uses a linearized variational technique for the analysis of total soil water content via the assimilation of screen-level observations (Balsamo et al., 2003), has been applied to the European data and also showed the importance of the soil moisture initialization for improving the skill of model predicted precipitation at large scale.

4. The role of the seasonal meteorological predictions and the ensemble prediction systems (EPS)

Until recent years, climate prediction had been viewed as a speculative and largely unproven venture. In fact, deterministic forecast lengths are limited by the Lorenz deterministic chaos theory to 2–3 weeks, and by the scarcity of data for the initialization to not more than 7 days, sometimes less. Thus, deterministic predictions of seasonal mean climate based on the mean anomaly are not always accurate for individual cases, even when a perfect atmospheric global circulation model and perfectly represented boundary conditions are used. More recently, when the meteorological models have been coupled with the oceanic models in order to better quantify the effects of interannual variability phenomena like the El Niño–Southern Oscillation – ENSO, seasonal forecasts of 3-month-average surface temperature or precipitation have been demonstrated to have some skill in particular seasons, regions, and circumstances (see for instance Shukla et al., 2000). This is the reason why some meteorological services and research institutes started few years ago to publish some climatic prediction maps referring to the future 1–3 months. However, as it is not yet clear why seasonal predictions succeed in some instances but fail in others. These forecasts are still regarded as research products and are not routinely used by the forecasters in their decision-making process, especially for the predictions related to the hydrological variables. On the contrary, their dissemination on the web sometimes causes some confusion for the public, who suppose that they have validity similar to the deterministic predictions (i.e. 5–7 days), and when these predictions fail the public tends to lose confidence in the quality of all meteo-hydrological predictions.

Thus, until now these products must still be regarded as a sort of very preliminary attempt to evaluate the performances of the models over so long a timescale, and, in my opinion, greater effort should be made in trying to explain to the general public that these products cannot be used in the same way as the short-term forecasts. From the point of view of the research, as our current knowledge about the relationships between boundary conditions and climate is still incomplete, many efforts will be required to fill this gap. For instance, Barnston et al. (2005) suggest that, for improving the quality of the climate predictions, a *two-tiered* prediction system is required: (i) to obtain a best prediction for the future boundary condition anomalies, and (ii) to specify the true probability distribution function corresponding to this boundary condition. Only when the accuracy of the seasonal climate predictions have increased, could they begin to be valued for meteorological and hydrological forecasts.

Regarding meteorological forecasts, one of the most recent and promising techniques in NWP is the development of the Ensemble Prediction Systems (EPS). These models have been recently developed in an attempt to overcome the problem of the limited temporal validity of deterministic forecasts (Ehrendorfer, 1997). As is known, these models generate an ensemble of predictions using the same model and different initial conditions, or, more recently, using the same initial conditions but different models or parameterizations (Buizza et al., 2005). To deal with the increased computational demand caused by running the models several times for the same prediction, normally EPS operate at a coarser grid scale compared with the deterministic model, but the advantage of this technique is that the generation of an ensemble of forecasts, rather than a single deterministic forecast, provides a way of quantifying the uncertainty about them, because it can be converted into a probability distribution function. The EPS has been part of the ECMWF operational suite since December 1992, when it has been gradually upgraded with the implementation of the Variable Resolution Ensemble Prediction System (VAREPS). This is designed to increase the ensemble resolution in the early forecast range and to extend the forecast range covered by the ensemble system to 1 month (Buizza et al., 2007).

At present, the vast majority of flood forecasting systems are deterministic in design and input, but this is changing fast, and new programmes are currently in development in Europe and in the US with the explicit aim of developing and producing ensemble forecasts that allow a reliable hydrological capacity (Demeritt et al., 2007). Examples in this sense are the expectation of an EPS for forecasting floods to be fully operational in the USA by 2013 (US National Weather Service, 2001), and the Hydrological Ensemble Prediction Experiment (Schaake et al., 2006) sponsored by the World Meteorological Organization (WMO).

5. The need of interfacing meteorological and hydrological models for flood forecasting

Flood forecasting is an important factor that can reduce the negative effects of flood events. A few years ago, the traditional method used to predict real-time flood events was to use a semi-distributed rainfall-runoff model (Bartholmes and Todini, 2005). Recently, to improve flood-forecasting capacity, more accurate methods are in process of development. For instance, the strategic plan for the US National Weather Service (2001) is investing strongly in the development of an Advanced Hydrologic Prediction System.

Nowadays, the continuous increase in computer power has allowed the utilization of distributed hydrological models with higher resolution (mesh size of 10–1,000 m). Also, meteorological models have increased their resolution in the recent years and now GCMs can run with a grid scale of 10–20 km or even less using nesting techniques. Mesoscale meteorological models can produce forecasted fields at a higher resolution (some hundred meters), but in normal operative conditions only for 2–3 days. Thus, since the aim of each meteo-hydrological system is to extend the forecasting horizon to 1 week or more, the GCMs are taken into consideration.

The gap between the resolutions required by hydrological models for the precipitation field and that furnished by the GCMs does not allow a direct coupling of the two models for an accurate prediction of the floods, especially for small- and medium-range basins. Cassardo et al. (2006) have indeed shown that it is possible to get some rough estimates of the river-flow increases for a large basin, such as that of the Po River in the Italian Piedmont region, during the exceptional flood of the October 2000, by taking into consideration the runoff evaluated by a SVAT (Soil Vegetation Atmosphere Transfer) scheme (in that case, the LSPM, Land Surface Process Model, has been used) coupled with a LAM (Limited Area Model). Nevertheless, the results reported by Ferraris et al. (2002) show that a direct coupling of meteorological and hydrological models might be considered not particularly useful, even in catchments showing scales that allow this direct coupling, because of the wrong location of high precipitation clusters in the meteorological model, which could have dramatic consequences on hydrologic modelling. Incidentally, the results of Cassardo et al. (2002b), who also studied the same flood episode in November 1994 in Piedmont, have shown some problems of the meteorological model in correctly locating the predicted precipitation cluster, which produced the flood in the Tanaro river basin.

Two techniques are currently used with the aim to improve the quality of the forecast. One is the Ensemble Prediction System (EPS – see Section 4), currently used by ECMWF and many other meteorological services, like the German DWD (see for instance Ferraris et al., 2002, or also Bartholmes and Todini, 2005), in order to better assess the spatial and temporal variability associated to the deterministic forecast. The second one is the downscaling of the rainfall data provided by the meteorological model prior to using the hydrological model. At this regard, one of the most used methods is the multi-fractal theory, that, according with Gupta and Waymire (1993), can be considered a very powerful approach to nonlinear and intermittent processes like precipitation, and can reproduce simultaneously the statistical properties of real rainfall in space and time.

Following the established success of EPS in weather forecasting, several initiatives are trying to promote the application of such techniques to hydrological modelling, and in particular to real-time flood forecasting.

6. The problem of the orography and the results of the MAP project

The Mesoscale Alpine Programme (MAP) was an international research initiative supported by the WMO (World Meteorological Organization). Scientists from 13 countries all around the world have been directly involved in MAP, which was devoted to the study of atmospheric and hydrological processes over mountainous terrains. MAP started officially in 1994 with the twofold aims (i) to understand and model the physical processes at the basis of the intense meteorological phenomena induced by the topography, and (ii) to refine the instruments to make short-term forecasts. Particular attention was devoted to the mesoscale phenomena, which are at the basis of the floods. The MAP projects and their scientific objectives were subdivided in the following way. Three projects were in direct support of the so-called "wet" part of MAP: the orographic precipitation mechanisms, the upper tropospheric potential vorticity anomalies, and the hydrological measurements for flood forecasting. Four projects came in direct support of the "dry" part of MAP: the dynamics of gap flow, the non-stationary aspects of foehn winds in large valleys, the three-dimensional gravity wave breaking, and the potential vorticity banners. Finally, one project, devoted to the planetary boundary layer structure, came in support of both wet and dry scientific objectives.

In 2007, when most results of MAP were already published (see the MAP website: http://www.map.meteoswiss.ch/), the MAP D-PHASE ('Demonstration of Probabilistic Hydrological and Atmospheric Simulation of flood Events in the Alpine region') was launched. This new project, which is a Forecast Demonstration Project (FDP) of the WWRP (World Weather Research Programme of WMO), aims at demonstrating some of the MAP achievements: in particular, the ability of forecasting heavy precipitation and related flooding events in the Alpine region, addressing the entire forecasting chain ranging from limited-area ensemble forecasting, high-resolution atmospheric modelling (km-scale), hydrological modelling, and nowcasting to decision-making by the end users. The MAP D-PHASE will thus summarize the state-of-the-art in the forecasting of precipitation-related high-impact weather, especially concerning the mountainous regions, where hazardous events are most common, most difficult to predict and often have the largest impacts.

7. The problem of communicating the flood risk

An important question is whether and how useful meteorologists and hydrologists actually find the new technologies in making the decision about whether or not to issue a flood warning. The problem of using the innovative technologies in the best possible way has been illustrated well by Morss et al. (2005). In practice, the problem is to find a balance between the desire for issuing a warning as early as possible (Harremoës et al., 2002) and the inevitable trade-off between false alarms and false negatives, that is, unforeseen flood events (see Hammond, 1996). A study by Demeritt et al. (2007) shows that flood forecasters welcome EPS predictions, but that they use mainly these data to confirm their deterministic models, and that, in the case in which there is a discrepancy between EPS and deterministic model predictions, they adopt the technique of "*wait and see*". This seems to indicate that the continuous development of techniques is welcome, but that it is also necessary to invest in explaining directly to the targeted users how to utilize these new technologies.

8. Conclusions

This paper has described some aspects of the currently most important and most recent developments in the field of the prediction of extreme and hazardous event. To pursue this end, a panoramic view of the techniques which are currently in development by the meteo-hydrological community has been described and summarized. The improvement of modelling skills in the prediction of natural hazards can be successful if several aspects of contemporary meteo-hydrological science are improved. Data availability is a key factor, as it allows the models to be tested and intercomparisons to be made. Model parameterizations, especially those related to the surface layer and to the soil surface, must be continuously developed to improve the physical mechanisms embedded in the code (as in the MAP experiment). Moreover, the quality of a deterministic meteorological forecast can be improved by running the same model but over a coarse scale in different configurations (EPS), which will give a better description of the model output.

References

Balsamo, G.P., Bouyssel, F. and Noilhan, J., 2003: A simplified bi-dimensional variational analysis of soil moisture from screen-level observations in a meso-scale numerical weather prediction. *Quarterly Journal of the Royal Meteorological Society*, **130A(598)**: 895–916.

Barnston, A.G., Kumar, A., Goddard, L. and Hoerling, M.P., 2005: Improving seasonal prediction practices through attribution of climate variability. *Bulletin of the American Meteorological Society*, **86**: 59–72.

Bartholmes, J. and Todini, E., 2005: Coupling meteorological and hydrological models for flood forecasting. *Hydrology and Earth System Sciences*, **9(4)**: 333–346.

Black, E. and Sutton, R., 2007: The influence of oceanic conditions on the hot European summer of 2003. *Clim dyn.*, **28(1)**: 53–66.

Buizza, R., Houtekamer, P.L., Toth, Z., Pellerin, G., Wei, M. and Zhu, Y., 2005: A comparison of the ECMWF, MSC and NCEP global ensemble prediction systems. *Monthly Weather Review*, **133**: 1076–1097.

Buizza, R., Bidlot, J.-R., Wedi, N., Fuentes, M., Hamrud, M., Holt, G., Palmer, T. and Vitart, F., 2007: The ECMWF variable resolution ensemble prediction system (VAREPS). *ECMWF Newsletters*, **108**: 14–20.

Cassardo, C., Ruti, P.M., Cacciamani, C., Longhetto, A., Paccagnella, T. and Bargagli, A., 1997: CLIPS experiment. First step: Model intercomparison and validation against experimental data. *MAP Newsletters*, 7: 26–27.

Cassardo, C., Balsamo, G.P., Cacciamani, C., Cesari, D., Paccagnella, T. and Pelosini, R., 2002a: Impact of soil surface moisture initialization on rainfall in a limited area model: A case study of the 1995 South Ticino flash flood. *Hydrological Processes*, 16: 1301–1317.

Cassardo, C., Loglisci, N., Gandini, D., Qian, M.W., Niu, G.Y., Ramieri, P., Pelosini, R. and Longhetto, A., 2002b: The flood of November 1994 in Piedmont, Italy: A quantitative analysis and simulation. *Hydrological Processes*, 16: 1275–1299.

Cassardo, C., Loglisci, N., Paesano, G., Rabuffetti, D. and Qian, M.W., 2006: The hydrological balance of the October 2006 flood in Piedmont, Italy: Quantitative analysis and simulation. *Physical Geography*, **27(2)**: 1–24.

Cassardo, C., Mercalli, L. and Cat-Berro, D., 2007: Characteristics of the summer 2003 heat wave in Piedmont, Italy, and its effects on water resources. *Journal of the Korean Meteorological Society*, **43(3)**: 195–221.

CSAEOS (Committee on Scientific Accomplishments of Earth Observations from Space), 2008. Earth Observations from Space: The First 50 Years of Scientific Achievements. The National Academies Press, Washington, D.C., 144 pages (available online at this site: http://books.nap.edu/openbook.php?record_id=11991&page=R1, checked on April 1st, 2009)

Demeritt, D., Cloke, H., Pappenberger, F., Thielen, J., Bartholmes, J. and Ramos, M.-H., 2007: Ensemble predictions and perceptions of risk, uncertainty, and error in flood forecasting. *Environmental Hazards*, 7: 115–127.

Ehrendorfer, M., 1997: Predicting the uncertainty of numerical weather forecasts: A review. *Meteorologische Zeitschrift*, Neue Folge 6: 147–183.

Ferraris, L., Rudari, R. and Siccardi, F., 2002: The uncertainty in the prediction of flash floods in the northern Mediterranean environment. *Journal of Hydrometeorology*, **3**: 714–727.

Grazzini, F., Ferranti, L., Lalaurette, F. and Vitard, F., 2003: The exceptional warm anomalies of summer 2003. *ECMWF Newsletter*, **99**: 2–8.

Gupta, V.K. and Waymire, E.C., 1993: A statistical analysis of mesoscale rainfall as a random cascade. *Journal of Applied Meteorolgy*, **32**: 251–267.

Hammond, K.R., 1996: *Human Judgement and Social Policy: Irreducible Uncertainty, Inevitable Error, Unavoidable Injustice*. Oxford University Press, New York.

Harremoës, P., Gee, D., MacGarvin, M., Stirling, A., Keys, J., Wynne, B. and Vaz Guedes, S. (Eds.), 2002: *The Precautionary Principle in the 20th Century: Late Lessons from Early Warnings*. Earthscan, London.

IPCC, 2001: *Third assessment report, intergovernmental panel on climate change.* WMO: Press release No 695.

Lorenz, E.N., 1963: Deterministic nonperiodic flow. *Journal of Atmospheric Science*, **20**: 130–141.

Meehl, G.A. and Tebaldi, C., 2004: More intense, more frequent and longer lasting heat waves in the 21st century. *Science*, **305**: 994–997.

Morss, R.E., Wilhelmi, O.V., Dowton, M.W. and Gruntfest, E., 2005: Flood risk, uncertainty, and scientific information for decision making: Lessons from an interdisciplinary project. *Bulletin of the American Meteorological Society*, **61**: 695–701.

Pasini, A., 2005: *From Observations to Simulations. A Conceptual Introduction to Weather and Climate Modelling*. World Scientific Publishers, Singapore.

Schaake, J., Franz, K., Bradley, A. and Buizza, R., 2006: Hydrologic ensemble prediction experiment (HEPEX). *GEWEX Newsletter*, **15(4)**: 10.

Shukla, J., Anderson, J., Baumhefner, D., Brankovi, C., Chang, Y., Kalnay, E., Marx, L., Palmer, T., Paolino, D., Ploshay, J., Schubert, S., Straus, D., Suarez, M. and Tribbia, J., 2000: Dynamical seasonal prediction. *Bulletin of the American Meteorolgical Society*, **81**: 2593–2606.

US National Weather Service, 2001: Strategic Plan for Weather, Water and Climate Services 2000–2005. National Oceanic and Atmospheric Administration, Washington, DC. Updated with changes approved as of December 2001, last accessed 19 January 2008 from http://www.weather.gov/sp/NWS_strategic_plan_01-03-05.pdf.

EXTENDING THE DANUBE FLOOD FORECASTING SYSTEM WITH THE USE OF METEOROLOGICAL ENSEMBLES

A. Csík[*] and G. Bálint

VITUKI Environmental Protection and Water Management Research Institute, Budapest, Hungary

Abstract. Flood forecasting schemes may have the most diverse structure depending on catchment size, response or concentration time and the availability of real time input data. The centre of weight of the hydrological forecasting system is often shifted from hydrological tools to the meteorological observation and forecasting systems. In lowland river sections, simple flood routing techniques prevail where accuracy of discharge estimation might depend mostly on the accuracy of upstream discharge estimation. In large river basin systems, both elements are present. Attempts are made enabling the use of an ensemble of short and medium term meteorological forecast results for real-time flood forecasting by coupling meteorological and hydrological modelling tools.

Keywords: Meteorological and hydrological ensembles, flood forecasting, likelihood-type data

1. Models

The hydrological modeling system NHFS GAPI/TAPI has been developed within the Hydrological Institute of the VITUKI Centre (Bartha et al., 1983; Szöllõsi-Nagy, 1982). The conceptual, partly physically-based GAPI model serves for simulations and forecasting of flow for medium and large drainage basins (Gauzer, 1990). The lumped system consists of sub-basins and flood-routing sections and a backwater module (Todini and Bossi, 1986).

The source of the meteorological forecasts was the European Centre for Medium Range Forecasting (ECMWF) archived forecast arrays. Operational use of the above system often revealed the uncertainty of QPF taken into consideration while calculating the expected Danube hydrographs. To test the feasibility of the use of meteorological ensembles, a forecasting experiment was designed. The aim of the investigation was also to assess how much prior estimates of uncertainty of can be given by the selected approach.

---

[*] To whom correspondence should be addressed. e-mail: csika@vituki.hu

2. Results

Figure 1 shows the main features of different sets of hydrological ensembles for the gauging station at Budapest. The specific Box-Whisker diagrams indicate observed hydrograph forecast arrays, showing minimum and maximum values of 50 element ensembles, while quartiles above and below the mean values are indicated by wider boxes.

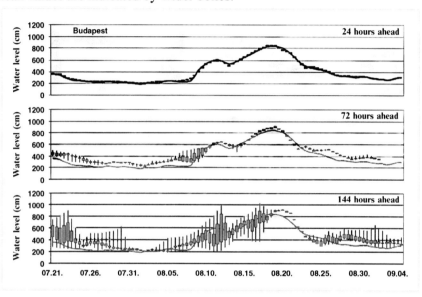

Figure 1. Hydrological ensembles indicating great dispersion on 7th and no dispersion on 15th August 2002, at the Budapest gauge

3. Conclusions

The forecasting experiment proved that the use of meteorological ensembles to produce sets of hydrological predictions increased the capability to issue flood warnings. The NHFS system can be used for such a purpose. However, for real-time use the linkage between meteorological and hydrological modules should be reviewed. The more than 5,000 model runs for the August 2002 extreme flood event were able to be performed within reasonable time.

Further important findings are:

– Appropriate decision support rules are needed to utilise the array of flood forecasts for flood management and warning purposes.
– The proper estimation of the contribution made to forecast error by the various modules in the system may help a better understanding of expected uncertainty of the forecast.
– Any future forecasting exercise should include longer periods of low flow or medium flow to have proper estimates of 'false warning' types of errors.

References

Bartha, P., Szőllösi-Nagy, A., Harkányi, K., 1983: Hidrológiai adatgyűjtő és előrejelző rendszer. A Duna. *Vízügyi Közlemények*, **65(LXV)**: 373–388.

Gauzer, B., 1990: A hóolvadás folyamatának modellezése (modeling of the processes of snowmelt). *Vízügyi Közlemények*, **72(LXXII)**: 273–289.

Szöllősi-Nagy, A., 1982: The discretization of the continuous linear cascade by means of state space analysis. *Journal of Hydrology*, **58**: 1–2.

Todini, E., Bossi, A., 1986: PAB (Parabolic and Backwater) an unconditionally stable flood routing scheme particularly suited for real time forecasting and control. *Journal of Hydraulic Research*, **24(5)**: 405–424.

WORKING GROUP III: THREATS DUE TO CLIMATE CHANGE – GLOBAL PROBLEMS IN A REGIONAL CONTEXT

Chair and rapporteur: **P. Robinson**
University of North Carolina Chapel Hill, USA

The expertise of the participants in the Working Group primarily encompassed the area stretching from Eastern Europe through the Caucasus to the states of Central Asia. This region is the focus of this report. The group was aware of the work and recent reports of the Intergovernmental Panel on Climate Change and used the information there as the global background to the regional analysis.

The region is one with a continental climate. There is a large annual temperature range, with cool to cold winters and warm to hot summers. Precipitation is concentrated in the summer, but commonly some precipitation occurs in all seasons. There is a general decrease in total annual precipitation from west to east. In addition to the strong seasonality, there are large inter-annual variations in temperature and precipitation. A major characteristic of the region is that all areas have snow accumulation in winter and a spring thaw. Some regions have permafrost, while higher elevations often have permanent glaciers. Topographically, the west has many flat open plains, while the east has many major mountain ranges interspersed with open basins.

Commonly, atmospheric water availability refers to the difference between precipitation and evaporation. In this region with annual snow accumulations and melting cycles, and with permanent glacial ice, the concern with atmospheric water availability must also incorporate this frozen water, taking into account the variable timings and amounts of storage and release. Thus, atmospheric water availability will change because of changes in the nature, intensity and frequency of atmospheric precipitation producing systems and as a result of changed evaporation and snow and ice regimes associated with changing temperatures. The impacts will also depend directly on changes in land cover and, more indirectly, on societal changes.

The group identified numerous issues of particular concern to the region, which were divided into three overlapping categories for analysis.

1. Trends in atmospheric water availability

Long-term trends in water availability are occurring, and will continue. However, the trends are likely to be in different directions in different areas within the region, giving no meaningful regional average. Similarly, the trends may not persist with the same magnitude, or even direction, during the next few decades.

The most significant actual or potential impacts identified were:

- Agriculture may directly benefit in some areas, but not in others
  o Impacts may change with time and initial benefits may turn to problems
- Ecosystem shifts are likely
  o Forest resources may be redistributed across the region
- Seasonal frost regime changes will affect soil moisture supply
  o Water-logging of soil may become more common
- Timing and rate of melting of glaciers will change and influence flood potential
  o Possibly more water may be available initially, but this may be a short-term effect. This may be marked in Central Asia
  o Mudslides may increase
- Potential transboundary and political problems
  o Potential for conflict (Water Wars) is likely to increase.

2. Periods of high or excess availability

Flood events will change in magnitude, timing and frequency because of annual snow melt changes, and long-term glacier retreat, as well as changes in atmospheric water availability. There have already been some suggestions of increased flood intensity in some areas:

- Increased flood magnitude leads to greater areas prone to flooding
  o Flow changes associated with land use changes may amplify this problem
- Increases in magnitude may be acute in urban areas
  o Urban infrastructure could become increasingly vulnerable
- A changed flood regime requires that agriculture adapt to the changed regime
  o Cropping patterns may need to be examined
- Current beneficial effects of some floods will likely be reduced
  o Floods nourish wetlands, clear channels and provide inundation irrigation
- Current dam operation, hydropower generation and irrigation strategies may become obsolete
  o New techniques must be investigated

3. Periods of low or deficit availability

Periods of low water availability usually last for a number of months, leading to low flows in rivers. More extended, or particularly intense, dry periods may lead to widespread drying of the whole landscape:

- Low flows directly create several problems for:
  - Irrigation and hydroelectric generation, and domestic water supply
  - Navigation and transportation using rivers
- Low flows are commonly associated with poor water quality
  - Public health issues and problems arise
- Low flows can induce coastal change
  - Beach resources become an issue
- The drying of large land areas decreases soil moisture
  - Soil becomes friable and susceptible to wind erosion
  - Irrigation water is often scarce and agricultural problems appear
- Peat drying creates a special set of problems
  - Increased risk of peat fires, air pollution and loss of resources
  - Increased movement of pollutants through soil to rivers, decreasing water quality
- Drying leads to shrinkage of lakes and wetlands
  - Changes of limnology and aquatic biology, with associated ecological changes on land
- Drying creates atmospheric aerosol
  - Windblown dust, soil and peat-fire particles become air pollution aerosol
  - Aerosols may include heavy metals and radio-nuclides among pollution types
  - Salt from lake beds (the Aral Sea problem) may become:
    - Deposits on agricultural crops, decreasing yield or killing plants
    - Incorporated into glacial ice, lowering melting temperature and enhancing thawing
    - Cloud condensation nuclei, modifying cloud formation and precipitation regimes

4. Action items

Not all of the effects noted above will occur simultaneously or in the same place. Nor are the impacts of equal magnitude, area, or importance, but they will vary in significance from place to place. But they are all possible in the region in association with climate change:

- Infrastructure
  - Adopt policies to adapt infrastructure, including urban design, to changing flood regimes
  - Ensure that land-use changes, however caused, are incorporated into planning activities
  - Plan domestic water supply systems with particular attention to wastewater treatment

- Operations & Management
  - o Develop strategies for assessing optimal agriculture
  - o Create better flood/drought forecast models
  - o Examine strategies for flood forecasting and management
  - o Review dam operation, hydropower generation and irrigation strategies
  - o Organize sub-regional consortia to manage effects beyond national boundaries
- Education
  - o Enhance flood-related education for managers, decision makers and the public
  - o Agriculturalists must be educated to ensure adequate response to changed conditions

This report has been prepared in the light of the papers presented at the NATO Workshop and through working group interactions between many experts. The experts note, however, that there is not enough knowledge to make definitive statements, particularly in assigning weights to the various climatic and human drivers of change. Further, there is a major need to ensure that the best data, quality-controlled, long-term and readily accessible, are used in analyses. Data archives should be freely open, and current and continuous monitoring should be encouraged.

PART IV:

PROTECTING AQUATIC ECOSYSTEMS

ASSESSMENT OF RISKS AND POSSIBLE ECOLOGICAL AND ECONOMIC DAMAGE FROM LARGE-SCALE NATURAL AND MAN-INDUCED CATASTROPHES IN ECOLOGICALLY VULNERABLE REGIONS OF CENTRAL ASIA AND THE CAUCASUS

A.N. Valyaev[1*], S.V. Kazakov[1], A.A. Shamaeva[1], O.V. Stepanets[2], H.D. Passell[3], V.P. Solodukhin[4], V.A. Petrov[5], G.M. Aleksanyan[6], D.I. Aitmatova[7], R.F. Mamedov[8] and M.S. Chkhartishvili[9]

[1] *Nuclear Safety Institute of the Russian Academy of Sciences (IBRAE RAS), 52b Tulskaya St., 115191 Moscow, Russia*
[2] *Vernadsky Institute of Geochemistry and Analytical Chemistry of RAS, 19 Kosygin St., 11999 Moscow, Russia*
[3] *Geosciences and Environment Center, Sandia National Laboratories, Cooperative Monitoring Center, USA*
[4] *Nuclear Physics Institute of National Academy of Sciences, 490000, Almaty, Kazakhstan Republic*
[5] *Technical University, 492000 Ust-Kamenogorsk, Kazakhstan Republic*
[6] *Yerevan State University, Alex Manoogian St., 0025 Yerevan, Armenia*
[7] *Institute of Physics and Mechanic of Rock Stones of National Academy Sciences, Bishkek, Kyrgyzstan Republic*
[8] *Geology Institute of Azerbaijan National Academy of Sciences, 29a H. Javid Av., AZ1143 Baku, Azerbaijan*
[9] *Scientific Center for Radiobiology and Radiation Ecology, Georgian Academy of Sciences, 51 Telavi St., 0103 Tbilisi, Georgia*

Abstract. Various threats to civilization have been recently observed to increase in number. They include natural and man-induced catastrophes, international terrorism, ecological imbalance, global climate change and many other hazards. According to UN and scientific forecasts, an increasing tendency for the catastrophe scale will obtain during the 21st century. Humankind has been facing most of the above issues for the first time and, therefore, there are no good suitable methods provided for solving them. A concerted effort is required by all States of the world community. This paper presents our international program which includes the participation of six countries: Russia, Kazakhstan, Kyrgyzstan, Georgia, Armenia and Azerbaijan. The program includes separate projects, developed by each participating country and briefly described here. Today, one of the possible methods is the assessment of risks and the possible ecological and economic damage resulting from catastrophes. We pay attention the most ecologically precarious regions and single out the typical factors that significantly increase

_____

[*] To whom correspondence should be addressed. e-mail: anvalyaev@mail.ru

the risk of natural and man-induced catastrophes. The prediction, prevention and assessment of the above mentioned threats are especially real and important for newly independent states, where the system of safety, frontier and customs control, strict visa control and other state safety measures have not yet been developed. Also, it is easier to implement attacks of terrorists in these regions. The consequences of such catastrophes will be followed by major human and huge material losses, and extremely negative irreversible environmental effects possibly of a global scale. Our Nuclear Safety Institute of the Russian Academy of Sciences (IBRAE RAS) has its own experience in these scientific directions (http://www.ibrae.ac.ru).

Keywords: Terrorism, ecologically vulnerable, safety measures

1. Introduction

Today, a steady increase in catastrophic events is apparent, which proves that the nature of their occurrence is far from being random. Not only the population has suffered, but also economic damage supported by negative or often irreversible environmental impacts are increasing abruptly and persistently. Risks of catastrophe have increased so much that it is becoming evident that none of the States is able to manage them independently. The situation is aggravated by the fact that a number of catastrophes may be caused by deliberate attacks of terrorists. Therefore, it is necessary to develop intensively and implement new special methodologies, techniques and approaches aimed at producing methods for forecasting large natural and man-induced catastrophes, as well as to prevent and eliminate their consequences. Also, it is important to realise that it is quite impossible to develop models of all the types of the possible major catastrophes and thus to predict their consequences, as may be done, for example, in physical experiments. In our opinion, one of the most effective ways to solve the issue can be the assessment of risks and ecological-economic damage from catastrophes. Opportunities for any country in this field are always limited, but integration with other States will allow not only assessment but also management of transboundary risks, which is an essential prerequisite for sustainable development of any country and its transfer to a higher management level.

The last greatest man-made disasters in Europe and Siberia, which have been due to the sources of radionuclides and toxins in northern marine environments, are presented in Fig. 1. For several years IBRAE RAS has been running systematic research on ecological risks, associated with environmental contamination and its health effects (http://ibrae.ac.ru). The methodology created is recommended by international organizations, such as WHO, UNEP, Russian Health Ministry and used presently in Europe and USA. Models are developed for the migration of contaminants into the environment and their entry into the human body, as well as methods for estimating risk factors, including comparative analysis for radiation and chemical risks. Methods are being developed to counter some kinds of radiation terrorism, which can be implemented by means of explosives (Valyaev and Yanushkevich, 2004). We pay great attention to the regions, where it is possible to stimulate natural calamities in this way or cause man-induced catastrophes with huge negative effects of an international global scale and we present the main details of our international program here. It includes the separate thematic projects, developed by nine scientific organizations from the six participating countries in Europe and Central Asia: Russia, Kazakhstan, Kyrgyzstan, Armenia, Azerbaijan and Georgia. These organizations are presented in the affiliation of this article. We combine our efforts and the experiences of our institutes in resolving of above mentioned problems. The brief special characteristics of these regions and some of our results on risk investigations are given below and in the thematic publications (Stepanets et al., 2005; Kazakov et al., 2004, 2005; Valyaev et al., 2004a, b and c; Passell et al., 2005; Tsitskishvili et al., 2005).

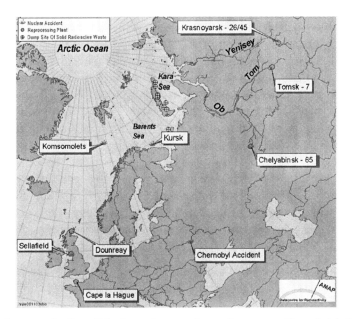

Figure 1. Sources of radionuclides in northern marine environments: Chernobyl in Ukraine, the reprocessing plants at Sellafield, La Hague and Dounreay, the dumping sites of nuclear waste in Kara Sea, Russian nuclear installations (Mayak [Chelyabinsk-65], Tomsk-7 and Krasjonoyarsk-26) releasing radionuclides to Russian Ob and Yenisey rivers and the sunken submarines Komsomolets and Kursk

2. Main results

2.1. CENTRAL ASIA

Practically, the entire 4 million square kilometers territory of Central Asian States, populated by nearly 50 million people, is an area of ecological risk on a global scale. We consider the two most hazardous regions:

1. The basin of the Naryn River, the main one in Kyrgyzstan
2. The industrial areas of Eastern Kazakhstan, located along the largest of the Republic's rivers, the Irtysh

The Kyrgyzstan Republic

The largest uranium, antimony and mercury deposits in the former USSR are located in Kyrgyzstan consisting of 49 tailing dumps, multiple burrows and slime basins containing about 100 million cubic meters of fine dispersed radionuclide waste, heavy metal salts and toxins covering c.1,200 ha. Highly radioactive waste of enriched-U products from Eastern Europe and China were also delivered here over the last decade. Central Asia's main headwater rivers, the Syrdarya and Amurdarya, have been contaminated and continue to be contaminated, along with transboundary contaminant transfer to the lands of Kazakhstan, Uzbekistan and Tajikistan, causing the growth of economic destabilization and political tension. The status of the majority of tailing dumps is extremely poor, as their dams have started to break with time under the impact of landslips, mudflows and ablation. Here we pay particular attention to the problems in the most dangerous geo ecological region of the Republic with U-tailing storages, located along the largest river, the Naryn, flowing into the Syrdarya river (Fig. 2). The Naryn River provides the whole country with energy supply, where the six major hydroelectric stations (HESs) and their huge water reservoirs are located. The largest, the Torkogul HES reservoir, impounds 20 billion cubic meters of water with the height of its dam c.200 m. The huge water mass in these HES reservoirs stimulates more frequent and intense earthquakes, for example, as happened recently in the Sysamyr ☆ (1992, M = 7.3 on the Richter scale) and Kochkor–Ata ☆ (1992, M = 6.2) earthquakes (Fig. 2). The detailed analysis showed the high concentrations of radionuclides, radon, radium and various heavy metals in the water and in the materials of the walls of the buildings and artificial constructions located nearby. In river water, pollutant concentrations exceeded the extreme limits of concentrations by 1–2 orders of magnitude. The worst situation is around 13 U-tailing storages, located along the Mailuu Su River and its tributaries. Earlier, the highly radioactive wastes from U-enrichment production in East Germany, Bulgaria, Czech and China were delivered here for secondary processing.

Today, the following element concentrations were recorded: 0.1–0.15% U; 0.11–0.18% Cu; 0.5–1.0% Mn; 0.026–0.036 G/kg Se; 0.02–0.034 G/kg Co. The total alpha activity of the waste ranges between 67,000 and 113,000 Bq/kg. The mean gamma intensity on tailing surfaces is 20–50 μR/h, but at some places its value reaches 250–2,000 μR/h. The mean radon concentration is reaches 400 Bq/m^3 in the walls of old buildings.

The huge water masses press on the surface of the Earth's crust and stimulate frequent and intense earthquakes. In addition, the dams of the reservoirs and tailing dumps, and Russian and American military air bases located in Kyrgyzstan are very attractive for controlled terrorist acts using explosives (Valyaev and Yanushkevich, 2004).

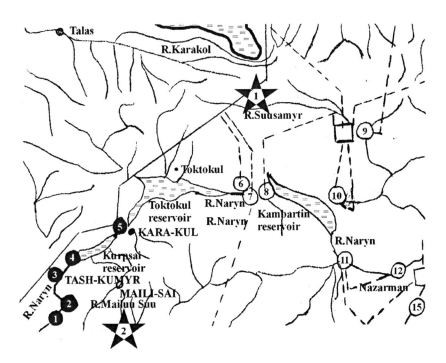

Figure 2. The fragment of objects in Tien–Shan mountains, located near HES cascade at the largest river in Kyrgystan Republic, the Naryn. • - Operative HES; ○ - projected HES; ▬ - operative high voltage lines; --- projected ones

Eastern Kazakhstan

This is also an area of major ecological risk, where multiple mines for extracting metals and minerals, as well as a number of industrial enterprises and their tailing dumps, including uranium, are located in the cities and their suburbs along the Irtysh, the largest river of the country. The river originates in China and flows via the lands of Kazakhstan and Russia, including such large cities as the East Kazakhstan capital, Ust-Kamenogorsk, Semipalatinsk,

Pavlodar, Omsk, and Tobolsk, and after its confluence with the Ob River it flows into the Arctic Ocean (Fig. 3). The largest plant is the Bukhtarma HES with a 100 m high dam has a sizeable reservoir, the basin length of which is above 300 km and the depth up to 100 m. The other is the Ust-Kamenogorsk HES with a 42 m high dam and a single outlet valve is located 15 km from the city.

The next is the Shulba HES (Valyaev and Yanushkevich, 2004) is located 180 km from Ust-Kamenogorsk. Errors in the design of these HESs have caused a number of different accidents. Unfortunately, the data on obser-vations of the HESs accident statistics are practically absent today. Ust-Kamenogorsk city with 330,000 inhabitants is one of the most contaminated towns in the world and located in the Irtysh mountain valley. It represents a unique urban system, oversaturated by different enterprises. The largest is the FSU Ulba Metallurgical Plant (UMP), which incorporates three separate works, producing enriched U for nuclear power plants, Be, Ta and their products. The UMP operational wastes storage, located in the city center, has accumulated c.100,000 t of wastes, contained U, Th and their decay products. It covers thousands of square meters with the depth of con-tamination of over 5 m. The level of gamma radiation at its surface reaches 360 µR/h and increases with depth up to 1,000 µR/h. Radioactive anomalies of 1,000 up to 6,000 µR/h are recorded at the region of UMP. Many other large plants operating in the city, such as lead–zinc, titanium–magnesium, ceramics, beryllium (Be) works, electrical capacitor plants, a nonmetalli-ferous group of enterprises and silk cloth enterprises, use various poisons and toxins in their technologies, while their wastes are also located within the city boundaries. For instance, the lead–zinc plant stores over 13 million tons of wastes and c.1,000 t of arsenic in the form of the highly toxic substances calcium arsenate and arsenite, which contain 7–10% As, in an opencast dump covering 17.5 ha in area. In Irtysh river basin, where over 40% of the hydroelectric power in Kazakhstan is generated, are also located large active non-ferrous pits, precious and rare-earth metals pits with their dumps. In the water of the Ulba River flowing into the Irtysh in the city toxin concentrations are: Cu (4.86–5.50) at the maximum permitted concentration (MPC), Zn 4.71–5.37 MPC, oil products 2.03–2.07 MPC, nitrite nitrogen 1.40–1.95 MPC.

Risks of ecologic catastrophes are increased, because the two above mentioned HES are located on the Irtysh upriver of the city (Fig. 3). The huge water masses in the man-made reservoirs press heavily on the base of the mountain surfaces, disturbing and deforming their initial natural states. We consider these factors have resulted to the increase in the frequency and intensity of the strong earthquakes, included catastrophic ones that already occurred not only near the city in 1990, but also in the Altai Mountains in Russia in 2003 and 2005. Such earthquakes may cause damage to HES dams, where in addition parts of the dams are in unsatisfactory conditions, especially the Ust-Kamenogorsk HES dam, which has operated for over 50 years. Also, any HES with its huge water reservoir is very attractive for possible controlled terrorist acts, including those using explosives. According to some initial estimates, the result of a dam burst would be a huge wave with its front height c.30 m which would destroy the city and its environs. All enterprises, their products and hazardous wastes in the storages would then be carried down the Irtysh River towards many other cities and after reaching the Irtysh-Ob river junction it would spread over a large region, including the Arctic Ocean through the Kara Sea (Figs. 1 and 3).

The Irtysh basin has accumulated 120 million cubin meters of various wastes, which amounts to 60% of the total pollution in the whole of Kazakhstan. This has resulted in a sudden worsening of water quality in all the cities: Ust-Kamenogorsk, Semipalatinsk, Pavlodar, Omsk, Tobolsk and in other many inhabited localities. The Irtysh-Karaganda man-made channel supplies water to Kazakhstan's central regions, such as Karaganda, the Kazakhstan capital Astana and their suburbs. The Irtysh pollution represents the most serious danger as a potential source of contamination of the world ocean through the Arctic Ocean.

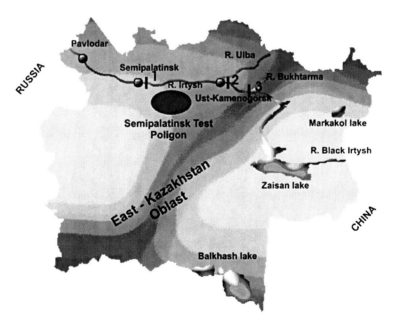

Figure 3. The scheme of the Irtysh River in Kazakhstan Republic. 1 – Shulba HES, 2 – Ust-Kamenogorsk HES, 3 – Bukhtarma HES

The following major pollutants were recently detected in soils and water: (1) toxin components due to sulfide non-ferrous ore processing: SO_4, NO_3, NH_4, Cu, Pb, Zn, Cd, Tl, Se, Hg, Sb, As, and also pH indicator; (2) complex components due to processing of the rare metal ores, Be and

Li; (3) complexes of detrimental substances: SO_4, Cl, NH_4, NO_3, F, Li, Be, Th, U with high general alpha and beta radioactivity and also the pH indicator.

Downstream of the Semipalatinsk Test Polygon (STP) [industrial complex] are deposited more than 18,000 t of radioactive waste with a total activity of 1,300,000 Ci, which significantly aggravates the situation. Radionuclides with a total activity over 10,000,000 Ci were accumulated during nuclear tests in the underground wells, located within 60 km of the Irtysh. As a result of multiple nuclear explosions (over 500) during the period 1949–1990, U-235 and Eu-152 are recorded as available over an area of more than 300 km². The Chagan River with very high radionulide concentration flows out of the man-made Atomic Lake (created as a result of the underground nuclear explosions) and flows into the Irtysh.

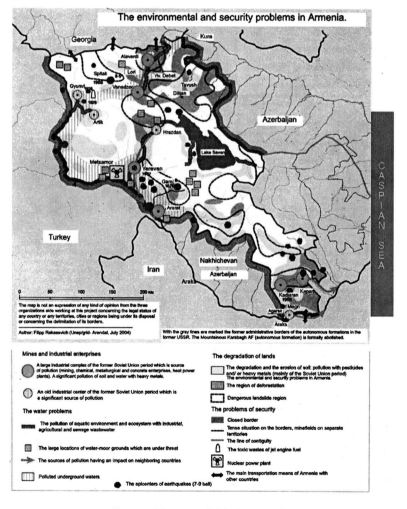

Figure 4. The map of risks in Armenia

3. The Caucasian region

The peoples of Armenia, Azerbaijan, Georgia, the North Caucasian part of the Russian Federation, Northeast Turkey and Northwestern Iran live on nearly 600,000 km^2. of the very seismically activity Caucasian Region situated between the Black and the Caspian Seas. The largest Caucasian rivers enter the Caspian and Black Seas.

Armenia

Armenia is located in the basins of the Araks and Kura rivers, which drain into the Caspian Sea, and the country has a mean height of 1,800 m above sea level. Severe earthquakes, frequent landslips, hail damage, droughts, strong winds and floods threaten human safety and cause considerable ecological and economic damage. The 880 MW Armenian Nuclear Power Plant (ANPP), where current protective measures do not meet today's requirements, is a potential man-induced source of hazard. In addition, the ANPP was designed in anticipation of the maximum seismic activity in its vicinity of grade 7 on the Richter scale. In fact, the activity can be 8–9 grades; in this case, it would affect a region populated by 20 million people. Other hazardous sites are the cascades of large Hydroelectric Stations (HES) and reservoirs, such as the huge Akhuryan reservoir with a 59 m height of dam and a water volume of 525 million cubic meters; multiple chemical and chemical-mining plants, their tailing dumps of toxins and radionuclides; and the gas-main pipelines and power lines. Overall, the above peculiarities are typical also of Georgia and Azerbaijan (Figs. 4–6).

Azerbaijan

Here is located a complete deathtrap that does not meet current operational standards. Multiple sites containing hazardous chemicals and radioactive substances with an average intensity of 500–600 μR/h have been created in the oilfields. Radium waters extracted jointly with oil create extra natural contamination. The country has 785 km of coastline along the Caspian Sea. About 100,000 hazardous radiation, chemical, biological, oil, and explosive production plants, and their technologies which deliver significant deterioration of the basic production waste deposits, create a major threat. During the period 1992–1998, more than 100 emergencies of a man-induced nature were registered. They were supported by explosions and fires causing irreversible ecological damage to the entire region, including the Caspian Sea (Fig. 5).

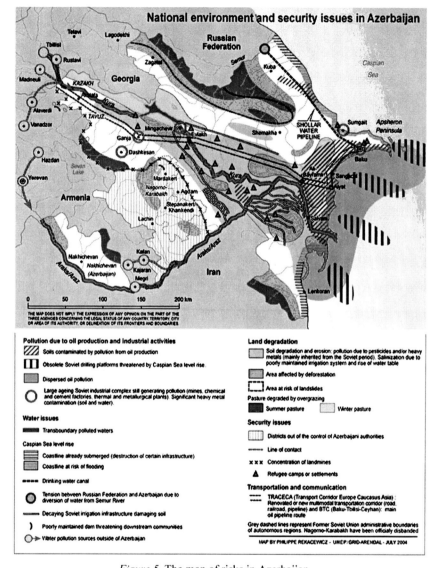

Figure 5. The map of risks in Azerbaijan

Apart from this, the majority of production processes leave sizeable reserves of hazardous chemical substances, such as chlorine, ammonia, hydrochloric acid, and others, and there are up to several thousand of tons of toxins at individual sites. Azerbaijan incurs huge extra losses due to flooding in the densely populated coastal area as a result of the water level rise of up to 3 m in the Caspian Sea. For the period 1978–1995, the total damage from such flooding amounted to US$12 billion.

At present the Kura-River, the largest in the Caucasus, is strongly regulated by the cascade of large hydroelectric stations (HES), such as the Mingechauri HES (with its 83 m high dam and 16 billion cubic meters of water in the reservoir), the Shamkir HES (with a 70 m dam and 2.7 billion cubic meters of water), and the Enikend HES (a 36 m dam and 158 million cubic meters of water). All the dams of the reservoirs are in an emergency state of repair. More than 150 kinds of hazardous chemical substances are recorded in the Kura-River's waters. The average annual concentration of water contamination exceeds the Maximum Permissible Concentration (MPC) several times over, and does so hundreds of times during emergencies; oil products and phenol concentrations exceed MPCs tens or hundreds of times.

About 3,800 km of oil and over 10,000 km of field pipelines are in operation at present, 60% of which are badly deteriorated. The areas of contaminated sites in Apsheron make up more than 30,000 ha with the radiation level up to 1,200 μR/h. In the case of a dam burst in any of the HES, numerous manufacturing products and their wastes will be spread by the flood wave over vast areas of land and the Caspian Sea. At the same time, the contaminants accumulated over decades in great quantities in the sediments and coastal soils will be entrained from the bed of the Kura-River and its tributaries and spread over large distances.

Georgia

Of Georgia's three major HESs, namely, the Zhinval, Shaor and Inguri HESs, the latter one represents the most hazard. It has one of the world's largest arch dams 271 m in height and impounds 12.1 billion cubic meters of water. A break in its dam would create a flood wave of the c.25 m height, which would be capable of washing off vast areas of land in Georgia and even reaching Turkey. Possible accidents at the two HESs being under construction on the Khudoni and Pari cascades (containing 700 million cubic meters and 200 million cubic meters respectively) upstream on the Inguri-River, will aggravate the severity of such a catastrophe, as they will double the front height of the wave. A dam break at the Shaor HES would flood the large cities of Kutaisi and Poti, and one at the Zhinval HES would cover Georgia's capital Tbilisi, and adjacent regions of Georgia and Azerbaijan. An accident at the Madneuli mining facility that uses cyanide technologies for extraction would cause severe water-poisoning of the Kura River and the Caspian Sea. Accidents at the Urev and Kvaiss mining facilities could contaminate, apart from major areas of land, the Black Sea basin with extraction products of arsenic and antimony. A burst in the tailing dump designed for toxic chemical compounds/radionuclides at Lilo in Eastern Georgia, as well as one in high toxic rocket fuel repositories in

Western Georgia would be followed by washing off contaminants into the basin of the Rivers Kura and Rioni, and further into the Caspian Sea and the Black Sea, respectively. Large ecological catastrophes are possible at the oil and gas terminals in the cities of Supsa and Batumi, as well as at sites of construction and operation of major oil pipelines (Baku-Jeikhan, Baku-Supsa, Novorossiysk-Abkhazia-Turkey), and gas pipelines (Baku-Erzrum, Vladikavkaz-Tbilisi-Yerevan), partly lying on the bottom of the Black Sea (Fig. 6).

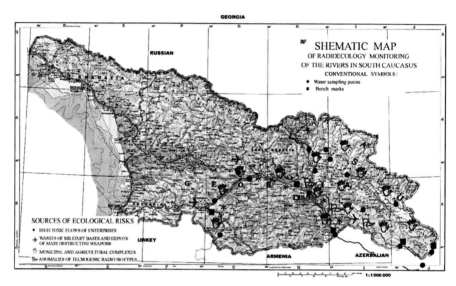

Figure 6. Schematic map of radioecology monitoring of the rivers in South Caucasus

Therefore, for all the regions mentioned, we may single out the following typical factors, that significantly increase a risk of occurrence of natural and man-induced catastrophes:

1. All the regions are located in the mountain lands that have a high seismic level from 5 to 9 grades on the Richter scale.
2. The largest mountain rivers have cascades of powerful HES with their sizeable water reservoirs and huge, high dams (over 100 m).
3. In the region's densely populated lands there are plenty of mines for extraction of metals and minerals, as well as industrial facilities and plants, including nuclear power ones, with U-tailing dumps and piles of many varied pollutants. The facilities use different radioactive, toxic and poisonous substances in their technologies.

4. The man-induced activity in the regions under review increases the probabilities for occurrence of not only severe man-induced catastrophes,
   but also natural ones.
5. An especially grave situation has been created in the transboundary lands
   of the States due to the lack of common ecological and geochemical
   monitoring systems, that increases political and economic tension between
   the countries and is generating of negative migration processes.
6. Risks and ecological-economic damage from catastrophes are not only
   regional, but also global by nature, since they entail contamination of
   vast lands, the basins of the Black, Caspian and Kara Seas, the Atlantic
   and Arctic Oceans and consequently the entire World Ocean.
7. Opportunity to perform deliberate attacks by terrorists with the using of
   explosives able to cause man-induced catastrophes and stimulate natural
   calamities (earthquakes, mudflows, landslips, etc.). It is easier for terrorists
   to implement controlled attacks there due to the intersection of main
   lines, an available border with current centers of international terrorism
   located in Chechnya, Afghanistan and some other nearby States. Especially
   great is the hazard for the newly independent States, where the system
   of safety, border and customs control, of strict visa control and other
   State safety measures have not yet been implemented. The consequences
   of terrorist attacks in the regions will be followed by major human and
   huge material losses, and extremely negative irreversible environmental
   effects of global scale.
8. Some special space observation technologies will be used in all regions
   for the current continuous observation and control for the most dangerous
   large items (Valyaev et al., online a, b and c).

4. The calculation of the total limited losses

Our method is the following. We consider the common case of any entity
for the fixed time interval under the following assumptions:

1. In the initial state, the object is in a normal non-accident situation.
2. Different kinds of accidents may occur as observed with $i = 2, 3,...$ m,
   where m is the total number of possible accidents (m = 1 corresponds to
   the normal regime).
3. Each accident may create different kinds of losses. Assume that j is the
   kind of loss with value a_j. Then $j = 1, 2, ...n$, where n is the total
   number of possible kinds of losses.
4. The realization of i accidents creates the loss of j kinds with P_{ij}
   probability. Thus, the matrix of loss probabilities is determined.

Then the total vector of limited losses $\vec{a}_{lim}$ may be determined according
to the following formula:

$$\vec{a}_{\lim} = P(1)\vec{a}_{1n} + \sum_{i=2}^{m} \hat{P}_{ij}\vec{a}_j \qquad (1)$$

where P(1) is the probability of loss formation under normal exploitation; $\vec{a}_{1n}$ is the vector of limited loss under regular exploitation. $P_{ij}a_j$ coordinate vector value in sum is equal the loss value of j kinds under realization of an i kind of accident.

In the absence of accidents, the second term in the right part of Equation (1) is equal to zero and then $\vec{a}_{\lim}$ the total vector of limited loss is determined by the first part of Equation (1):

$$\vec{a}_{\lim}n = P(1)\vec{a}_{1n} \qquad (2)$$

The main problem in this calculation is in the determination of the loss probability matrix. As one possible method, we propose to use the method of expertise estimates.

The plan of calculations of the total losses includes the following main elements. At first, for every object or installation we have to point out and develop the classification of the main types of possible accidents. For example, in the case of an HES disaster we have to take into account the following possible kinds of accidents:

1. Total damage or break of one or some HES dams
2. Partial damage of HES dam
3. Destruction of water lock
4. Stopping of HES turbines

Let us consider the most dangerous first accident, investigate the extreme cases developing the worst catastrophes and analyze the possible scenario of their realization connected with damage of two HES, located upriver on the Irtysh near Ust-Kamenogorsk sity (Fig. 3):

(a) Bukhtarma hydro electric station (HES)
(b) Ust-Kamenogorsk HES
(c) Both HES simultaneously

Here we have to take into account that the total damage of Bukhtarma HES dam with the height c.100 m will probably stimulate the total destruction of the Ust-Kamenogorsk HES with the height c.40 m. Then it is necessary to evaluate the parameters of catastrophic submergence in every scenario:

(a) Maximum possible height and speed of break-through wave
(b) Estimated time of wave crest coming and the front of the wave crest arriving in the town area
(c) The boundaries of the possible submergence zone in the vicinity

(d) The maximum depth of submergence for every fixed locality and the length of time of its submergence

(e) To identify all the main enterprises that will be swamped

For these estimates and calculations we shall use computer modeling taking into account the real profiles of the local land surface, mountains and valleys (including its rock and soil materials), the HES with its own and other water bodies, such as lakes and rivers, and any other natural features.

In the case of a possible HES disaster near U-tailing storages (Fig. 2) our analysis will include the following:

In order to analyze the possible scenes of realization of situations with pollutant migration from tailing storage:

(a) Constant pollutant migration without damage of tailing storage dams

(b) Similar migration with the partial damage of tailing storage dams, for example, under landslide or earth flow

(c) Pollutant migration under complete destruction of dams, for example, as the result of earthquake

(d) Pollutant migration in result of: (1) partial outflow; (2) total leakage

In the case of the last two scenes, the following two cases of development of the catastrophic situation are possible:

(1) All tailings are washed out by the river over a few days.

(2) All tailings are washed out by the river instantly.

The last situation is the most extreme and dangerous, because it will cause the maximum pollution with maximum losses both for the environment and the population.

For all cases, it is necessary to take into account the following kinds of possible losses:

1. Caused by death or injury to people or harm to the population's health

2. Caused by pollution of a wide area of land with subsequent losses in the forest, agricultural and fish industries

3. Caused by severe pollution of buildings and constructions

4. Resulting from pollutant migration throughout basins of the largest rivers

In risk evaluations, it is necessary to take into account the possible chemical nuclear reactions and transformations of pollutants in the soil, water and air. For example, transport calculation will be done for the decay chain $238U > 234U > 230Th > 226Ra$.

The project's objectives are as follows:

1. Selection for each country a site for which the risk of occurrence of one or several catastrophes is maximum and where the potential damage is the greatest

2. Development of scenarios for implementing possible catastrophes for the selected site
3. Estimation of risks and possible ecological and economic damage under various scenarios of catastrophe development
4. Suggestion of some recommendations on risk reduction and actions to eliminate the effects of accidents and catastrophes

5. Conclusion

For all the regions discussed, the following typical factors significantly increase the risk of natural or man-induced catastrophes:

1. These regions are located in mountainous lands with a high level of seismic activity (5–9 grades on the Richter scale).
2. The largest mountain rivers have cascades of large hydroelectric stations with their huge dams (>100 m).
3. In the regions' densely populated lands there are plenty of mines for extraction of metals and minerals, industrial facilities and plants with U-tailing dumps and piles of various pollutants resulting from different radioactive, toxic and poisonous substances used in their technologies.
4. The man-induced activity here increases the probabilities of occurrence of natural as well as severe man-induced catastrophes.
5. An especially grave situation has been created in transboundary lands between the States, due to the lack of common ecological and geo-chemical monitoring systems, which increases political and economic tensions between the countries and generates an outward migration processes.
6. Risks and ecological-economic damage from catastrophes are not only regional but also global, since they entail contamination of vast lands, the basins of the Black, Caspian and Kara Seas, that of the Arctic Ocean and, consequently, the entire World Ocean.
7. The opportunity to perform deliberate terrorist attacks using explosives, which are able to cause man-induced catastrophes and stimulate natural calamities. It is easier to implement terrorist attacks there due to the inter-section of main lines of communication, an available border with current centers of international terrorism, located in Chechnya, Afghanistan and some nearby countries. Especially great is the hazard for the newly independent States, where the system of safety, frontier and customs control, of strict visa control and other State safety measures have not yet been implemented. The consequences of such attacks will be major human and material losses, and extremely negative, irreversible environ-mental effects potentially on a global scale.

The program's results will be used in the following way:

1. Evaluation of levels of risks, resulting from the possible natural or man-made catastrophes for the most dangerous enterprises, so to develop estimates using a methodology or strategy to regulate and manage risks in emergencies.
2. Mapping risk estimates throughout the various lands.
3. Developing a common system for emergency prevention or elimination. To formulate the preventive counter measures and for decreasing of the level of risks, their preventions and ameliorating or mitigating their results.

The results obtained will have a universal character and may be used for analysis of similar events and situations in other countries.

This program will promote the realization of a concept of substantial development with the growth of economic cooperation and stability, decreasing political stresses not only for the countries that are participants in this program, but also for all countries on the continent.

References

Kazakov, S.V., Valyaev, A.N. and Petrov, V.A., 2004: Estimations of risks, ecological and economical losses under catastrophic damages of hydro stations in East Region of Kazakhstan Republic. *Proceedings of International Russian-Kazakhstan Scientific – Applied Conference*, October 5–6, Ust-Kamenogorsk, Kazakhstan, Part 1, pp. 330–333.

Kazakov, S.V., Utkin, S.S., Linge, I.I. and Valyaev, A.N., 2005: Categorization of water media and water bodies by the level of radioactive contamination". *Abstracts of International Conference "Radioactivity after Nuclear Explosions and Accidents"*, pp. IV-20–IV-21, December 5–6, Moscow, Publishing House: St. Peterburg, GIDROMETIZDAT.

Passell, H.D., Barber, D.S., Kadyrzhanov, K.K., Solodukhin, V.P., Chernykh, E.E., Arutyunyan, R.V., Valyaev, A.N., Kadik, A.A., Stepanets, O.V., Alizade, A.A., Guliev, I.S., Mamedov, R.F., Nadareishvili, K.S., Chkhartishvili, A.G., Tsitskishvili, M.S., Chubaryan, E.V., Gevorrgyan, R.G. and Pysykulyan, K.A., 2005: The international project of radiation and hydrochemical investigation and monitoring of general Caspian Rivers. *Abstracts of 5th International Conference "Nuclear and Radiation Physics"*, September 26–29, Almaty, Kazakhstan, pp. 487–489.

Stepanets, O.V., Borisov, A.M., Ligaev, A.R., Vladimirov, M.K. and Valyaev, A.N., 2005: Estimation of parameters of radiactive environmental contamination in places of a burial objects in shallow bays of Archipelago Novaya Zemlya on Data 2002–2004. *Abstracts of 5th International Conference "Nuclear and Radiation Physics"*, September 26–29, Almaty, Kazakhstan, pp. 463–464.

Tsitskishvili, M.S., Kordzakhia, G., Valyaev, A.N., Kazakov, S.V., Tsitskishvili, N., Aitmatov, I.T. and Petrov, V.A., 2005: Insurance of risk assessment and protection from distant transportation and fall out of pollutants under large anthropogenic damages on nuclear power stations due to mountainous regional peculiarities. *Abstracts of 5th International Conference "Nuclear and Radiation Physics"*, September 26–29, Almaty, Kazakhstan, pp. 460–461.

Valyaev, A.N. and Yanushkevich, 2004: Using acoustic techniques for detection of explosives in gas, liquid and solid mediums. NATO Science Series II "Detection of Bulk Explosives: Advanced Techniques against Terrorism" (Mathematics, Physics and Chemistry), vol. 138, pp. 175–183 and June 16–21, 2003, Kluwer, The Netherlands. Proceedings of NATO Advanced Research Workshop, St. Petersburg, Russia.

Valyaev, A.N., Kazakov, S.V. and Petrov V.A., 2004a: "Estimations of risks, ecological and economical losses under catastrophic damages of hydro stations in East Region of Kazakhstan Republic. *"Proceedings of International Symposium Complex Safety of Russia - Investigations, Management, Experience."* May 26–27, Moscow, Publishing House: Informizdatcenter, pp. 348–353. (in Russian)

Valyaev, A.N., Kazakov, S.V., Aitmatov, I.T. and Aitmatova, D.T., 2004b: Estimates of risks, ecological and economical losses from the emergence of natural and man-made catastrophes in uranium tailing storages in Tien-Shyan mountains. *Proceedings of International Symposium "Complex Safety of Russia – Investigations, Management, Experience."* May 26–27, Moscow, Publishing House: Informizdatcenter, pp. 353–358. (in Russian)

Valyaev, A.N., Kazakov, S.V., Aitmatov, I.T. and Aitmatova, D.T., 2004c: Earth from space – some actual problems and their possible decisions with using of space monitoring. *Proceedings of International Russian-Kazakhstan Scientific – Applied Conference*, October 5–6, Ust-Kamenogorsk, Kazakhstan, Part 3, pp. 196–199.

Valyaev, A.N., Kazakov, S.V., Aitmatov, I.T. and Aitmatova, D.T., online a: Problems of ecologic safety under displacement of rock stones, controlled from space, at ecological dangerous regions of Tien-Shyan Mountains" *Proceedings of the First International Conference "Earth from Space – the Most Effective Solutions"* http://www.transparentworld.ru/conference/presentations/operative.htm tyan_shyan_prsnt.zip

Valyaev, A.N. and Kazakov, S.V., online b: Earth from space – some actual problems and their possible decisions with using of space monitoring. *Proceedings of the First International Conference "Earth from Space – the Most Effective Solutions"*. http://www.transparentworld.ru/conference/presentations/operative.htm

Valyaev, A.N., Kazakov, S.V., Stepanets, O.V., Malyshkov, Yu.P, Solodukhin, V.P., Aitmatov, I.T., Aitmatova, D.I., online c: Using space technologies for monitoring of large river basins and prediction of large-scale natural and man-induced catastrophes in ecology-hazard regions of Central Asia and Caucasus. In: *Proceedings of the Second International Conference "Earth from Space – the Most Effective Solutions" Section "Space Monitoring in Problems of Management of Territories"; http://www.transparentworld.ru/conference/2005/thesis.pdf*

ENVIRONMENTAL MANAGEMENT OF INTENTIONAL OR ACCIDENTAL ENVIRONMENTAL THREATS TO WATER SECURITY IN THE DANUBE DELTA

L.-D. Galatchi[*]
Department of Ecology and Environmental Protection, Ovidius University of Constanta, Mamaia Boulevard 124, 900527 Constanta-3, Romania

Abstract. An ecosystem-based approach is essential for a successful conservation of Danube Delta. The most important threats to the Danube Delta are changes in the quantity and quality of water reaching the delta from its upstream catchments. Therefore conservation approaches will only succeed if they are accompanied by an integrated approach to catchment management. Such an approach will require the establishment of appropriate authorities, integrating the interests of the various stakeholders and taking biome needs into account when considering water allocations. Since the Danube Delta has catchments that are shared between many countries, this approach will also require a high degree of international co-operation. The value of the Danube Delta for biodiversity is broadly recognized in the level of protection afforded to them through international and national legislation. The Danube Delta has been designated as a Ramsar site. Other designations recognize the value for biodiversity. Of course, designation as protected areas alone does not guarantee the protection of those areas' biodiversity. Appropriate administrative structures to provide integrated management, appropriate management plans, and adequate resources for the implementation of management plans are all needed. It is pointed out that the main constraints on appropriate management are not, in the first place, scientific information, but personnel, the level of training, equipment, and last but not least, funds.

Keywords: Danube River, pollution, biological diversity and conservation, resources, threats, sustainable development

1. Danube river basin

Having its legal basis established since 1930, when the first law to protect natural monuments was passed, the study and protection of the natural assets in Romania saw a steady development until the 1960s, thereafter falling into neglect and being completely abandoned in the 1980s. A few

---

[*] To whom correspondence should be addressed. e-mail: galatchi@univ-ovidius.ro

years after 1990, conservation activity experienced a revival. The Danube Delta Biosphere Reserve was established, several international conventions regarding the conservation of the biological diversity were entered into, new environment protection projects and plans were drawn and the declaration of new protection areas was made. From 1993, though, the idea of conserving biodiversity has once again been abandoned and, as a result, no progress has been made in the field. Moreover, the objectives of the National Strategy for the Conservation of Biodiversity, published in 1996, have not been achieved. Therefore, it is highly likely that, due to the characteristics of the current period (confusing and contradicting legislation and overexploitation) and to the total lack of interest on the part of the institutions with responsibilities in the field, the situation of biological diversity in the Danube Delta has worsened. It is difficult to know to what extent, as there is not enough information in this respect.

The Danube hydrographic basin of 817,000 km^2 stretches across 17 countries could be a classical example of what can happen when the natural balance is perturbed. Of them all, Romania is the largest tributary to the river, its drainage surface being 228,580 km^2, almost 30% of the total Danube drainage area. Once in the Romanian Plains, there were huge holms (islets) and sandbanks next to the Danube, with great environmental importance, being natural filters for any waste or polluting substances carried by the tributary rivers. These have huge importance for fishing and wildlife, which extends far beyond the needs of the local population. Unfortunately, between 1964 and 1985, the communist régime completely drained this area to create more farmland. The result was a disaster: within just a few years the soil was dry, salty, filled with chemical pesticides used to protect the crops, and the farming on the 435,000 ha drained became an illusion. The fish population and the wildlife were lost forever; the dams are continuously eroded by the strong current of the Danube or by floods and require a huge budget for maintenance and reinforcement. Across the whole Romanian Plains, the drying up of the underground water reserves is obvious, leading to the death of the Plains' remaining forests and to the continuous decrease in crop productivity. This is a classic example of what can happen when the natural balance is perturbed.

Today, the Danube carries 80 million tons of suspended substances every year to the Black Sea, mainly bicarbonate, calcium, sulphide, sodium and potassium, chlorine, magnesium, but also nitrogen, phosphorus, oil by-products, and metal residues, the average water mineralization being 324 mg/dm^3. Another significant value is the biological oxygen demand (BOD 5), which is on average 1–3 mg/dm^3 (Chirila, 2003; Galatchi and Vladimir,

2007). The result is a marked decrease in the river's fish population, both in quantity (100 times less today than in 1960) and in quality (the size of fish and number of species).

Regarding this drained area, it is probably more profitable and logical to flood the drained areas once again, in order to restore the natural balance of the region and the large Romanian Plains. Regarding the Danube's polluting factors, it is most urgent to stop direct pollution of both the main river and its tributaries – if not, the lower basin of the Danube will be lifeless within 20 years.

The embankment works executed in the 1960s on approximately 800 km along the Romanian border, in order to obtain new land have practically led to the disappearance of the flooding areas. As happens in large systems, the ensuing effects appeared much later and were obviated by:

- Eutrophication of Danube Delta waters, and partly of those in the northwest Black Sea, due to the elimination of the filtering effect caused by intensive agriculture and non-filtered streams from towns along the rivers
- Changes in the specific diversity of fish and the dramatic decrease in fish populations with great economic value, especially carp, due to a lack of shallow water zones in the flooding areas, which is needed for their reproduction (Galatchi and Tudor, 2006)

Another factor, whose detrimental effects were not initially considered, has been the building of dams and water reservoirs for electric power supplies. Their appearance has led to changes in flood patterns and a fall in the quantity of alluvium carried by the Danube, due to water abstraction. It has also caused major distortions in the Romanian coastal ecosystems. Another effect caused by dams is disruption of migration paths for the breeding of the valuable sturgeon.

The Danube Delta is the largest wetland in Europe. It is 75 km long E–W and 75 km wide N–S. Its total surface area is 564,000 ha and until 1940 all belonged to Romania. Today, 442,000 ha are Romanian land and 121,700 ha are Ukrainian. Most of the surface is covered by water (65% permanently and up to 90% at flood time). The only firm lands are the two big islands, but there are many major sandbanks, some of them with unique habitats, like the famous oak forest on Letea Sandbank. The diversified structure of the natural ecosystems, most of which were unaffected by human activity, have made the Danube Delta, together with the Razelm-Sinoe lagoon complex, a Biosphere Reserve. It has also been included on the list of International Natural and Cultural Heritage, and in the Ramsar Convention.

As a direct effect of the disappearance of the flooding areas, the Danube Delta's capacity to retain nutrients has increased sharply since the 1980s. It is currently affected by eutrophication, which has led to the reduction or loss of low water macrophytes, to changes in the range of periphytic and epiphytic algae, and to the spreading of rival species sustained by the excess nutrients, for example, the blue-green bacteria – *Cyanobacteria* (Galatchi, 2006).

Apart from the losses caused by eutrophication, the Danube Delta's biodiversity was and still is affected by the changes in, or destruction of, the habitats, by changes in the hydrological patterns. The latter are due to the creation of man-made channels or meander interruption, by transforming large areas into agricultural or fishing zones, or by the change in water quality, like the transformation of the Razelm lagoon into a fresh water lake.

The reed-covered surface in the Delta is the biggest in Europe and other species are remarkable: oak, Turkistan ash trees, aspen trees, crawling and climbing lianas. But the fame of Delta is due to its wealth of fish species, most importantly the sturgeon family, and water birds. Between 1878 and 1945 Romania invested in considerable channel and dam works, so that navigation between the Black Sea and the Danube stream should be continuous.

After 1950, the Romanian communist régime came up with a plan for total and swift exploitation of the Delta, thus nearly destroying it:

1. The first idea was of exploiting the reed, *Phragmites australis*, for a paper factory built especially for that. The only result was the final and permanent destruction of 60,000 ha of reed-covered surfaces and the knowledge that nobody can make paper using only reed.
2. The second idea was that of changing a great part of the Delta into agricultural land. Many works have been achieved and in 1990 there were dams and drying installations covering 50,000 ha. The result was pitiful.
3. The third ecological attack on the Danube Delta involved changing part of the lakes and marshes into fishing ponds by surrounding them with dams. Exploiting them intensively in order to get huge fish production proved to be a disaster because the fish population did not have time to recover and the biological and chemical pollution of the waters increased to dangerous levels.

All these actions had important results, such as: (1) a considerable decrease in fishing production by 33%; (2) the quality of the fish also decreased, first diminishing the valuable species of fish (it is significant that sturgeon production was reduced between 1900 and 2000 by more than 99%). So, today's image of the economic–ecological situation of the Danube Delta is not satisfactory.

In the current period, the main problems connected to the biological diversity and conservation in Danube Delta are:

(a) *Specific diversity*
- Although the number of botanical species is known, for the fauna there are blank areas with respect to the number of species living in the Danube Delta, due to the important number of insect species. In both categories, only limited data are available on the current status of the natural populations of species.
- The lack of information on the present state of most of the wildlife does not allow for the evaluation of the extent to which these species are threatened or nearly extinct and this is what makes them so difficult to protect.

(b) *Ecosystem diversity*
- There are not enough data available on the existing types of natural ecosystems, their distribution or their abundance in the whole territory of Danube Delta. The programme of ecological zoning carried out in the last decade was concluded at the level of the secondary eco-regions and therefore there are some data about the most endangered species.

(c) *Conservation*
- Since the 1990s, when there were 23 protected plant species and 24 protected animal species in the Romanian part of the Danube Delta, no further species have been placed under protection.

(d) *Legislation*
- There are numerous legal contradictions in the Environment Act, the Forestry Code, Hunting Law and the Bern and Bonn Conventions, ratified by Romania.
- International conventions on biodiversity conservation ratified by Romania, which, according to the Constitution, become an integral part of national law, are neither observed nor put into practice.

The Danube Delta is a dynamic, changing structure. In its natural state, it is continually shaped and re-shaped by three different factors: sediment input by the river, wave action from the sea and lakes, and energy of tidal currents.

Its position as a link between the Black Sea and the mainland gave
Danube Delta a major function in transport and trade. Its high biological
productivity also made it an important source of food. The Danube Delta is
by nature a highly open system. It is greatly influenced by human interference in the upstream sections of the river. Its sensitivity to external
influences forces those responsible for management of the Danube Delta to
take various restrictions into account, and sometimes to deal with hazards
resulting from actions beyond their control.

Danube Delta ecosystems sustain a rich and specific flora and fauna,
which are in decline as long as the human interference continues. Over the
past decades, the specific value of the Danube Delta as a natural biome has
become a point of increasing concern for those involved in delta management and development. Today, management has to deal with many, often
contradictory objectives, of which nature conservation is only one. How
can biodiversity and naturalness be reconciled with the conservation of
these delta resources? To put it in simpler words: *How to deal with nature
in the Danube Delta?*

This review is based on a questionnaire distributed to key wetland
experts familiar with deltas.

The review focuses on the following main themes:

- Values and resources utilization in Danube Delta
- Threats to biodiversity and naturalness
- Monitoring and evaluation
- Management and conservation

2. Values and resources utilization

The values of Danube Delta can be divided into products, functions and
services. Each of the following aspects gives an overview of the relative
importance of specific values in the Danube Delta.

Let's look at products first. A wide variety of plants and animals, as
well as mineral products and water, are harvested from the Danube Delta,
providing considerable economic returns to various stakeholders.

Water-bird hunting occurs over almost the entire delta. The hunting is
mainly recreational, although it also provides food for local people and
considerable economic returns. For example, approximately 150,000 ducks
are shot each year, increasingly by syndicated hunters from outside the
area. Tourism also contributes to degrading the Delta especially by unrestricted
hunting, which has led to some rare species of birds nearing extinction.

The hunting of mammals occurs on a smaller scale and is considered less important. Fish is generally considered to be an important product. Extensive fish farming in the natural wetland areas occurs at several sites in the Delta. The whole fishing policy until now proved to be wrong and it must be reconsidered ecologically.

Exploitation of the vegetation occurs through:

- Grazing, which occurs everywhere and is considered relatively important.
- Cutting the reeds, particularly for thatching, is less widespread, but is important.
- Exploiting the forests is a real ecological attack, intensified especially after 1990, when the new regime gave the Delta forests to some private companies and not to Romsilva, an unique exception in the country. These companies undertook massive deforestation without any concern for the future or for the natural balance.

The exploitation of the water resources is by far the most significant non-wildlife product, particularly water used for irrigation, but also domestic water supply. This exploitation is a major threat to natural areas, due to unsustainable levels of abstraction.

Secondly, let us consider the functions of Danube Delta. The Delta has a wide range of functions, including groundwater recharge and discharge, flood control, sediment, toxin and nutrient retention, shoreline stabilization and storm control. The most important of these are considered to be groundwater recharge, toxin and nutrient retention, shoreline stabilization and sediment retention.

Now let us consider services in the Danube Delta. It also provides a wide range of services important to people. Amongst those identified, the importance of supporting biodiversity stands out. Other services are the recreation and tourism values, although eco-tourism is rather poorly developed in the Delta. Its importance for research is also significant. The Delta has important culture or heritage values, reflecting the long history of settlements in the area. The Delta provides important services for transport, trade, and human settlements.

3. Threats to biodiversity and naturalness

The threats to biodiversity can be divided into two groups: one that results from human action outside the delta area, and the other due to unsustainable human action within the delta.

3.1. WATER QUANTITY

Humans have modified the river discharge to the Delta, mainly by hydro-technical works, particularly dams, in upstream river sections. Such interventions were generally considered the most important threat recorded. These changes affect many aspects of the functioning of the Delta, but are particularly important for migratory fish species, and also flora and fauna that depend on specific flooding patterns.

Increased drainage intensity and decreased water storage in the catchment cause increased amplitude and a shortened duration of floods. Storage lakes for hydropower production cause a leveling-off of the flooding régime in the Delta, especially of the minor floods outside the main flood season (Compania Nationala "Apele Romane", 2006).

3.2. SEDIMENT SUPPLY

Regulation of the upstream sections of a river tends to reduce the sediment transported to the delta. Most of the bedload and part of the suspended sediment are trapped in storage lakes. Sediment transported to the Danube Delta was reduced by about 30% in past decades by the construction of the Iron Gates Dam on the Danube and by numerous minor dams at the apex of the Delta.

3.3. SEA LEVEL RISE

Not only the upstream conditions influence the development of deltas – downstream conditions such as sea level rise have an influence as well. In fact, a relative rise in the sea level is a precondition for the development of a delta at a river mouth. But a rapid acceleration in the rise of sea level increased the wave action, causing erosion on the Delta's marine margin. That is why the very famous and profitable beaches of the Romanian Black Sea coast have narrowed over the last 2 decades, affecting the ecological equilibrium.

The Biosphere Reserve's Black Sea coastline, with a length of almost 200 km, is currently affected by an intensive and continuous degradation caused by marine erosion. The shoreline is shrinking at an annual rate varying from a few meters to 15–20 m for the coastline of the Delta between Sulina and Cap Midia, and around 0.2–0.5 m for the coastline with sea cliffs south of Constanta.

Erosion along the coastline is both natural and man-made. From among the human activities of great impact on the coastline one may mention the hydro-technical works on the Danube and its main tributaries, the ports, and coastal engineering works. The hydro-technical changes performed on the Danube and its main tributary have caused a fall in the input of sediments to the coastal area by over 50%, as compared to the values registered before the building of the dykes. Thus, a great sedimentary imbalance has been created in the coastal zone, which has in turn initiated the erosion process. The port facilities and other coastal engineering works, such as the protection dykes for the Sulina navigation channel, the breakwater piers of the Midia, South Constanta and Mangalia ports, and the coastline protection works on the tourist beaches, also cause major environmental imbalance along the littoral.

Concerning these tendencies, I should say that considering global climatic changes and the general rise in the sea level, as well as the regional geo-ecological conditions that characterise the Danube – Danube Delta – Black Sea geo-system, one can estimate that the medium-term erosion process will be at least as active as in the past 2 decades. The long-term predictions reveal an extension of beach erosion, especially because of the continuous decrease of sand material in the coastal area, because of the persistent rise in sea level and the increasing energy level of the hydro-meteorological factors – storms.

Many countries consider the problem of losing land by erosion of their coastline to be of national importance. Beach erosion leads to territory loss, but it especially affects the tourist industry, causing significant losses to national economies. The process of erosion also disturbs the ecological state of the coastal area almost irreversibly.

3.4. CHANGES IN THE LAND USE

One of the greatest threats to the biodiversity and naturalness of the Danube Delta Biosphere Reserve has been the appropriating of natural areas for other purposes, principally for agriculture, industrial zones, forestry, fish farms etc. The area of natural habitat, particularly wetlands, in the Delta has been greatly reduced in the past. Forty five percent of the Delta remains as natural habitat, mainly due to strict protection of the scientifically interesting areas. Now, I can express optimism that the international designation of the Delta under the Ramsar Convention and European Directives will reduce the risk of further losses.

4. Concluding remarks

1. *An ecosystem-based approach is essential for successful conservation of the Danube Delta.* The most important threats to the Delta are changes in the quantity and quality of water reaching it from upstream catchments. Therefore, conservation approaches will only succeed if they are accompanied by an integrated approach to catchment management. Such an approach will require the establishment of appropriate authorities,

integrating the interests of the various stakeholders and taking biome needs into account when considering water allocations. Since the Delta catchments are shared between many countries, this will also require a high degree of international co-operation.

2. *Management within the Delta should be undertaken within the framework of integrated land use and management plans.* The intense pressures for development in the Delta require that the conservation of biodiversity and naturalness be put within the framework of an integrated land-use plan covering the whole delta. Within this, appropriate management decisions for the natural areas require the establishment of appropriate authorities on whole delta level, including the areas in Moldavia and Ukraine. Because it is a transboundary delta, special mechanisms are needed to ensure international co-operation.

3. *Adequate resources must be available for long-term and sustainable management plans.* There are numerous examples of management plans that have been developed for wetland areas, including deltas, which have failed to be implemented due to the lack of available resources in terms of training, personnel and equipment. A key priority in the Delta should be to review those constraints and to assess how they can be addressed both at the national level and through international co-operation. Also, mechanisms for the financial involvement and responsibility of the private sector should be promoted.

4. *A coordinating mechanism should be established to exchange ideas and best practice between delta management experts in Europe.* European deltas share many common values, threats and management constraints. Therefore delta experts and managers can gain considerably from sharing expertise and information. Several initiatives to promote international co-operation between deltas already exist, of course. Some kind of coordinating mechanism would be helpful to integrate these initiatives. Activities would focus on training programs, international exchanges, specialist workshops, modern communications media, plus joint project initiatives. The exchange of ideas and learning from each other by bringing people together has to be very effective.

5. The fifth point is short but clear. *Not planning and decision-making for local people, but **with** local people.* Ultimately, protecting nature in the Delta will only be successful if there is a strong commitment on the part of local communities and stakeholders to the planning and decision-making process.

Unfortunately, Romania currently faces many transitional problems. It is not easy, but we are optimistic, as local communities accept the saying that to deal with the nature in Danube Delta Biosphere Reserve is doing more, with less.

References

Chirila, E., 2003: *Analiza poluantilor*. Brasov, Editura Universitatii Transilvania, pp. 1–30.

Compania Nationala "Apele Romane", 2006: Internal report of the "Romanian Waters" National Company, Bucharest.

Galatchi, L.D., 2006: The Romanian National Accidental and Intentional Polluted Water Management System. In: Dura, G. et al. (Eds.), *Management of Intentional and Accidental Water Pollution*, Springer, New York, pp. 181–184.

Galatchi, L.D. and Tudor, M., 2006: Europe as a Source of Pollution – The Main Factor for the Eutrophication on the Danube Delta and Black Sea. In: Simeonov, L. and Chirila, E. (Eds.), *Chemicals as Intentional and Accidental Global Environmental Threats*, Springer, New York, pp. 57–63.

Galatchi, L.D. and Vladimir, A.N., 2007: *The Sustainable Development Plan of the City of Tulcea (Romania). Case Study: Analysis of the Danube River Water Quality Parameters*, Ovidius University Annals of Chemistry, **17(2)**: 242–247.

PREVENTING DISASTER ON FRENCH CREEK, ONE OF THE MOST BIOLOGICALLY DIVERSE RIVERS IN NORTH AMERICA

J.O. Palmer and E. Pallant[*]
Department of Environmental Science, Allegheny College, Meadville, Pennsylvania, USA

Abstract. Disasters created by chronic perturbations are as debilitating as acute disasters such as tsunamis and oil spills. French Creek is the most biologically diverse river in Pennsylvania but is threatened by invasive species, poorly managed farm animals, urban and agricultural runoff, leaky septic systems, and loss of riparian buffer zones. For chronic, non-point threats, prevention is the best protection and depends on education and action by watershed residents. Creek Connections at Allegheny College engages teachers and students throughout the watershed with biological and chemical research of the Creek and threats to its integrity. Through independent research, stream restoration projects and public presentations, students become an educated citizenry as vital to ecosystem protection as any regional or national response plan for more acute disasters.

Keywords: Aquatic diversity, French Creek, invasive species, non-point source pollution, stream restoration, watershed education

1. Introduction

Disasters created by low-level chronic perturbations are every bit as debilitating as attention-getting, georegional, acute disasters such as tsunamis, hurricanes, floods, tornadoes, radiation, toxicity, and oil spills. In fact, disaster preparedness is largely geared toward responding to emergencies. Yet, rivers are subject to devastating impacts of the slow and creeping variety, which often take years to manifest: invasive species, falling water tables, excessive nutrient inputs, sedimentation, and increases in impervious surfaces within a watershed.

French Creek is a 188-km long river within a watershed of 328,930 ha. Its headwaters are in western New York State, but 91% of the watershed is located in northwestern Pennsylvania. Approximately one half the watershed is forested and one half is agricultural; the human population is nearly half a million. French Creek is the most biologically diverse river in Pennsylvania and one of most diverse in the USA (Smith and Crabtree, 2005). There are

---

[*] To whom correspondence should be addressed. e-mail: epallant@allegheny.edu

89 species of fish and 27 species of freshwater mussels; only one mussel from the historical record has become extinct (Smith and Crabtree, 2005). By comparison there are eight species of mussels in continental Europe.

Within the French Creek watershed, federal, state, and local authorities are charged with planning for acute disasters such as a toxic spill from a point source, but just as hazardous to the Creek are invasive species, poorly managed farm animals, urban and agricultural runoff, leaky septic systems, and the loss of riparian buffer zones.

2. Threats from non-point and chronic sources

Streamside vegetation offers a first line of protection against the input of exogenous chemicals and sediments. Trees, shrubs, and dense, native grasses slow the flow of runoff allowing infiltration into porous riparian soils and the recharge of groundwater. Forested buffers greater than 20 m in width provide the most flood control, followed by shrub buffers and grass buffers (Osborne and Kovacic, 1993).

In the state of Pennsylvania, for example, 40% of the stream kilometers have less than 30 m of riparian forest buffer (Novak and Woodwell, 1999). Forests have been cleared for timber, to accommodate housing, business, and retail developments, and for agriculture. Other forests have been compromised by the introduction of invasive and exotic species.

Agriculture is the number one industry in Pennsylvania accounting for 15% of the degraded streams in the state (Novak and Woodwell, 1999). In some farms livestock, particularly dairy cattle, trample stream banks. This causes sedimentation that is debilitating to benthic organisms. Animal waste also contributes to excess nutrient loading. Cropland within watersheds contributes soil, manure, fertilizers, and pesticides to regional waterways, and in some cases is responsible for contaminating drinking water supplies.

Municipal sources of pollution contaminate approximately ten percent of Pennsylvania's rivers (Novak and Woodwell, 1999). Inadequately functioning sewage treatment plants allow untreated sewage to periodically enter rivers while some industries are responsible for the release of solids, metals, nutrients, pesticides, other chemicals, and heat. Poorly functioning septic systems release untreated sewage and nutrients into rivers. Municipal solid waste and construction site runoff also contributes to water pollution in some parts of the state (Novak and Woodwell, 1999). Smaller, though still significant, sources of pollution include the discharge of industrial chemicals and the accumulation of roadside and urban chemicals that reach rivers via urban runoff and storm sewers, e.g., salt, sand, oil, and grease.

Though small in terms of total area polluted the massive accumulation of discarded automobile tires have become dangerous not just to rivers where they contaminate soil and water, but to human populations as well.

Tire dumps provide habitat for rodents and mosquitoes that act as vectors for West Nile virus, meningitis, and encephalitis.

A major threat to the integrity of Pennsylvania's rivers is the invasion of non-native species (Pimentel et al., 2000). Many kilometers of native riparian vegetation have been replaced by Japanese knotweed (*Polygonum cuspidatum*), Multiflora rose (*Rosa multiflora*), Purple loosestrife (*Lythrum salicaria*), and Taratarian honeysuckle (*Lonicera spp.*). Eurasian watermilfoil (*Myriophyllum spicatum*) grows densely in North American waterways. In addition, several aquatic species have entered North American rivers in the ballast water of ocean-going vessels. Among the most pervasive of these species are Zebra mussel (*Dreissena polymorpha*) and Northern snakehead (*Channa argus*) (Ricciardi et al., 1998; Strayer, 1999; MacIsaac, 1996). For these chronic, non-point threats to French Creek, prevention is the best method of protection.

3. Creek connections and French Creek

Integral to any plan to protect the river is education of the citizens within the watershed. Creek Connections, a watershed education program run by Allegheny College, is one such effort. "Kids in Creeks" is an elegantly simple concept at the heart of Creek Connections that captures its core activities and enormous environmental learning, action and stewardship possibilities. Since 1995, Creek Connections has engaged K–12 schools in western Pennsylvania and New York turning waterways into outdoor laboratories and helping teachers bring watershed education and scientific research into their classrooms. Creek Connections annually involves over 5,000 students from more than 45 schools, emphasizing a hands-on, inquiry-based investigation of local waterways. Allegheny College interns from all majors act as human resources and mentors to assist the ongoing watershed studies.

Students conduct water chemistry and biological studies, design research projects, and attend an annual Student Research Symposium. With the opportunity to practice real science using a natural resource located in their own backyards, students strengthen quantitative skills and come to appreciate the connection and impacts they have on their waterways. Sharing their knowledge, research, and concerns with others in their watershed community cultivates scientific literacy and an ethic of environmental stewardship for students at all levels. Training thousands of students per year, who in turn, bring their newfound knowledge home to their families is one of the most effective tools available in the French Creek watershed for the prevention of non-point sources of pollution and contamination.

A Creek Connections school year begins with a professional development workshop for teachers focusing on a specific topic such as mussel ecology, stream bank restoration, wetlands or groundwater, and often includes a field trip to French Creek or another waterway. Generous grant support provides each teacher with equipment and academic standards-based activities to conduct water quality tests in the field and the lab. At least eight times during the school year, students visit their local field site to collect data and report it to the project web site: http://creekconnections.allegheny.edu. Thus, 13 years of Creek Connections data is accessible to all students, government agencies and the general public. Many teacher and student resources, and watershed curricular activities, are also available at the site. Classes also design independent research, art, music and skits reflecting watershed ecology and natural history themes. Increasingly, classes put their new knowledge to work by implementing stream bank restorations in collaboration with regional conservation districts, state agencies and land trusts.

Celebrating a year of hard work, the annual Student Research Symposium convenes in April at Allegheny College or a Pittsburgh location offering a forum for students to share their research findings with each other and the public. Displays, oral presentations, educational displays from environmental organizations and agencies are featured as well as focus group activities for hands-on learning about various watershed topics.

Additional environmental security is provided at these symposia by the participation of governmental and non-governmental agencies. In one location, at one time, students, their teachers, and the public view displays and collect materials prepared by the likes of the Carnegie Science Center, the Pennsylvania Department of Environmental Protection, the Erie National Wildlife Refuge, and more than 20 additional organizations.

One teacher speaks of the impact of Creek Connections on her science class: "Our students live in a low socio-economic community which struggles to find role models to make a difference in their lives. Our Allegheny intern was an excellent role model for our students. She spoke with them about education as well as environmental issues. We spent a day building stream restoration structures with the Conservation District and our Allegheny intern. The students living in the community took ownership of the project and began to see themselves as a change agent." This is just one example of how low income residents become active protectors of water quality.

Creek Connections exemplifies the educational synergies that can result from collaborations between communities, their schools, and their colleges. Beyond the science and stewardship, getting kids into their creeks again is a significant step toward "saving our children from nature-deficit disorder" (Louv, 2006).

For 13 years several thousand students per year have studied the unique biological and chemical characteristics of French Creek, its tributaries, its downstream watersheds, and threats to its integrity. High school and middle school students do independent research projects and make presentations in their schools, communities, to their families, and to residents of the French Creek watershed. In addition they complete environmental restoration projects such as installation of streambank fencing to keep dairy cows out of waterways, planting trees in the riparian zone to lessen erosion, and transferring thousands of discarded tires from dumps to be recycled. The Creek Connections approach to French Creek is as vital to ecosystem protection as any national response plan for an emergency. The long-term health of this unique and vital ecosystem requires awareness – and an education plan – for coping with the insidious, low-level threats to diversity just as much as preparedness for a more newsworthy disaster.

References

Louv, R., 2006: *Last Child in the Woods: Saving Our Children from Nature-Deficit Disorder*, Chapel Hill: Algonquin Books.

MacIsaac, H., 1996: Potential abiotic and biotic impacts of zebra mussels on the inland waters of North America. *American Zoologist*, **36**: 287–299.

Novak, J. and Woodwell, W. (eds), 1999: A Water*shed Primer for Pennsylvania*, Pittsburgh, CA: Pennsylvania Environmental Council.

Osborne, P. and Kovacic, D., 1993: Riparian vegetated buffer strips in water-quality restoration and stream management, *Freshwater Biology*, **29**: 243–258.

Pimentel, D., Lach, L., Zuniga, R., and Morrison, D., 2000: Environmental and economic costs of nonindigenous species in the United States, *BioScience*, **50**: 53–65.

Ricciardi, A., Neves, R., and Rasmussen, J., 1998: Impending extinction of North American freshwater mussels (Unionidae) following the zebra mussel (Dreissena polymorpha) invasion, *Journal of Animal Ecology*, **67**: 613–619.

Smith, T. and Crabtree, D., 2005: *Freshwater Mussel (Unionidae) and Fish Assemblage Habitat Use and Spatial Distributions in the French Creek Watershed: Reference for Western Pennsylvania Unionid Protection and Restoration.* Unpublished July 30, 2005 Final Report submitted to the Pennsylvania Fish and Boat Commission iv + 180 pp.

Strayer, D.L., 1999: Effects of alien species on freshwater mollusks in North America. *Journal of North American Benthological Society*, **18**: 74–98.

APPRAISAL OF METHODOLOGY OF ECOLOGICAL RISKS ASSESSMENT ARISING FROM POLLUTION OF THE RIVERS OF THE UKRAINE

V.D. Romanenko, S.A. Afanasyev and A.I. Tsybulskiy[*]
Institute of hydrobiology NAS of Ukraine, 12 Prospect Geroiv Stalingrada, 04210 Kyiv, Ukraine

Abstract. A methodology for the assessment of ecological risks arising from pollution in water bodies has been developed and applied in the Dnipro basin countries within the framework of the UNDP-GEF Dnieper Program during 2002. The methodology was developed considering Dnipro's basin peculiarities and directed for processing of data, collected according to a specially developed scheme of field investigations, based mainly on the biological parameters. This ensured taking into account local and regional peculiarities of water quality origins; accumulated human load on the water body, not just for separate pollutants. This is the fundamental difference between ecological standards of surface water quality and standards of ecological safety of water use, e.g., maximum permissible concentration. Hydrochemical and hydrophysical data have been used only as indicative while carrying out field investigations and as testing in risk assessment calculation. In calculations we have not distinguished biological and hydro-chemical parameters, but during data analysis biological parameters were preferable as ecological effects indicators.

Keywords: Ecological risk, ecological risk assessment, ecological risk indicators, hydrochemistry, hydrobiology

1. Introduction

The main goal of this work was to appraise a methodology of assessment of ecological risks, arising after the impact of major point sources of pollution on different types of river ecosystems. It was undertaken in the countries sharing the Dnipro basin within the framework of the UNDP-GEF Dnieper Program during 2002 (Afanasyev and Grodzinskiy, 2004). The methodology was developed according to a scheme of field investigations described by Afanasyev (2001), based mainly on the biological parameters.

---

[*] To whom correspondence should be addressed. e-mail: acybula@ukr.net

J.A.A. Jones et al. (eds.), *Threats to Global Water Security*,
© Springer Science + Business Media B.V. 2009

The basic concepts are defined as follows:

Ecological risk – The possibility of development of ecosystem changes, which bring about its deterioration, depletion or transformation to a state which can threat the health of the population health and/or loss of its economical value.

Ecological risk assessment – Scientifically based estimation considering possibility and magnitude of adverse changes in the ecosystem or its components given a certain human load upon it.

Ecological risk probability – Probability of an event arising, which is considered as undesirable for the ecosystem and will damage it. Probability values are estimated for a certain time scale, e.g., 1 , 3, 10 years etc.

Risk indicators – Characteristics of ecosystem and its elements, which are used as means for judgment about risk probability and magnitude.

2. Materials and methods

As models, some rivers in Steppe and Polissya zones of Ukraine as well as in the Carpathian foothills have been chosen.

As an example, realization of ecological risk assessment methodology on the Dnipro river near Kherson city have been considered. Field investigations, sample collection and biological assessment of impact of the point source of pollution (Kherson city) on the Dnipro's hydroecosystems were carried out at different distances: immediately downstream of wastewater discharge (tens of meters from the real point discharge), within a 1 km zone, within a 3 km zone, etc., considering discharge magnitude and diluting capacity of river, up to the zone where the structure of biotic communities shows no visible deviations from background.

Distribution of turbidity, temperature, pH, and dissolved O_2 within the river has been studied visually and with instruments. In the selection of sampling points we obligatory consider their landscape-biotopic similarity. In the biotopes upstream, point pollution source samples have been collected and communities have been described as a control. Investigations were also carried out directly in the ecologically sensitive and ecologically valuable zones, within water areas having protected status and situated within 15 km of the source of point pollution.

General patterns of biotic communities were described on the basis of development and distribution of macroforms (aquatic plants, filamentous algae, macroinvertebrates). Samples of zoobenthos and zooperiphyton were collected, as well as samples of phyto- and zooplankton where possible. The species composition of ichthyofauna was estimated by analyzing amateur catches and interviewing the local population. The following background abiotic characteristics have been determined: concentration of dissolved oxygen, pH, temperature of water (using portable oxymeter Oxi 315i/SET), electrical conductivity, mineralization (using portable field device DurOx 325), transparency (using Secchi disk), chromaticity (according the chromaticity scale ШЦВ-000Т0), velocity of flow (visually). At the same time, samples were collected for laboratory processing to determine the forms of mineral nitrogen and phosphorus, BOD_5 and some other hydrochemical parameters.

Samples were collected considering visually defined units, using light diving equipment if needed. Laboratory processing of samples was realized using standard methods. Calculation and statistical treatment of data were undertaken using WACO software, developed in the Institute of hydrobiology of the Ukrainian National Academy of Sciences.

3. Results of the investigation

The reach of the lower Dnipro near the city of Kherson belongs to the Low Dnipro wetland geobotanical district. Wetland with high reed vegetations and floodplain meadows are the main features of its landscape. Downstream, the Kherson main channel is divided into numerous arms and creates a delta with floodlands, islands and water bodies of different shapes and sizes. This district is peculiar with its well preserved woodlands and wetlands with high species diversity, including many rare, relict and endemic species. The Dnipro delta is a habitat for numerous species of aquatic and amphibian fauna. It is used by migrant birds for seasonal habitation. This territory is considered a Ramsar wetland (3UA009), including stream stretches, arms and channels, shallow floodwater bodies, reed vegetations, island massifs, and the Dnipro liman.

Most part of the area is used for intensive economic activity: cattle pasture, haymaking, fishing, hunting, large groups of summer cottage settlements, resorts and boarding houses, and there are also some fishpond farms, etc.

On the stretch from Kherson to the liman headwater there are some features with strict and partial protection – the Bakayskiy state and Dniprovskiy fish reserves, Krasniukovske, Prorezhanske and Limanske hunting holdings. Within the Dnipro delta are the main habitats for spawning, hibernation and feeding of valuable food fish (bream, pike perch, etc.). This part of the delta and the adjacent part of the Dnipro liman is the most important region for the protection of ichthyofauna and restoration of fish stock.

Average indices of general mineralization of the Dnipro water upstream and downstream of Kherson during 1995–2000 did not exceed 0.40 g/l (National report about state of environment of Kherson oblast, 2001). Values of pH and dissolved oxygen concentration within the same period varied insignificantly.

Discharge of wastewaters from Kherson is carried out through the chain of sludge ponds and channels to the River Verevchyha and then to the

River Koshevaya, which then flows into the Dnipro. The scheme of the discharge is rather complicated. After a cascade of ponds, situated in the floodland of River Verevchyha, through the system of arms water passes through the reed vegetation for a considerable area. Downstream of the confluence of all the arms, about 3 km downstream of the water release from the system of ponds, the River Verevchyha flows through summer cottage settlements, where its water is used for irrigation, recreation and fishing.

At the point of wastewater discharge from the sludge ponds system, the hydroecological situation was the follows. The outlet channel was 1.5–2.0 m wide, 0.5–0.7 m deep, and the velocity of flow – 1.2 m/s. Temperature +24.2°C; pH – 8.0; dissolved oxygen concentration – 8.4 mg/l; and mineralization – 1.4 g/l. There is no aquatic vegetation because of an unsuitable substrate, concrete walls are covered with intensive growth of green filamentous algae. At the outlet of water from the channel: width is 2.5–3 m, depth 1–1.5 m, and a high velocity of flow. On the water surface oil spots and foam were recorded, the temperature was +22.7°C and the water was muddy, Secchi transparency – 0.3 m; pH – 8.7; dissolved oxygen concentration – 8.8 mg/l; mineralization – 1.4 g/l. These are the only points in our investigations with a lack of macroinvertebrates.

In the system of channels there is a gradual restoration of the biotic communities. After their confluence, 3 km downstream of the discharge, the River Verevchyha is 1.5–2.5 m wide, 0.5–1.5 m deep, and the velocity of flow – 0.6–0.8 m/s. Later on, the river bed widens significantly and after passing though a mass of vegetation transforms into an elongated bay (width – 10 m; depth – 3–4 m, with a low velocity of flow, backwater under proper wind direction or discharge from Kakhovka HPS is not excluded).

Already in 0.5 km downstream of the confluence of the channel arms, the Secchi transparency increases up to 1 m, dissolved oxygen concentration increases to 11.5 mg/l, mineralization decreases to background values 0.4 g/l. Within the stream section, the low riverside land is disturbed because of the presence of summer cottage settlements. The *Trent Biotic Index* (TBI) (Woodiwiss, 1964) is equal to 6; according to trophic indices river is characterized as a polyhypertrophic water body. 6 km downstream of the discharge point, near the point where the River Koshevaya flows into the Dnipro, all the parameters examined are close to background and the influence of Kherson's wastewater discharge is no longer apparent, particularly as the highest phytoplankton saprobity indices were recorded here – 1.55–2.94 (up to the α-mesosaprobic zone).

The water quality of the River Verevchyha, estimated on the basis of bottom invertebrates, agrees with the α-mesosaprobic zone classification (2.83), and in the Dnipro stream with an oligosaprobic classification (1.50).

In particular, the difficult environment for macrozoobenthos was recorded in 1975, where bottom invertebrate fauna was practically absent or minimal. Installation of the wastewater treatment plant and stopping the discharge of untreated wastewater into the River Koshevaya has contributed to environmental sanitation on these streams.

Saprobic indices and biotic index TBI indicate significant improvement of water quality and water body status from the wastewater discharge up to the point of inflow into the River Dnipro (Table 1).

TABLE 1. Values of biotic indices of water body status

Site	Name of water object and distance from discharge point	Indicator		
		TBI benthos	S of aquatic vegetation	S of bottom fauna
1	Kherson wastewater discharge	0	_[a]	_[a]
2	Intermediate discharge in cascade of sludge ponds of Kherson waste water treatment plant (0.65 km upstream)	0	_[a]	_[a]
3	River Vervchyha downstream of Kherson wastewater discharge (2.7 km)	2	2.05	3.6
4	River Vervchyha downstream of Kherson wastewater discharge near bridge (2.4 km)	1	2.05	3.6
5	River Vervchyha downstream of Kherson wastewater discharge (2.1 km)	3	2.05	3.6
6	River Verevchyha mouth (3.4 km)	6	1.86	2.75
7	River Verevchyha arm (6.0 km)	7	1.86	1.85
8	Dnipro, Kherson biological station (background) (6.2 km upstream)	7	1.77	1.92

[a] "_" means lack of this group at the given point.

Hydrochemical analysis on the chosen points as well as on the Dnipro near Kherson hydrobiological station (as a background) has been carried out. It was stated, that pH values in all the water bodies investigated was comparable with those during 1995–2000.

An increase in pH values in the mouth of the River Verevchyha and in control points (sites 6 and 8) was shown by the intensive growth of blue-green algae. Comparatively low dissolved oxygen concentrations at the discharge point were related to high stream turbulence and water mass aeration. After passing through the mass of reed vegetation (points 4 and 5), dissolved oxygen concentration gradually decreases, and then increases as the water flows into the mouth of the River Verevchyha (algae bloom). General mineralization in the investigated water bodies is rather high (especially at points 1 and 3) compared with the Dnipro. This indicates salt pollution of the waste waters. It is worth noting that there is a high content of ammonia and the nitrite form of nitrogen at the discharge point. The content of phosphates in the water of all the water bodies was also high. Concentrations of nitrogen and phosphorous mineral forms decrease with distance from the wastewater discharge point.

Thus, the results of analysis of general mineralization and nitrogen and phosphorous mineral form content in the water bodies investigated make it possible to conclude that there is considerable inflow of polluted wastewaters into the River Verevchyha. But after passing through the natural purifying and rehabilitating area of reed vegetation in the River Verevchyha flood plain, pollution levels as indicated by biogenic substances at the junction of the River Verevchyha and the Dnipro's Koshevaya arm were similar or even lower than in the control.

Analyses of the features at risk and received stresses and hazards due to Kherson makes it possible to determine the types of possible ecological risks, as well as to evaluate their level of probability and magnitude (Table 2).

TABLE 2. List of ecological risks types, evaluation of their reality level and magnitude

Risk index	Ecological risk type	Level of reality	Magnitude evaluation
HWR	Risk of river waters pollution	High	Extremely undesirable
HWS	Risk of bottom sediments pollution	High	Extremely undesirable
HBP	Risk of aquatic plants communities deterioration	Moderate	Undesirable
HBSF	Risk of bottom fauna communities deterioration	High	Undesirable
HBPHPL	Risk of phytoplankton community deterioration	Moderate	Undesirable
HBZPL	Risk of zooplankton community deterioration	Moderate	Undesirable
HBZBT	Risk of zoobenthos community deterioration	Moderate	Undesirable
EUF	Risk of flood lands and first terrace impoundment by polluted waters	Low	Undesirable

Referring to this, bottom fauna and aquatic thickness should be considered as the most threatened by pollution components of ecosystem. Comparison of probability evaluation and magnitude of every ecological risk type makes it possible to state those of them, requiring preliminary evaluation (Table 3).

TABLE 3. Ecological risk type requiring preliminary assessment

Risk index	Ecological risk type
HWR	Risk of river waters pollution
HWS	Risk of bottom sediments pollution
HBP	Risk of aquatic plants communities deterioration
HBSF	Risk of bottom fauna communities deterioration
HBPHPL	Risk of phytoplankton communities deterioration
HBZPL	Risk of zooplankton communities deterioration
HBZBT	Risk of zoobenthos communities deterioration

4. Preliminary assessment of risk probability

The quantitative reference data needed for calculating levels of probability for risk types *HWS, HBP, HBSF* are summarized in the Table 4. As risk indicators for *HWS, HBP, HBSF* consequently *TBI, S* of aquatic vegetation, *S* of bottom fauna were used.

Arithmetical and quadratic means deviation for every risk indicator were calculated according to data from Table 1: for *HWS* 3.44 and 2.877 consequently; for *HBSF* – 2.81 and 0.794 consequently.

Available data are sufficient for calculation of ecological risks probabilities for different points of around Kherson. Calculated values are summarized in the Table 4.

Values of ecological risk probabilities and general probability of river water pollution for every of pollutants are listed. The probability of water pollution at least with one of measured indicators can be used as a overall indicator of river water pollution probability.

TABLE 4. Probability of ecological risks occurring

I^a	Number of sample point							
	1	2	3	4	5	6	7	8
Chemical								
NH_4^+	0	0	0	0	0	0	0	0
NO_2^-	0.47	0.49	0	0.49	0.47	0.47	0.49	0.48
NO_3^-	1	0.82	0.96	0.82	0.65	0.44	0.44	0.45
PO_4^{3-}	0.98	0.59	0.96	0.59	0.74	0.32	0.34	0.33
O_2	0.08	0.32	0.10	0.55	0.58	0	0	0.28
Biological								
HWS	1	0.91	1	0.98	0.80	0.04	0	0
HBP	0	0	0	0	0	0	0	0
HBSF	–	0.77	–	0.77	0.77	0.38	0.07	0.08
P^b	1	0.97	0.99	0.98	0.98	0.80	0.81	0.86

[a]Indicators.
[b]General probability.

Table 4 clearly shows that even according to the hydrochemical parameters investigated, the risk of river pollution with at least one of them is rather high and even at the control point amounts to 86%. This is a illustration of the generally unfavorable status of the river and that it is at "risk". On the basis of all the data from all the points investigated, it can be stated that the ecological status of the River Verevchyha returns to background levels (control point parameters) 3.4 km downstream of the wastewater discharge point. This is evidence of a high degree of self-purifying potential in the given watercourse, which passes through a large mass of reed vegetation, where polluted water purifies quickly and effectively.

Considering appraisal of methodology of ecological risks assessment in general, it can be concluded:

The following types of ecological risks have a high level of probability: risk of river water pollution, risk of aquatic plant communities deterioration, risk of bottom fauna community deterioration (Table 5).

The risk of pollution by at least one of the hydrochemical parameters considered is rather high in all the rivers investigated (except the Dniester) even at the control points, which illustrates the overall unfavorable and risk status of rivers.

Practically in all the risk probabilities in the rivers return to control level after 1–10 km. The exception is discharge of cold technogenic waters from the Dniester hydropower plant, the risk from which can be measured at a distance of more than 130 km.

TABLE 5. Appraisal of methodology of ecological risks assessment on the rivers of the Ukraine

Water body (details below table)	Ecological risk (value)				
	Control	Point of stress	500 m	1,000 m	Distance at which risk is equal to control (risk value)
1. River Dnipro	0.86	1.00	0.97	0.98	3.4 km (0.80)
2. River Goryn'site 1	0.07	0.99	0.68	0.11	4.8 km (0.07 for biotic parameters)
	0.76	1.00	0.99	0.99	4.8 km (0.94 for chemical parameters)
3. River Goryn' site 2	0.87	1.00	0.99	0.93	3.2 km (0.91)
4. River Samara	0.88	1.00	0.98	0.98	23 km (0.96)
5. River Desna	0.16	0.84	0.69	0.16	1.0 km (0.16 for biotic parameters)
	1.00	1.00	0.99	0.83	7.2 km (0.40 for chemical parameters)
6. River Dniester	0.66	1.00	1.00	1.00	130 km (0.70)

5. Conclusions

In all the water bodies investigated, results were found which cannot be derived by other methods or which are difficult to obtain by other methods. On the River Desna, it was observed that the total pressure of diffuse and small point sources of pollution (uncontrolled flood plain development and construction of summer cottage settlements) leads to more risk of decreasing water quality than the city of Chernigiv wastewater discharge. For the mountain rivers investigated, the impact of flood waters was many times higher than the risks from point pollution sources. In general, we came to the conclusion that pressure from major point sources of pollution can be noticeably low compared with the background of small and diffuse pollution sources or other sorts of pressure.

1. River Dnipro, mouth area of large river in the steppe zone, city of Kherson, Vodokanal wastewater treatment plant; 2. River Goryn', Ustia, mouth area of small river in the steppe zone, Rovnoazot JSC; 3. River Goryn', mouth area of small river in the steppe zone, Lutskvodokanal wastewater treatment plant; 4. River Samara, small river in the steppe zone, city Novomoskovsk wastewater treatment plant; 5. River Desna, medium river in Polissia, city of Chernigov, Chernigivvodokanal wastewater treatment plant; 6. River Dniester, medium river in Carpathian foothills, Novodniestrovska HPP.

Data collected from rivers of different types make it possible to state that, in many cases, for the evaluation of the impact of pollutants on the environment it is not necessary to know the real source and quality of pollution. Moreover, often we deal with integral pressure from different sources and different pollutants. The ecological effects that occur after the releasing and accumulation of pollutants into hydroecosystems are extremely important to know.

On the basis of the analytical data, it is possible to conclude that analysis and assessment of ecological risks are especially effective in cases where:

1. There are considerable gaps in reference data about the anthropogenic loads on the ecosystems and the status of the ecosystem itself.
2. Where responses of ecosystems to these loads are not identified and have a probabilistic character.
3. Where possible future use of ecosystems assumes a range of alternative scenarios.

This work is executed within the framework of the program "Development of technology of minimization of the ecological risks related to human-induced (technogenic) and biological contamination of superficial waters with the purpose of improvement of the environment of man" (State registration No. 0104U007822 within the framework of the Government program the "Newest medical-biological problems and environment of people").

References

Afanasyev, S.A., 2001: Development of European approaches to biological assessment of hydroecosystems in the monitoring of the rivers of Ukraine. *Hydrobiologicheskiy J.*, **3(5)**: 3–18. (In Russian.)

Afanasyev, S.A. and Grodzinskiy, M.D., 2004: Methodology of evaluation of ecological risks, occur after pressure of pollution sources on the water objects. Kyiv.: АйБи, 59pp. (In Russian.)

National report about state of environment of Kherson oblast, 2001. (In Ukrainian.)

Woodiwiss, F.S., 1964: The biological system of stream classification used by the Trent River. *Board Chem. Ind.*, **11**: 443–447.

EMERGENCY RESPONSE AND WATER SECURITY OF THE BTC PIPELINE IN ECOLOGICALLY SENSITIVE AREAS OF GEORGIA

M. Devidze[*]

Tbilisi State University, Sokhumi Branch, 9 Jikia St., 0143 Tbilisi, Georgia

Abstract. The BTC Pipeline crosses several sensitive areas in Georgia, including surface and groundwater bodies. The territory that is crossed by the pipeline is seismically active and there is risk of oil spillage. Any oil spilled from a damaged pipe will quickly mix with the river water. It is planned to construct permanent facilities equipped with five permanent tanks and six block valves on six tributaries of the rivers. Nevertheless, the risk of oil spillage remains dangerous for water ecosystems in these sensitive areas. A specific methodology of bio-indication and remediation of water ecosystems which would be used for emergency response in case of oil spills was tested in these sensitive areas. Laboratory experiments on raw oil show differences in sensitivity for a range of phytoplankton species. It is recommended to use indigenous *Cyanobacteria* compounds for sensitive areas in the Borjomi region for remediation of oil spills.

Keywords: *Cyanobacteria*, bioindication, bioremediation, sensitive area, oil spill, pipeline

1. Introduction

The total length of the Baku-Tbilisi-Ceyhan (BTC) pipeline is 1,760 km. The pipeline will provide crude oil from the Caspian Azeri-Chirag-Gunashli oil deposits to the terminal at Ceyhan on the Mediterranean. The length of the pipeline across the territory of Georgia is 249 km, with a diameter of 115 cm. The territory which is crossed by the pipeline is seismically active and there is risk of oil spills. Also, the pipeline crosses several sensitive areas in Georgia including ground water deposits. Apart from the above, the pipeline corridor crosses numerous small rivers and springs of surface water with substantial seasonal flow fluctuations. There are a number of wetlands in the high mountainous volcanic areas and a large volume of high quality underground water (e.g. Dazbashi spring) and curative mineral waters (e.g. Borjomi).

The Borjomi Gorge, which is covered with large areas of forest and wetlands, has a diversity of ground, surface, potable and mineral waters. Also, following from the climatic conditions of the gorge and the very

---

[*] To whom correspondence should be addressed. e-mail: mdevidze@caucasus.net

unstable hydrological profile of the mountain rivers expressed by intense
flooding, any oil spilled from a damaged pipe will quickly mix with the
river water; it is impossible to calculate precisely the concurrence time of
an accident and peak moment of the flood.

It is planned to construct permanent facilities equipped with five permanent tanks and six block valves on six tributaries of the rivers. The
structure of each facility is designed in the following way: (1) the riverbed
will be consolidated and an accumulating tank will be constructed to
dampen the flood; (2) in order to prevent filtration from the accumulating
tanks, its bottom and sides will be covered with water-resistant membranes,
while the riverbanks will be consolidated with gabions.

Despite construction of these pipeline safety facilities, the risk of oil
spills still exists, which is more dangerous for water ecosystems in sensitive
areas. In our research, we carried out a specific methodology of bio-
indication and remediation of water ecosystems which would be used for
emergency response in case of an oil spill in the sensitive areas along the
pipeline. The water ecosystems are characterized by a wide diversity of
biota, but priority should be given to study of phytoplankton algae species
and their response to oil spills, as they are basic for primary production
(Nelson-Smith, 1977).

2. Assessing water contamination

One of the popular methods of assessment of water contamination is monitoring and bio-indication of phytoplankton species quantity and diversity.
Laboratory experiments on raw oil show differences in sensitive for a range
of phytoplankton species. The species *Ditylum brightwelli, Coscinodiscus
granii* and *Chactoceros curvisetus* perish within 24 h after oil treatment in
100 mcl/l concentration (Mironov, 1970). Experiments show delay of reproduction of phytoplankton species *Nitzschia closterium* with oil concentrations
of 25% (Galtsoff et al., 1935). Other experiments show gathering of large
numbers of active forms of *Infusoria* around an oil spot.

We carried out laboratory experiments on aquarium fish (*Lebistes
reticulates*). They were kept in an oil polluted environment. Two aromatic
fractions of Mirzaani oil showed mutagenic effect in chronic tests. The
level of chromosomal aberrations was increased to 7–8% per cell compared
with control cells (0.2%). The same experiments were carried out on laboratory mice, where aromatic fractions of oil were shown to have a mutagenic
effect (Devidze, 1986).

The second part of our methodology for emergency response on oil
spills is bioremediation, as a natural process that uses microorganisms to
transform harmful substances into non-toxic carbon dioxide, water and
fatty acids. We studied the microbial diversity of Cyanobacterial species in

the mountain mineral waters of the Borjomi region, which is crossed by the pipeline. *Phormidium, Oscillatoria, Mycrocystis, Gloeocapsa,* and *Synechocystis* cyanobacterial morphotypes were dominant in this area. Laboratory research showed that in oil-contaminated cultures cyanobacterial communities of *Phormidium tenuissimum, Synechocystis minuscula* and *Synechococcus elongates* are edificatory. In small compounds of these communities, physiological groups of microorganisms are present that activate the process of degradation of the hydrocarbon components.

From the northern Caspian was picked out cyanobacterial community *Phormidium gelatinosum Ph. tenue, Oscillatoria pseudogeminata, Osc. tambi, Osc. Amphibian* (Soprunova, 2005). In our experiments it was shown that from these species the most resistant to oil contamination is *Oscillatoria pseudogeminata* as bacterial and fungi associates are present in its glicocalics and they are able to oxidize oil hydrocarbons.

Upon introduction into an oil spill contaminated environment, the microbes will germinate and become active instantly. In good conditions, these special microbial blends multiply exponentially within minutes. These begin the bioremediation to digest environmental pollution and wastes as long as all the necessary ingredients of water, oxygen and a food or waste source are present. Normal effective temperature range for these bioremediation microbes is between 50°F to 100°F.

The underlying idea is to accelerate the rates of natural indigenous populations of microbial bacteria which can be stimulated through the addition of nutrients or other materials. Exogenous microbial populations can be introduced in the contaminated environment. But we do not recommend using exogenous microbial populations in sensitive areas as they will have negative impact on ecological balance in these ecosystems.

We recommend using indigenous *Cyanobacteria* compounds for sensitive area in the Borjomi region for remediation as emergency response to oil spills. It was shown that the introduction of *Cyanobacteria* associations in oil polluted ecosystems creates conditions for the development of polyfunctional associations (Cyanobacteria microalgae, bacteria, fungi), which degrade pollutants and increase the effective biological rehabilitation of contaminated areas.

In conclusion, our methodologies of bio-indication and defining Cyanobacterial natural compound are being developed for bioremediation of oil contaminated waters, which will be used especially for sensitive areas as improved safety methods for emergency response for water security.

References

Devidze, M., 1986: Study of mutagenic activities of Mirzaani and Samgori oil in bone marrow cells of laboratory mice. *Inform. Acad. Sci. Georgia,* **123**(3): 605–608.

Galtsoff, et al., 1935: Last call for cultural methods. *Science*, **81(2098)**: 272–273.

Mironov, O.G., 1970: The effect of oil pollution on the flora and fauna of the Black Sea. *FAO Tech. Conf. Mar. Pollut.*, pap. E–92, Roma.

Nelson-Smith, A., 1977: *Oil Pollution and Marine Ecology*. London: Publisher "Progress", Moscow.

Soprunova, O.B., 2005: Особенности функционорования альго-бактериальных сообществ техногенных сообществ техногенных экосистем. *Автореферат*.

PART V:

INFRASTRUCTURE – TECHNICAL INNOVATIONS AND FAILURES

EFFECTS OF RESERVOIRS ON STREAMFLOW IN THE BOREAL REGION

Ming-Ko Woo* and **R. Thorne**
School of Geography and Earth Sciences, McMaster University, Hamilton, Ontario, Canada

Abstract. The circumpolar boreal region is generally well endowed with water resources relative to its low population. The most common natural river flow pattern is the nival regime in which snowfall accumulated during the long winter is rapidly melted in the spring, releasing large meltwater fluxes that cause the hydrograph to peak. Human development has altered this flow rhythm through water use and consumption. Reservoirs constructed for hydro-power production change the timing of flow while water withdrawal for consumptive purposes causes changes in both the timing and the amount of discharge. The most profound effects come with river diversion which alters the annual flows and re-distributes the monthly discharges, with the donor river losing water and the recipient river and the connecting channel gaining flows. Examples are drawn from Canada and Siberia to illustrate the effects of reservoir operation on streamflow. Flow regulation produces hydrological impacts that have local to regional consequences.

Keywords: Reservoir, streamflow, water use, hydropower, floods, flow diversion, boreal region

1. Introduction

The boreal region covers about 900 million hectares and lies between 50° and 70°N. Rivers of this region flow mainly to the Arctic Ocean. Although the region is sparsely populated compared to the lower latitudes, there are many northern communities that depend on the rivers for water supply and are afflicted by their floods. River discharge also brings considerable amounts of freshwater to the polar seas and is important in influencing sea ice formation and decay, and the heat and water circulations of the polar ocean (Aagaard and Carmack, 1989; Manak and Mysak, 1989). Knowledge of the flow patterns of boreal rivers has applications to the water supply and flood problems of the region, and permits proper assessment of river discharge to the Arctic.

---

* To whom correspondence should be addressed. e-mail: woo@McMaster.ca

J.A.A. Jones et al. (eds.), *Threats to Global Water Security*,
© Springer Science + Business Media B.V. 2009

While there are many local situations where water is withdrawn or diverted from rivers, such as for individual farms or industrial plants, major projects involve river impoundment and accompanying reservoir operations. Most large rivers in Siberia are thus affected, as are many Canadian rivers that drain to Hudson Bay. This paper examines the effects of human interference with natural river regimes, using case studies to highlight the hydrological changes associated with reservoir and flow management.

2. The boreal region

The boreal region is bounded by the treeline in the north. The southern boundary may be the temperate forests or grasslands. Within the region are flat-lying to undulating plains and plateaus that may be occupied by extensive wetlands, rolling terrain underlain by Precambrian Shields, and rugged mountain ranges that often rise high above the alpine treeline (Fig. 1). Lakes and wetlands are common on the plains and on the Shield (National Wetland Working Group, 1988; Zhulidov et al., 1997). One unifying factor is the cold temperate to subarctic climate, with long (five or six months) cold winters and warm summers accompanied by long daylight hours. The region experiences recent warming (ACIA, 2005) and large inter-annual variations in its climate (Kistler et al., 2001), and this can have effects on streamflow (e.g. Burn and Hag Elnur, 2002; Dankers and Christensen, 2005; Déry and Wood, 2005). Boreal forest is the dominant vegetation cover and the principal tree species include spruce, larch, pine, birch and poplar, in various proportions depending on location.

Streamflow regime is the average seasonal rhythm of discharge. The most common seasonal pattern of streamflow is the nival regime, so termed by Church (1974) to signify the dominance of snowmelt in producing runoff. An example is provided in Fig. 1 which shows the long-term daily mean and the standard deviation of daily flow for Upper Liard River (60.05°N 128.90°W, area 33,400 km^2) in Canada. The annual peak flow occurs in the snowmelt season. After the snow is depleted, the spring high flow recedes rapidly. Summer low flow is occasionally interrupted by hydrograph rises due to rainfall or summer snowfall, but the summer peaks are generally lower than those in the spring (Kane et al., 1992). There are modifications to the nival regime due to the presence of glaciers, lakes or extensive wetlands in the basins (Woo, 2000) but the common consideration is the dominance of snowmelt contribution in terms of total river flow.

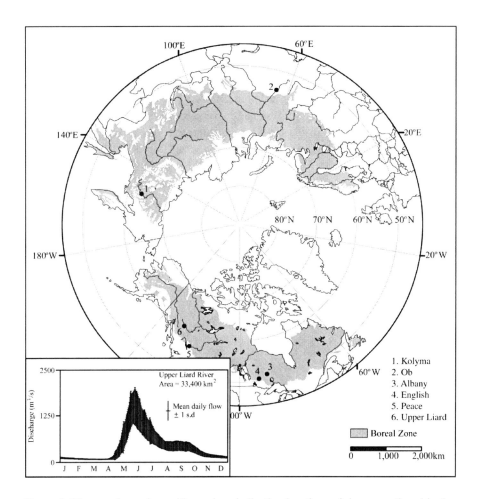

Figure 1. Circumpolar region, with numbers indicating locations of rivers mentioned in the text. Inset provides an example of nival regime discharge (Upper Liard River, Canada) that features high flow during the snowmelt period and low flow in winter

3. Flow modifications through water use

Many boreal rivers are subject to several forms of flow regulation for hydro-power generation and for water supply. Three types of flow regulation are recognized: alteration of the timing of flow, withdrawal of water for consumptive use and flow diversion from one drainage system to another.

3.1. REGULATED FLOW

Hydro-power generation alone does not diminish the water flow (unless it involves large reservoirs that lead to increased evaporation loss) but the timing of discharge can be significantly altered to meet electricity demands.

One much studied example is the Peace River in western Canada (56.02°N, 122.21°W) where a reservoir was built for the generation of hydropower. Since the reservoir went into operation after 1967, there has been a substantial increase in winter flows accompanied by a reduction of spring peaks, rendering the annual discharge to be far more uniform than the natural nival regime (Fig. 2; Woo and Thorne, 2003). Compared with the natural flows, Peters and Prowse (2001) found that even at 1,100 km downstream of the reservoir, average winter flows became 250% higher, annual peaks (for the 1-day, 15-day and 30-day highs) were 35–39% lower, and overall variability in daily flows decreased, though inflows from tributaries below the dam add back some of the natural variations of the nival rhythm.

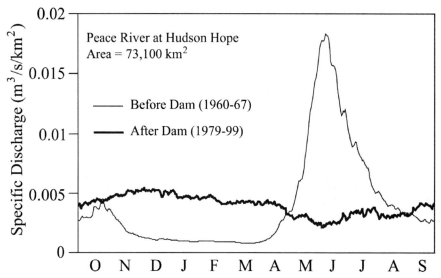

Figure 2. Example of flow regulation for hydro-power generation, Peace River in British Columbia, Canada, showing the mean daily flow before and after the construction of the Bennett Dam. The after-dam line shows the elimination of spring freshet but augmentation of winter flow due to reservoir operation (After Woo and Thorne, 2003)

A change in the flow regime in the boreal region can greatly impact the formation and breakup of river ice which causes blockages of hydro-power facilities and intensifies floods that can lead to serious economic losses (Prowse and Beltaos, 2002). Beltaos et al. (2006) noted that as a result of flow regulation of the river, the freeze-up stage has increased while together with changing climatic patterns, the spring flow has diminished. Both these tendencies inhibit the occurrence of mechanical breakup (in contrast to thermal breakup which is less intense), contributing to less frequent occurrence of ice-jam flooding. This is not without negative consequences, however, because wetland ecosystems depend on the annual floods for storage replenishment and floods can transport sediments effectively for the maintenance of river deltas. A weakening of river and lake ice will also restrict winter traffic to remote locations as winter roads constructed over-ice cannot be reliably maintained.

3.2. CONSUMPTIVE WATER USE

Reservoirs have been built to store water for consumptive use by agriculture, industries and municipalities. Many reservoirs on boreal rivers in Siberia serve this water use function. River flow undergoes losses during period of high water demand but storage often allows a steadier water release during the period when the natural flow is low. Below a reservoir built along the main valley of the Kolyma River in Siberia, the Ust'-Srednekan station (area of 99,400 km^2; located about 1,500 km below the dam) experienced a large winter (December to April) flow increase of about 200 m^3 s^{-1} and a reduction in June peak flow by 1,330 m^3 s^{-1}, after the reservoir was filled between 1986 and 1990. The monthly hydrograph changed significantly: summer flows became lower but winter flows increased, thus moderating the seasonal differences (Fig. 3a).

The upper Ob River in Siberia also loses flow downstream as water is withdrawn for consumptive use. Figure 3b shows the mean monthly flows at three stations: Fominskoye (52.45°N, 84.92°E, area of 98,200 km^2), Barnaul (53.4°N, 83.92°E; area of 169,000 km^2) and Novosibirskaya (54.8°N, 82.95°E; area of 232,000 km^2). Comparing the discharges of these stations between the period 1960–1979 and 1980–1999, there was a progressive increase in the loss of flow downstream during the growing season. This suggests a tendency that consumptive water use was on the rise.

3.3. FLOW DIVERSION

Rivers have been diverted for the purposes of hydro-power generation, water supply for irrigation and industries, and flood control. The flow regimes of both the donor and the recipient rivers are affected. As an example, Fig. 4 shows the flows of English River and Albany River which drain into James Bay in northern Ontario (Woo and Waylen, 1988). English River flows westward to Lake Winnipeg and had natural flow up to 1928. Then it was regulated for hydro-power generation and to control the water level of the Lake of the Woods downstream. Since 1958, it received water diverted from the Albany drainage and thereafter, there was a substantial flow increase, as demonstrated by the discharge at Ear Falls (50.38°N, 93.12°W, with a natural flow area of 26,400 km^2). The coefficient of variation of annual flow was kept at around 25% so as to maintain a steady level of water supply from year to year. The range of annual floods was also reduced, with a long

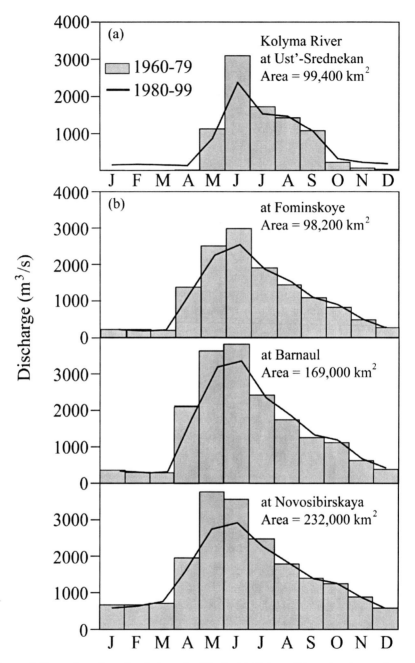

Figure 3. Examples of regulated rivers in Siberia (**a**) Kolyma River showing an increase in winter flow but reduction in spring discharge after reservoir construction in 1986; and (**b**) Upper Ob River showing increasing water loss to agricultural and industrial use as it flows downstream

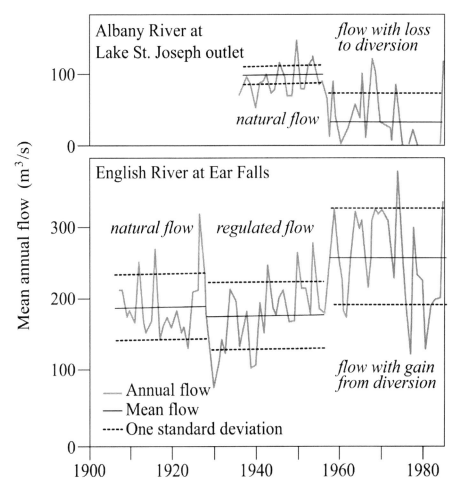

Figure 4. Example of flow diversion in Northern Ontario, Canada, with English River gaining flow from Albany River after 1957 (After Woo and Waylen, 1988)

term coefficient of variation of about 30%. On the other hand, the exporting Albany River experienced a flow reduction because an average of 75 m³ s⁻¹ was diverted to the English River, and the variability of flow increased at the outlet of Lake St. Joseph that flows toward the lower Albany River. Low flows are notably affected, sometimes with zero flow during periods of water export.

Although flow regulation is less intense compared with the temperate latitudes, there are many cases of streamflow regulation in the circumpolar boreal region. In general, water withdrawal for consumptive use is more advanced in Eurasia than in North America. In Canada, flow regulation mainly serves the purpose of hydro-power production. A notable example is the Baie James project which involves an extensive water diversion scheme in boreal Quebec.

A general consequence of flow diversion schemes is that the rivers that export water undergo a loss of flow. When diversion is in operation, the peak flow for the exporting river becomes more variable from year to year because very high floods may still be permitted to pass through the original channel to avoid flooding of the importing basin, but the lower floods may be released to the importing stream to supplement its water needs. In terms of annual low flows, the exporting basins are often responsible for maintaining a guaranteed minimum discharge to satisfy the requirement of the recipient basin. Thus, compared with the natural regime, the donor river experiences an increased probability of zero or very low flows. Unlike individual floods or drought events, flow diversion leads to long term changes in the hydrological environment that has lasting effects on river ecology and the economy of water use.

4. Discussion

At present, boreal rivers are not as heavily utilized as those in the temperate latitudes. Most of the water use is for hydropower production and to meet agricultural, industrial or community demands in limited areas. The natural flow regimes will be altered through climate change and increasing pressure on the water resources due to economic development (Yang et al., 2002, 2004; Ye et al., 2003). With climate warming, the ice-jam breakup floods are expected to be less severe and the snowmelt freshets will arrive earlier (Woo et al., 2007). Total annual flow may or may not change because enhanced evaporation may be compensated by higher precipitation, depending on location. The change in streamflow regime due to natural causes will be considerably less significant compared with possible human interference of the drainage systems. Within the boreal region, economic development will not only lead to higher consumptive use of water, but the pristine state of the water will unlikely to be preserved as agriculture, industries and population centres will discharge more wastes into the drainage system.

With moderate precipitation, low evaporation loss and meager consumptive exploitation, the boreal region is richly endowed with water at the present. However, there has been and will continue to have pressure for diverting water from the boreal rivers southward to satisfy the thirsts of the agricultural, industrial and urban needs in the temperate areas. These activities can deprive the boreal ecosystems of their needed water and can affect the discharge into the northern oceans. Judicious development of water resources in the boreal region is therefore critical to the well being of the people and the environment of the North.

References

Aagaard, K. and Carmack, E.C., 1989: The role of sea ice and other freshwater in the Arctic circulation. *Journal of Geophysical Research,* **94(C10)**: 14 485–14 498.

ACIA, 2005: *Arctic Climate Impact Assessment.* Cambridge University Press, Cambridge, 1042 pp.

Beltaos, S., Prowse, T.D. and Carter, T., 2006: Ice regime of the lower Peace River and ice-jam flooding of the Peace-Athabasca Delta. *Hydrological Processes,* **20**: 4009–4029.

Burn, D.H. and Hag Elnur, M.A., 2002: Detection of hydrologic trends and variability. *Journal of Hydrology,* **255**: 107–122.

Church, M., 1974: Hydrology and permafrost with reference to northern North America. *Proceedings, Workshop Seminar on Permafrost Hydrology.* Canadian National Committee, IHD, Ottawa, 7–20.

Dankers, R. and Christensen, O.B., 2005: Climate change impact on snow coverage, evaporation and river discharge in the sub-arctic Tana Basin, Northern Fennoscandia. *Climatic Change,* **69**: 367–392.

Déry, S.J. and Wood, E.F., 2005: Decreasing river discharge in northern Canada. *Geophysical Research Letters,* **32**: L10401. (DOI: 10.1029/2005GL022845).

Kane, D.L., Hinzman, L.D., Woo, M.K. and Everett, K.R., 1992: Arctic hydrology and climate change. In: Chapin, F.S., Jefferies, R.L., Reynolds, J.F., Shavers, G.R. and Svoboda, J. (eds.) *Arctic Ecosystems in a Changing Climate: An Ecophysical Perspective.* Academic, San Diego, CA, 35–57.

Kistler, R., Kalnay, E., Collins, W., Saha, S., White, G., Woollen, J., Chelliah, M., Ebisuzaki, W., Kanamitsu, M., Kousky, V., Van Den Dool, H., Jenne, R. and Fiorino, M., 2001: The NCEP–NCAR 50-year reanalysis: monthly means CD-ROM documentation. *Bulletin American Meteorological Society,* **82**: 247–267.

Manak, D. and Mysak, L.A., 1989: On the relationship between arctic sea ice anomalies and fluctuations in northern Canadian temperature and river discharge. *Atmosphere-Ocean,* **27**: 682–691.

National Wetlands Working Group, Canada Committee on Ecological Land Classification 1988: *Wetlands of Canada.* Ecological Land Classification Series No. 24, Environment Canada, Ottawa.

Peters, D.L. and Prowse, T.D., 2001: Regulation effects on the lower Peace River, Canada. *Hydrological Processes,* **15**: 3181–3194.

Prowse, T.D. and Beltaos, S., 2002: Climatic control of river-ice hydrology: a review. *Hydrological Processes,* **16**: 805–822.

Woo, M.K., 2000: Permafrost and hydrology. In: Nuttall, M. and Callaghan, T.V. (eds.), *The Arctic: Environment, People, Policy.* Harwood Academic Publishers, Amsterdam, The Netherlands, 57–96.

Woo, M.K. and Waylen, P.R., 1988: Effects of interbasin transfers on the regimes of three northwestern Ontario rivers. *Canadian Geographer,* **32**: 319–327.

Woo, M.K. and Thorne, R., 2003: Streamflow in the Mackenzie Basin, Canada. *Arctic,* **56**: 328–340.

Woo, M.K., Thorne, R., Szeto, K.K. and Yang, D., 2008: Streamflow hydrology in the boreal region under the influences of climate and human interference. *Philosophical Transactions of the Royal Society B,* **363**: 2251–2260. DOI: 10.1098/rstb.2007.2197.

Yang, D., Kane, D.L., Hinzman, L.D., Zhang, X., Zhang, T. and Ye, H., 2002: Siberian Lena river hydrologic regime and recent change. *Journal of Geophysical Research,* **107**: 4694. (DOI: 10.1029/2002JD002542).

Yang, D., Ye, B. and Shiklomanov, A., 2004: Discharge characteristics and changes over the Ob River watershed in Siberia. *Journal of Hydrometeorology,* **5**: 595–610.

Ye, B., Yang, D. and Kane, D.L., 2003: Changes in Lena River streamflow hydrology: human impacts versus natural variations. *Water Resource Research,* **39(7)**: 1200 (DOI: 10.1029/2003WR001991).

Zhulidov, A.V., Headley, J.V., Robarts, R.D., Nikanorov, A.M. and Ischenko, A.A., 1997: *Atlas of Russian Wetlands: Biogeography and Metal Concentrations.* Environment Canada, Saskatoon.

HYDROGEOLOGICAL FACTORS AND THE SAFE OPERATION OF HYDRAULIC STRUCTURES

R. Minasyan[*] and S. Zeizafoun
Department of Geophysics, Yerevan State University, Yerevan, Republic of Armenia

Abstract. Numerous cases of accidents involving hydraulic structures are known from many countries. These accidents can cost lives and cause considerable economic damage. The breakdown of these structures can be a sequential process, whereby one may identify a number of phenomena and processes in causally linked relationships. The origin of such accidents is often explained by the complicated geological–tectonic and engineering–geological structure of areas where they are constructed, and, in particular, by the influence of the water. The influence of the latter is especially evident on earth dams. The classification of these accidents shows that the hydrogeological factor conditioned by percolation processes was the main reason for many of the dam failures.

Keywords: Dam failure, technogeneous earthquakes, natural risks, seepage, reservoirs, earth dams

1. Introduction

A dam may cause a severe hazard should failure occur, but may present a small risk if there is no hazard to life or property within the inundation zone of the dam.

The risk factors that can cause dam failure are translated into high risks when people or property are threatened. These risk factors can be classified and grouped into one of four categories: natural factors, structure factors, human factors and operating factors. Natural risks such as earthquakes, natural floods, floods caused by dam failure, and landslides are the most crucial risk factors.

The consequences of a flood due to an earthquake which causes dam failure usually considerably exceed the consequences of an expected natural flood at maximum level. Therefore, residences and businesses that would escape natural flooding can be at more extreme risk from dam failure flooding.

---

[*] To whom correspondence should be addressed. e-mail: hydroscope@netsys.am

2. Structural factors

The condition of the dam and its foundation, the abutment conditions, seepage potential, dam material characteristics, the characteristics and conditions of spillway structures, water outlets, inlets and gates, cracking, settlement and slides are the more common signs of impending structural failure of a dam. All sorts of other human and operational factors should be included in risk analyses, including population and properties downstream and the seismicity of the territory (cp. Selim et al., 2002).

For different parts of a dam, the most characteristic problems can be identified as (Laffitee, 1996):

• *Crest problem*: ruts, along crest, excessive settlement, sinkhole, longitudinal cracks, transverse cracking.

The greatest weight is assigned to the following:

• *Upstream slope problem*: local settlement, inadequate erosion problem, too steep a slope, a rapid drawdawn of the reservoir.
• *Downstream slope problem*: seepage through the dam, settlement of the embankment, loss of embankment material strength from excessive seepage, internal erosion, seepage water existing from a water boil, standing water at the downstream toe.
• *Outlet problem*: inoperable low-level outlet, deterioration of concrete, conduit in poor condition, corrosion, inoperable gate.
• *Spillway problem*: eroded channel, blocked channel, inadequate erosion protection, too steep a gradient.

The process of filling and emptying the water behind the dam causes disturbance of the regime of the geodynamic field in the areas adjacent to it, decreases the stability of the rocks and creates the natural conditions for the well-known "technogeneous earthquakes". This phenomenon is especially dangerous in seismically active zones and requires careful monitoring, in particular, of the regime of groundwater and its temperature, as well as changes in the discharge of springs and geochemical characteristics, and the tectonic stresses in the vicinity of the reservoirs caused by excessive seepage losses from the dam. In earth dams, this system supports the growth of holes or tunnels generated by water pressure. These tunnels grow to the point where dam failure takes place.

(*Editor*: the authors are here referring to piping failure of earth dams, first studied by the Karl Terzaghi, the father of soil mechanics, which he likened to the "boiling" of soil made weightless by the pressure of seepage water emerging on the dam toeslope. Subsequent research has identified the importance of highly expansive or dispersive clay species used in dam construction, the water content and degree of compression applied during construction, and the ionic balance between the soil and the seeping reservoir water. Reference: Sherard, J.L. and Decker, R.S. (eds), 1977: *Dispersive clays, related piping, and erosion in geotechnical projects.* American Society for Testing Materials, Special Technical Pub. 623, 486 pp.)

3. Summary

As proposed by the International Commission on Large Dams (1982), permanent observations and the installation of measuring devices are necessary on dams, as well as remote geophysical, seismological and other methods in the general vicinity of the dams, in order to predict developments and guard against possible disasters. These measures are especially important in the case of dams constructed of earth materials.

References

International Commission on Large Dams, 1982: *Automation in monitoring the safety of dams.* Paris, France, ICOLD Bulletin, No. 41.
Selim, M.M., Imoto, M., Hurukawa, N., 2002: Statistical investigation of reservoir-induced seismicity in Aswan area, Egypt. *Earth Planet Space*, **54**: 349–356.

EVALUATION OF THE EFFECT OF A WATER HAMMER ON THE FAILURE OF A CORRODED WATER PIPELINE

G. Pluvinage[1*], C. Schmitt[1] and E. Hadj Taeib[2]

[1] *Laboratoire de Fiabilité Mécanique (LFM), Ecole Nationale d'Ingénieurs de Metz, Ile du Saulcy, 57045 Metz, France*
[2] *Applied Fluid Mechanics Dept., Ecole Nationale d'Ingénieurs de Sfax, BP W 3088 Sfax, Tunisia*

Abstract. A numerical model has been created to simulate the propagation of pressure waves in water networks. The model takes into account the structure of the water network and the pressure loss. When several valves are closed simultaneously, pressures can exceed the admissible pressure and cause failure. In this case, the severity of a defect such as a corrosion crater (pit) has been estimated by computing a safety factor for the stress distribution at the defect tip. To investigate the defect geometry effects, semi-spherical and semi-elliptical defects are assumed to exist up to one half thickness of the pipe wall. The results have been input into the Structural Integrity Assessment Procedure (SINTAP) Failure Diagram Assessment (FAD) to obtain the safety factor value. Conventionally, it is considered that the failure hazard exists if this safety factor is less than two.

Keywords: Water network, water hammer, pipe, failure, corrosion, safety factor

1. Introduction

Most water distribution systems in the world are older than 30 years. The outdatedness of the installations and the poor quality of the pipe materials, which are often corroded or damaged, are responsible for failures. Consequently, a large amount of water is lost by leaks, often more than 30%. Leakage occurs in many components of the distribution system: pipes, service connections, valves or joints. Leaks and failures are a major economic problem. When internal pressure decreases, leaks can also serve as pathways for contamination by harmful organisms from outside the pipes and pathogen intrusions can have health consequences.

Hence, programs have been developed to evaluate the degradation state and the risk of failure of pipe systems and to plan their rehabilitation. Preventive and predictive maintenance strategy programs with systems for inspection and control are developed (Sægrov et al., 1999; Hunaidi et al., 2000; Rajani and Kleiner, 2002). Leakage and cracking of pipes results from the initiation and propagation of defects. Initial defects can be assumed to be:

---

* To whom correspondence should be addressed. e-mail: pluvina@univ-metz.fr

J.A.A. Jones et al. (eds.), *Threats to Global Water Security*,
© Springer Science + Business Media B.V. 2009

(a) corrosion pitting or (b) mechanical damage, e.g., scratches and gouges, occurring during pipeline construction or during in-service maintenance. Variations in internal pressure may be caused by soil-movement induced bending or repeated loading from road traffic. Failure occurs when the defect has reached its critical size.

2. Water grids and water pipes failures

2.1. POLITICAL, SOCIAL, HUMAN AND ECOLOGICAL ASPECTS

Since prehistory, mankind has used artificial pipes and channels to transport water to settlements and fields. In the Middle Ages wooden tubes were used. Some are still in service in some districts of Paris and London. Water distribution systems must be constantly expanded to satisfy the needs of the population. In many countries, population is more and more sensitive to the fact that water problem is also a sanitary problem. Some disease have appears by introduction of bacteria into drinkable water through the orifice of the leak.

2.2. ELEMENTS OF A WATER GRIDS AND WATER NETWORK DENSITY

Pipelines are complicated three-dimensional structures that include straight pipes, nozzles, pipe-bends and different welded joints. In Western Europe, the distribution of drinking water has been practically completed over the last 20 years and now the problem of renewing the grids is considered as an urgent matter in terms of money and time. Average networks loose about 30% by leaks or ruptures. In some Mediterranean countries, more than 80% is lost by leaks, breaks and illegal withdrawing. Studies made in North America indicate that each year 10.4 billion litres of drinking water never reaches the tap. The most important part is lost by leak. The annual cost of the lost resources is estimated to US$3.6 billion in the USA and CAN$625 million in Canada. By reducing leak and rupture of the water pipes which cause about two thirds of the losses considerable economy can be made. In these countries, it has been pointed out that water distribution companies or city water offices needs new methods and technologies in order to reduce rapidly, efficiently and economically the water losses.

2.3. WATER PIPE MATERIALS

The materials used for water piping are grey cast iron, ductile iron, PVC, cement-based materials and sometimes steel. Grey cast iron has been the main standard material for the past 150 years: the majority of distribution piping installed in North America and Western Europe, beginning in the late 1800s up until the late 1960s, was manufactured from this material.

Grey cast iron is among the most common material used in the water distribution pipes of developed countries: about half of the networks in North America and more than a third in France. Grey cast iron is a problematical material because it is relatively brittle and susceptible to corrosion. Because of this and their age, grey cast iron pipes have the highest number of failures per kilometre per year, and they must be replaced as soon as possible. Ductile cast iron pipes were introduced in the late 1960s in Western Europe and in the late 1970s pipe in North America to replace grey cast iron pipe. Ductile cast iron piping is stronger and more resistant to corrosion than grey cast iron. Its lifespan is estimated at 100 years. Steel pipes have been used since the 1930s in Western Europe. The older ones are joined with lead and have to be replacing as soon as possible. The lifespan is estimated at 100 years for recent pipes. Polyvinyl chloride (PVC) pipes have been used in Western Europe and Australia since the early 1960s and in North America since the late 1970s.

Cement-based materials include reinforced or pre-stressed concrete, cement-mortar linings and asbestos-cement. Two general components of cement-based materials include the aggregates and the binder. Asbestos cement water mains were installed in North America, Europe, and Australia from the late 1920s to the early 1980s to replace the failed cast iron pipes and still form a significant component of water distribution networks of many cities.

2.4. OPERATING CONDITIONS AND PIPE FAILURES

The operating conditions of the piping can be quite severe, with internal pressure, bending, and thermal and cyclic loading combined with the influence of internal and external corrosive environments. The potential synergy of such parameters can lead to an increase in the risk of damage and unexpected fracture of these structures. Corrosion reduces the thickness of metal pipes and thus their ability to resist failure. The primary stress is due to internal pressure which produces a uniform circumferential tension stress across the wall. Pipes can also be subjected to secondary loadings, due to the soil, frost, bending loading due to the vehicular loading, and thermal stress. Bending produces an axial compression and tension stress while soil pressure and the changes in temperature induce axial and circumferential stresses.

3. The water hammer

3.1. DEFINITION OF A WATER HAMMER

A water hammer is a pressure surge or wave caused by the kinetic energy of a fluid in motion when it is forced to stop or change direction suddenly. For example, if a valve is closed suddenly at the end of a pipeline, a water hammer wave propagates in the pipe. Water in a pipe has kinetic energy proportional to its mass times the square of the velocity of the water.

The water hammer phenomenon over-pressurises the pipe due to transient changes in flow rate (Samani and Khayatzadeh, 2002; Boulos et al., 2005). It occurs during starting and stopping of pumps or rapid valve closure. The pressure wave variations propagate along the pipes and induce stresses within the pipes and associated equipment. Figure 1 shows a typical pipeline failure arising from the combination of a water hammer and corrosion.

Figure 1. Typical pipeline failure

3.2. EXAMPLE OF WATER HAMMER ASSESSMENT

The present example assesses the water network connecting Lassifer, Guellala and Zarzis reservoir tanks on Djerba Island, Tunisia (Fig. 2).

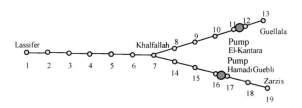

Figure 2. Schematic illustration of a "Y" type water pipeline network in Tunisia

The compressive wave of the water hammer provides instantaneous high pressure within the pipeline. The maximum stress can be readily calculated via thin-walled hollow cylinder assumptions: $\sigma_{\theta\theta} = PR/e$, wherein, P, R and e are internal pressure, cylinder radius and pipe wall thickness, respectively. In Table 1, the maximum stress for various closing and opening operations is summarized, including node number, pipe diameter, pipe thickness and available maximum pressure at corresponding nodes. It can be seen that the maximal stress is produced at node 12, downstream of the pump at El Kantara.

The maximum pressure value is assumed to occur with two simultaneous closing operations with the maximum stress resulting at node 11. Water hammer failure hazard is traditionally considered from the point of view of a safety factor, which in this example is greater than two. The semi-elliptical and semi-spherical defects including a defect depth of half the pipe thickness were applied to determine the potential failure (Table 2). The results show that semi-elliptical defects behave in a more brittle manner than semi-spherical ones, and their safety factor is much lower. Semi-elliptical defects are more critical than semi-spherical defects for a given defect depth, pipe geometry and loading.

TABLE 1. Maximal pressure and stress due to two simultaneous fast closures at nodes 13 and 19

Action	Cast iron pipe		Node	P_{max} (bar)	Maximum stress (MPa)
	Diameter (mm)	Thickness (mm)			
Closing of valves at node 12	800	11.7	2	6.10	21.25
	600	9.9	10	13.73	42.41
	500	9.0	11	14.40	40.77
	450	8.6	13	11.82	31.52
Closing of valves at node 17	800	11.7	2	4.02	14.00
	600	9.9	7	9.09	28.07
	500	9.0	11	7.33	20.75
	450	8.6	13	10.24	27.31
Superposition of closing of valves at 12 and 17	800	11.7	2	6.99	24.36
	600	9.9	10	16.74	51.70
	500	9.0	11	17.21	48.73
	450	8.6	13	16.88	45.01

TABLE 2. Safety factor for semi-elliptical (SE) and semi-spherical (SS) defects via SINTAP

Pipe type	Pressure	Safety factor	
		S-E	S-S
D = 450 mm, t = 8.6 mm	16.88 bar	2.99	4.72
D = 500 mm, t = 9.1 mm	17.21 bar	2.99	4.14
D = 600 mm, t = 9.9 mm	16.74 bar	2.58	3.56
D = 800 mm, t = 11.7 mm	6.99 bar	4.64	6.73

4. Conclusion

A recent UN report has pointed out that water distribution companies or authorities need new methods for reducing water loss rapidly, efficiently and economically. These technologies can generate new economic activities. System renewal can only be done at the scale of the whole network, but it is expensive and it is important to estimate the lifespan of the materials.

References

Boulos, P.F., Karney, B.W., Wood, D.J. and Lingireddy, S., 2005: Hydraulic transient guidelines for protecting water distribution systems. *American Water Works Association Journal*, **97**: 5.
Hunaidi, O., Chu, W., Wang, A. and et Guan, W., 2000: Detecting leaks in water-distribution plastic pipes, *Journal AWWA*, **92(2)**: 82–94.
Rajani, B. and Kleiner, Y., 2002: Toward pro-active planning of water supply systems. International Conference on Computer Rehabilitation of Water Network CARE-Germany.
Sægrov, S., Melo Baptista, J.F., Conroy, P., Herz, R.K., LeGauffre, P., Moss, G., Oddevald, J.E., Rajani, B. and Schiatti, M., 1999: Rehabilitation of water networks. *Urban Water*, **1**: 15–22.
Samani, H.M. and Khayatzadeh, A., 2002: Transient flow in pipe networks. *Journal of Hydraulic Research*, **40**: 637–644.

SEWER SYSTEM CONDITION, TYPE OF SEWERS AND THEIR IMPACTS ON ENVIRONMENTAL MANAGEMENT

Š. Stanko[*] and I. Mahríková

Department of Sanitary and Environmental Engineering, Slovak University of Technology Bratislava, Slovak Republic

Abstract. The new trends of draining dry weather flow in flat areas require use of the pressure sewerage system, which is more economical from the investment view, very easily applied in a relatively short time and can cope with a faster expansion of the urbanized areas. But here is a question: Is this system safe considering water sources, which are very often present especially in flat areas? Is it only a modern trend or a necessity? Which is better to build: a gravitation sewer system, a pressure or a vacuum sewer system? The costs influence choice. On the other hand, the gravitation sewer system is safe, especially for draining dry weather flow. What impacts we can expect on the local environment? The paper shows when and where it is better to use the pressure, vacuum and gravitation sewer system and illustrates the failures in existing sewer systems, and how we can plan to reconstruct the existing sewer system and to decrease negative impacts on the environment.

Keywords: Sewer, gravitation system, pressure system, vacuum system, recon-struction, rational formula, population

1. Introduction

For water experts two significant days are important, the World Water Day (22nd March) and the World Population Day (11th July). These Special Days show the importance of water at world scale. They remind us that more than 50% of the world population will live in cities from 2008.

According to this trend, Slovakia as the small country with no more than 49,035 km^2, population 5.4 million, density 109.7 inhabitants per square kilometer, elevation range from 94 to 2,654 m elevation above sea level, and the population concentrated in the cities in both flat and hilly areas and the villages surrounding the cities. The largest city holds 10% of country's population with 0.5 million people. The remaining municipalities contain 40% of the population.

---

[*] To whom correspondence should be addressed. e-mail: stefan.stanko@stuba.sk and ivana.mahrikova@stuba.sk

Because of the country's relief, Slovakia has a really serious problem
with water supply and achieving waste water drainage. The Operational
Program of the Environment for years 2007–2013 in Slovakia has designated
water supply and waste water drainage as priority number one. The program manual shows that the Operational Goal 1.2 is to build 1,211 km of
new sewer systems with EU financial support by 2015. This goal has some
restrictions. It must cover the predefined municipalities and solve the problems for communities with no more than 2,000 PE. The financial support
is limited. From this point how to build the sewer system is a key question.

In 2004, a new Water Act came into force in Slovakia. It is in line with
the requirements of Directive Nr. 91/271/EEC. The harmonisation of wastewater treatment in Slovakia with the requirements of this Directive will
require a substantial amount of funding for construction of new wastewater
systems and reconstruction of existing ones (Mahríková, 2007).

2. Which sewer system?

The answer to this question must consider the financial aspect. If we decide
to expand the gravity sewer system, the investment costs rise. If we decide
to build the pressure system, it decreases the investment costs. Or if we
decide to build the vacuum system, the investment depends on the amount
of investment in the vacuum station and other equipment related to this
system. We have to consider the following aspects:

- Investment and operational demands
- Environmental impacts
- The terrain possibilities
- Operational impacts concerning the costumers, the finance and practical
  view
- The revenue aspects for the operational and assets company

2.1. INVESTMENT AND OPERATIONAL DEMANDS

When we decide to build the sewer system, we need to prepare a cost-
benefit analysis. Depending on funding, we have to decide which is the most
effective sewer system we can build. For this reason, there is frequently a
conflict between financial and technical considerations. The financial considerations put constraints on the feasible size of the sewer system, but on
the other hand the technical aspects enforced by the engineers, who want to
build the perfect and long-lasting system, often take no consideration of the
financial aspects. The investors in Slovakia are very often the water companies, together with some city and village authorities, and they try to reduce
costs, because their own capital sources are exhausted and the financial

possibilities are only bank loans, which constrain the investors' decisions. The banks need some guarantee, but the investors have only exhausted credit. The State is committed to the European Union to build a promised number of new sewer systems and waste water treatment plants by 2015. The solution is to build sewers as cheaply as possible, which has an impact on the quality.

2.2. ENVIRONMENTAL IMPACTS

These solutions have environmental impacts. A wrong decision on the building technology has a negative impact on underground waters: for example, building an inadequate pressure sewer system in good slope conditions, or building a gravity system with no guarantee that it is watertight in areas that also serve as sources of water supply. The sewer systems are very often built by the local people, who have none of the requisite experience, and they can build a leaky waste water system. From this reason, the engineers must to decide to use an adequate sewer system, which is foolproof for the inexpert workers. Very often the problem is due to separate seals in sewer pipelines. A better solution is to use sewer pipes with built-in seals.

The next problem is using the same excavation for the water supply pipe and waste water pipe. There is some danger that the failure of pressurised water pipes will cause water to drain from the leaking sewer pipe into the groundwater. These can impact on groundwater pollution, the impact on the wastewater treatment process worsening and a danger of underground cavern creation (*Editor*: referring to the development of soil piping, cf. comments on Minasyan and Zeizafoun, this volume).

2.3. THE TERRAIN POSSIBILITIES

The terrain either allows or prohibits the construction of some types of sewer systems. Slovakia's terrain is very varied country. The hill country has large boulders at shallow depth. Excavation is very expensive and the pipe material is very often endangered by the sharp stones. The better solution is using the pressure sewer system, but it needs to be set in bedding designed to protect it against freezing.

The research at the Department of Sanitary and Environmental Engineering of Slovak University of Technology shows that in the mountainous landscape we very often witness many failures of the sewer system caused by incorrect bedding of the sewer system.

The flat landscape country offers all three possibilities for sewer system building. The gravity sewer system is suitable in areas with a suitable soil. It requires clay soil, not sandy soil. Sandy soil is absolutely inappropriate for building gravity sewer systems, because making excavations in the sandy is very dangerous at depths of 1 m or more. Excavation in "liquid sands" is very expensive and technologically very complicated. There is a great danger of burying the workers.

From these reasons, we suggest using the vacuum or pressure sewer systems for waste water drainage, especially for dry-weather sewer systems (STN, 2003a). Storm water drainage is suitable only in open trenches, with a suitable surface adjustment (STN, 2003c).

2.4. OPERATIONAL IMPACTS CONCERNING THE COSTUMERS, THE FINANCE AND PRACTICAL VIEWPOINT

The operational conditions for the costumer very often decide the investor's choice. Systems with a sophisticated operation require a "wise" costumer, who has some professional education in waste water drainage with regular maintenance. The vacuum and pressure system require this type of general costumer awareness. The pressure systems contain special equipment in a manhole in the house – the pressure pumping station and valves, which need maintenance from time to time. The vacuum system contains valves in the home, which operate automatically, like a pressure pumping station, but it needs to be inspected of the costumer. The gravity system does not assume any special knowledge about the sewer system operation. The finance aspect is important too. The operation of a pressure sewer system, which has a pumping station in costumer's house that usually belongs to the costumer, requires higher operational energy costs and maintenance. Not every costumer understands the need for regular maintenance. The pressure pumping station's lifetime is also limited.

2.5. THE REVENUE ASPECTS FOR THE OPERATIONAL AND ASSETS COMPANY

The right choice of sewer system has a big influence of revenues. Not every system is operationally effective. The factors described above are the main influences on the revenues of water companies, which need to generate profits from the investment as well as building an operationally sound system. These factors affect the "water price", comprising both water supply and wastewater drainage prices. Slovakia's water prices are regulated by the regulator, which is the Regulatory Office for Network Industries. The system of water price regulation is based on analysis of statistical data, which is not very suitable for water price determination. The water companies' opinion is that the water price is too low. The present water price is not enough for revenue generation, and for new sewer system building, reconstruction, rehabilitation concerning not only the sewer system, but the wastewater treatment plant too.

Sewer system investments in Slovakia are very often decided by the financial costs. Together, the financial costs and the legal Directives determine the amount and type of new sewer systems being built.

3. Reconstruction and rehabilitation

Slovakia's sewer systems are old systems and a high percentage of them need reconstruction (Stanko and Mahríková, 2006). The lifetime of sewer pipes is usually no more than 50 years (STN, 2003b). The present state in the big cities is very inadequate. Table 1 gives an indication of the amount of reconstruction that is necessary.

TABLE 1. Present state of the sewer system pipe composition by material and diameter

		Percentage of all pipes
Pipes needing rehabilitation/reconstruction		9.1
Composition and size of pipes		
Material	Concrete. reinforced concrete	95.5%
	Stoneware	3.7%
	AZBC. PVC	0.8%
Diameter	<=DN 700	84.4%
	DN 800 – 1,400 (600/800 – 1,400)	13.2%
	>DN 1,400	2.4%

TABLE 2. Reasons for inadequate sewer conditions in Slovakia

Lifetime overrun	46.1%
High water leakage in lifetime	39.9%
Insufficient capacity	10.4%
Disturbed by static overload	3.6%

Table 2 shows the necessity of sewer system reconstruction, which amounts to 473 km. Sewers that have exceeded their design lifetime are responsible for the highest percentage of the sewers that need reconstruction, because many of the sewers in the big cities are more than 50 years old. Only 10.4% of total amount of reconstruction is caused by sewers that have insufficient capacity. From these statistics, we can say that the population increase over the last 50 years has no significant influence on the need for sewer reconstruction, and we can conclude that the old calculation methods based on the rational formula gave a good estimate for the design capacity of the sewers for a long time forward (Stanko et al., 2006).

4. Conclusion

Sewer system condition, the type of sewers and their impacts on environmental management depend on various factors. Only a knowledge of local conditions, the economic level of the country, living standard and the water companies, covering both water supply and waste water operations, can determine what is the most suitable sewer system for a specific region. The greatest influence on sewer system building is the financial aspect. The technical aspects do not have a significant role in every case. The Slovakian experience concerns not only new sewer system building, but also the reconstruction and rehabilitation of sewers system. The solution very often depends on engineers, but the complex knowledge of the designers can have a positive influence on the right choice.

References

Mahríková, I., 2007: Disposal of waste waters from small urban areas. In: *Proceeding of NATO Advanced Research Workshop Dangerous Pollutants (Xenobiotics) in Urban Water Cycle*, Lednice, Czech Republic, 2.-6.5.2007.

Stanko, Š., Mahríková, I. and Gibala, T., 2006: Development of a complex system for pipeline design in Slovakia, In: *Proceeding of NATO Advanced Research Workshop, Security of Water Supply Systems: from Source to Tap*, Murter, Croatia, ISBN 1-4020-4563-8. Dordrecht: SPRVE, 137-147.

Stanko, Š. and Mahríková, I., 2006: Implementation of fiber optic data cables in sewage system. In: *Proceeding of NATO Advanced Research Workshop, Integrated urban water resources management*, ISBN 1-4020-4684-7. Dordrecht: SPRVE, 171–180.

Stanko, S., 2007: Reconstruction and rehabilitation of sewer systems. In: *Proceeding of NATO Advanced Research Workshop Dangerous Pollutants (Xenobiotics) in Urban Water Cycle*, Lednice, Czech Republic, 2–6 May, 2007.

STN EN 13380 (2003a) *General requirements for components used for renovation and repair of drain and sewer systems outside buildings.*

STN EN 13566-(1-4) (2003b) *Plastics piping systems for renovation of underground non-pressure drainage and sewerage networks.*

STN EN 13689 (2003c) *Guidance on the classification and design of plastics piping systems used for renovation.*

Tóthová, K. and Mahríková, I., 2006: Security of water supply and sewerage systems in Slovakia - present state. In: *Proceeding of NATO Advanced Research Workshop, Security of Water Supply Systems: from Source to Tap.* Murter, Croatia. ISBN 1-4020-4563-8. Dordrecht: SPRVE, 155–167.

PART VI:

RESTORING THE WATER RESOURCES OF THE ARAL SEA BASIN

.

THE ARAL SEA: A MATTER OF MUTUAL TRUST

Y. Kamalov[*]

NGO Union for Defense of the Aral Sea and Amudarya. 41 Prospekt Berdakh, Nukus, Uzbekistan

Abstract. The matter of saving and restoring the Aral Sea is not currently on the agenda of any conference, project or scientific research supported by governmental bodies. There is a sort of silent agreement on the impossibility of saving the Sea. Meanwhile, there have been no attempts to save the Sea, not even one single approach has been taken. There is a demand for new arguments for saving and restoring the Sea.

Keywords: Aral Sea, water, upstream, downstream, rights, economics

The water crisis in the Aral Sea catchment area is tending to get worse due to many reasons. Low water efficiency, primitive approaches to irrigation, slow reforms in economics and lack of public awareness are among them. The main reason is reluctance or maybe inability to switch economics from agricultural to industrial. Obviously, under present circumstances the natural ecosystems are the last to receive water, after irrigation, industry, and water supply for human needs.

It could be asserted that there are no ecosystems in the Aral Sea basin that are not affected by the human-made water deficit, except some mountain ecosystems. International conventions intended to save nature have been approved, but this does not solve the problem. Economics is dominating. Moreover, there is a lack of effective legislative and economical tools even for equitable and agreed water delivery among business enterprises along the rivers.

The mountain countries are trying to implement a rule by which downstream states should pay for access to water coming from upstream. They recognize water as a mineral resource, like petrol, coal, etc. This contradicts international practice, but could be justified from other points of view. The downstream states should not pay for water itself, but for a guarantee of a secure and timely water flow. This type of payment will oblige upstream states to provide all possible measures to annually renew water resources, including, amongst other things, the monitoring of glaciers, cleaning up of river beds, saving of forests, and preventing floods and droughts. It is very important to keep hydropower reservoir dams safe from collapsing.

---

[*] To whom correspondence should be addressed. e-mail: yusup.kamalov@yahoo.com

If upstream countries do not meet their obligations, they should have to compensate for the losses endured by downstream countries. This rule would oblige upstream states to allow representatives from downstream countries to check the adequacy of those measures. Unfortunately, today upstream economic interests do not recognize their obligations to downstream stakeholders.

What may lead us to positive changes? First of all we need examples of good practices of economic development and saving ecosystems, i.e., an integrated water resources management practice. Such case studies are eminently possible from areas where rivers or tributaries are situated within one single country. Additionally, in those countries an emergency institution is able to react to natural disasters immediately along the river as a whole. Maintenance of this kind of body in the case of transboundary rivers has to be based on the upstream states' responsibility. It should push them to be more careful with threats of both natural and human-made disasters.

The downstream countries should pay for the above mentioned "guarantee", in as much as they are taking water for irrigation and industry, but not for direct human needs. It will stimulate them to save water in the rivers so that more water will flow towards the Aral Sea and to natural ecosystems. This interdependence could provide a basis for future integration of the economics of the countries sharing the rivers.

Another reason to save the Sea is dust storms on the former bottom of the Sea. The storms are bringing millions of tons of salts and dust up to the higher layers of the atmosphere, due to predominantly North winds and the two deserts, which have convective air currents above them. These aerosols clearly have a great impact on weather both at regional and global scales, being equal to the activity of several volcanoes. The bottom of the Sea is a brand new surface for winds to work on, compared with other sources of dust on the Earth. This matter should be further investigated in a deep and careful way.

The negative impact of this dust and salt on agricultural fields and glaciers in the mountains of the basin of Aral Sea is well known. The only single way for now of stopping this dangerous erosion from millions of hectares of the bottom is to cover it with water. Taking into consideration that water efficiency in the basin of the Sea is no more than 10%, it is possible to save water to cover at least half of the former seabed.

It is also well known that the productivity of soils is decreasing due to "secondary salinisation" that is a consequence of over-irrigation. We lost at least 50% of crop productivity compared with the time before World War II. The "secondary salinisation" is happening because underground waters raised due to irrigation are bringing salts to the surface of the soils. These salts are located everywhere at depths of several meters in sediments laid down on the bottom of a giant, ancient ocean that existed millions of years ago in the territory of contemporary Central Asia. To clean up the soils, more volumes of water are used that is raising the underground water table further. A vicious circle is locked in!

Implementing water saving technologies in agriculture will lead to increased productivity by avoiding salinisation. Moreover, investing in water saving means investing in economics and investing in security, because using less water in agriculture is demonstrating preparedness for future sustainability.

The next argument for saving the Sea comes from the moral dimension. As for many international cases, the Aral Sea issue is a matter of inter-relations between upstream and downstream people and states. Like in many other cases, the upstream states and people are not respecting the rights of downstream ones very much. In the case of the Aral Sea, the whole natural body is going to be destroyed without any agreement with the owners of it. It seems like people living around the Sea do not have rights to own the Sea, because upstream states should develop their economics by using up all the water in the rivers. Of course, nobody will deny formally those rights, but in reality nobody is taking care about the provision of the water to somebody downstream. Moreover, in all cases rivers are used as a hidden, free and convenient transport to send pollutants, waste and sewage waters far away from some counties, states and regions.

There is a belief among authorities at all levels of governance and among ordinary people that water in the river is the inalienable property of the country irrespective of to where the river is flowing. The same rights for downstream and upstream people are not included in any legislation of upstream countries. Is it worthy for the human race in the 21st century to have this kind of behaviour? Of course not, and we should prove it by collaboration to save the Sea. If we are not able to cooperate in saving a relatively small natural feature like the Sea, what can we expect in the process of saving the Earth?

The Aral Sea must be saved and restored!

EXTREME AND AVERAGE GLACIER RUNOFF IN THE AMUDARYA RIVER BASIN

V. Konovalov[*]

Institute of Geography, Russian Academy of Sciences, Moscow, Russia

Abstract. Glaciers located in the headwaters of the Amudarya basin cover nearly 78% from the total glaciated area in the Aral Sea basin. The long-term regime of seasonal glacier meltwaters during 1935–1993 was computed by the author's model REGMOD. For this computation, glaciers in the region were divided into 138 groups. Empirical integral function probability distributions of seasonal melting were estimated for each group and averaged for all groups for each year. Finally, high and low meltwater years were determined. The spatial distribution of melting was described in extreme and average years. It was revealed that the spatial variability of meltwater volume V_M is most variable in average years and not significant in minimal and maximal years. This is most important for stabilizing the balance between inflow and outflow in reservoirs supplying hydropower and agriculture. Regional dependencies of V_M in extreme years are closely related to the spatial distribution of the firn boundary averaged for glacier groups.

Keywords: Amudarya river basin, glaciers runoff, minimal and maximal values, spatial variability

1. Introduction

The main sources of runoff formation in the Amu River are melting of seasonal snow and long-term storage of ice and firn in the Pamir glaciers. At the beginning of the 1960s, the area of glaciers in the upper watersheds of the Panj and Vakhsh, which form the Amudarya was estimated as 11,600 km^2 or nearly 78% from the total glacier area in the Aral Sea basin. Combined analysis of long-term variability of the Panj and Vakhsh riverflow components during May–October showed that relative contributions of glacial runoff and total melting increase in low flow years and decrease in high flow ones. This peculiarity of glacial runoff is very important for water supply to agriculture and hydropower in the Central Asian states, because it provides natural regulation of the intra-seasonal distribution of runoff. However, the stabilizing role of glacier runoff in the Amudarya basin is becoming less effective due to shrinkage of glacier area by 2,324 km^2 over the period 1961–2000. This is very significant and undoubtedly affects the sustainable availability and utilization of river runoff in the Aral

---

[*] To whom correspondence should be addressed. e-mail: vladgeo@gmail.com

J.A.A. Jones et al. (eds.), *Threats to Global Water Security*,
© Springer Science + Business Media B.V. 2009

Sea basin, especially in low water seasons. Temporal variability in seasonal volumes of glacier melt were estimated previously by Konovalov (1979, 1985, 2006) for the Vakhsh and Panj river basins for the period 1935–1994. It was revealed that a characteristic feature of the temporal range in both basins is a decreasing linear trend of meltwater output. The present paper analyses the spatial variability of meltwater volumes V_M for May–October in low and high water years.

2. Computation of total glaciers melting

The Vakhsh and Panj river basins contain several thousands of glaciers (Inventory 1971–1978). This set is not enough for computing V_M volume, where V_M is a function of meteorological variables and empirical parameters, due to the very sparse meteorological network in the mountains. Thus, the glaciers were grouped into 138 quasi-homogeneous groups (Fig. 1) (cp. Konovalov, 1979, 1985, 2006). Each group is characterized by: geographical coordinates; areas of glaciers and solid moraine; altitudinal distribution of area; altitude of glacier head, end and firn boundary; upper limit of solid moraine cover; depth of solid moraine on the end of glacier; slope and azimuth of the glacier surface.

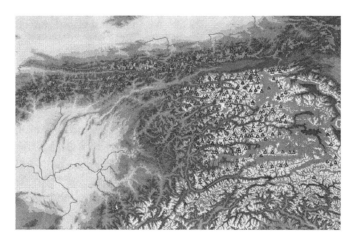

Figure 1. Location of glacier groups (blue triangles) within of Pamir-Alai and Hindu Kush mountain areas. Bold green lines are boundaries of main river basins

A set of computer programs and data on climate variables were developed for computing long-term series of glacier hydrological regimes using REGMOD. The formula used in REGMOD for calculating total volume of glacier melting v_m in the moment t, has the form:

$$v_m(t) = M_c(\tilde{z}_{im},t)S_{im} + M(\tilde{z}_i,t)S_i + M(\tilde{z}_f,t)S_f + M(\tilde{z}_{ws},t)S_{ws} + M(\tilde{z}_{ss},t)S_{ss}$$

$$(1)$$

Here M is intensity of melting for open ice or snow, $M_c = M \cdot f(h_c)$ is intensity of ice melting under solid moraine cover (*im*) of depth h_c, i is bare ice, f is old firn, *ws* is winter snow, *ss* is summer snow, $f(h_c)$ is a function of extinction of ice melting under moraine cover of depth h_c, $\tilde{z}$ is mean weighted altitude for the certain S area. To get total melt volumes V_M from Equation (1):

$$V_M = \sum_{d_{bp}}^{d_{ep}} v_m(t) \qquad (2)$$

where d_{bp} and d_{ep} are dates of the beginning and end of the calculation period. Computations of V_M according to Equations (1–2) were based on using several numerical methods, described in detail in Konovalov (1979, 1985, 2006). REGMOD model and its main subroutines have been successfully tested. The most recent was computating long-term glacier runoff for large glacierized river basins (Konovalov, 2007). Total river runoff computed by means of the water balance equation showed very close coincidence with measured data. The model was used for the selected 138 groups of glaciers between 1935 and 1993.

3. Spatial variability glacier melting

Computing empirical probabilities of $X > x_i$ is done by the following procedure (Alexeev, 1971):

$$p_i(x_i) = \frac{m(x_i) - 0.25}{N_i + 0.5} * 100, \qquad (3)$$

where $m(x_i) = 1, 2, ..., N_i$ are ordinal numbers of the x_i values after their disposition in the descending order. All temporal ranges of V_M volumes in the 138 groups of glaciers were processed according to formula (3) for the river basins in average and extreme years with statistical probability $P(V_M)$. Years when $7\% \geq P(V_M) \geq 93\%$ were considered extreme and when $45\% \leq P(V_M) \leq 55\%$ were considered average. Here $P(V_M)$ is the mean for 138 values in each year. Table 1 presents the results. During 1935–1993 we had 4 extreme and 10 average years.

TABLE 1. Characteristics of average and extreme years

Statistical parameters	Limits $P(V_M)$ %		
	Extreme years		Average years
	min	max	
Average	93.87–94.59	3.33–4.85	46.59–55.56
Minimum	75.21–86.97	1.26	11.34–31.51
Maximum	98.74	14.71–16.39	70.17–90.34
Standard deviation	2.67–4.84	2.63–3.76	6.51–19.25

Spatial variability $p_i(V_{Mi})$ (i = 1,...138) in average years is much more significant than in minimal and maximal years. This result is important for hydropower, because an even spatial distribution of V_M volumes in low water years has a stabilizing role for the balance of input and output of reservoirs. It was suggested that small or large spatial variability $p_i(V_{Mi})$ in extreme and average years could be influenced by the variability of air temperature and precipitation. However, a definite tendency was found only with greater spatial variability of air temperature in high water years.

Analysis of the spatial distribution of V_M volumes and $p_i(V_{Mi})$ was continued by presenting these variables as one-dimensional functions of geographical coordinates in low and high water years. This analysis was performed for 1941 and 1972, which were chosen as examples of low and high water cases. Figures 2–3 shows relationships with longitude and latitude for V_M and $p_i(V_{Mi})$. Trend lines in Figure 2a, b show clear dependencies of $p_i(V_{Mi})$ in 1941 and 1972. This type of regional variability in $p_i(V_{Mi})$ is due to the combined influence of climate and relief on glacier melting. It is known that V_M depends on the altitude of the firn boundary Z_{fg}. Therefore, this was analysed further. It was confirmed by the spatial relationships mean altitude of Z_{fg} in glacier groups from geographical coordinates (see Figure 2c, d). For preparation dependencies on Figure 2c, d, values of Z_{fg} in

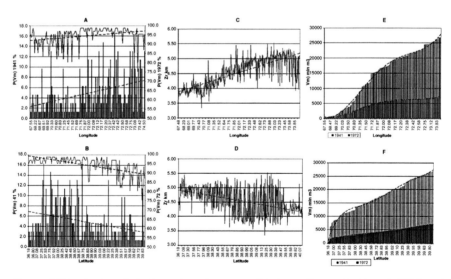

Figure 2. Distribution of glacier regimes by geographical coordinates. (**a, b**) probabilities of total melt volumes for May–October in high (1941) and low (1972) years; (**c, d**) – average f firn line in glacier groups; (**e, f**) – integral function of total melt volume distributions in high and low water years. Dotted line on graphs (**a–d**) shows linear trend related to geographical coordinates. Dotted line on graphs (**c, d**) shows adjusted empirical equation

408 groups of glaciers located within the Amu Darya headwaters were used. As suggested by Schetinnikov (1998), decreasing Z_{fg} along degrees of latitude is related to the prevailing (more than 76%) concentration glaciers in the north of the region ranging between 3.7 and 4.4 km above sea level, but in the south the largest proportion of glaciers (62%) are located between 4.0 and 5.0 km above sea level. Increasing Z_{fg} along degrees of longitude in the same part of the Pamir-Alai mountain region was revealed earlier by Kotlyakov et al. (1993). Most probably it should be considered as the combined result of changing relief and precipitation along the longitudes. Seasonal air temperature is not significant in this case, because this depends on latitude.

4. Conclusions

1. Regional determinations of glacier regimes are necessary for solving the problems of water consumption and forecasts of runoff.
2. Empirical functions of spatial distribution V_M volumes related to geographical coordinates have the following forms for high and low water years:

$$V_M \text{(high)} = 7E{-}06 \cdot lat^5 - 0.0028 lat^4 + 0.4261 lat^3 - 28.796 lat^2 + 976 lat \qquad (4)$$
$$V_M \text{(low)} = -0.0001 \cdot lat^4 + 0.037 lat^3 - 3.8553 lat^2 + 186.9 lat \qquad (5)$$
$$V_M \text{(high)} = -1E{-}06 long^5 + 0.0009 long^4 - 0.1825 long^3 + 13.899 long^2 -$$
$$115.36 long \qquad (6)$$
$$V_M \text{(low)} = -1E{-}06 long^5 + 0.0005 long^4 - 0.0801 long^3 + 5.1975 long^2 -$$
$$32.441 long \qquad (7)$$

These equations give estimates of total glacier melt in high water years of 26,888 and 7,108 km^3 in low water years.

3. Intra-seasonal distribution of total melting in glacier areas is closely connected with the type of water year. Glaciers feeding the Vakhsh and Panj rivers in maximal and average years concentrated in July–August when winter-spring accumulation of snow has been exhausted outside of glaciers area.

Acknowledgments This research was supported by the Russian Fund of Basic Research grant No. 05-05-64296.

References

Alexeev, G.A., 1971: *Objective methods of smoothing and normalization of correlation dependencies.* Leningrad: Hydrometeoizdat Publishing House, 362 pp. (In Russian).

Inventory of USSR glaciers, 1971–1978: Vol. 14, issue 3, parts 7–12. Leningrad, Hydrometeoizdat Publishing House (In Russian).

Konovalov, V.G., 2007: Long-term change of the water balance components in the river basin of snow and ice melted feedings. *Meteorology and Hydrology*, **8**: 77–89 (In Russian).

Konovalov, V.G., 2006: Regional model hydrological regime of glaciers (REGMOD). In: Kotlyakov V.M. (Ed.) *Glaciation of Northern and Central Eurasia in the current epoch*, **1**, Moscow: Nauka, 488 pp. (In Russian).

Konovalov, V.G., 1985: Melting and glacial runoff processes in the Central Asian river basins. Leningrad: Hydrometeoizdat Publishing House, 236 pp. (In Russian).

Konovalov, V.G., 1979: Computations and forecasts of melting and runoff of the Central Asian glaciers. Leningrad: Hydrometeoizdat Publishing House, 230 pp. (In Russian).

Kotlyakov, V.M. et al., 1993: Glaciers of Pamir-Alai. Moscow: Nauka, 256 pp.

Schetinnikov, A.S., 1998: Morphology and regime of the Pamir-Alai glaciers. Tashkent: SANIGMI Editorial Board. 219 pp. (In Russian).

FUTURE OF THE ARAL SEA AND THE ARAL SEA COAST

V.A. Dukhovny*, A.I. Tuchin, A.G. Sorokin, I. Ruziev and G.V. Stulina
Scientific Information Center of Interstate Commission for Water Coordination in Central Asia, Tashkent, Uzbekistan

Abstract. Heads of State of the Central Asian Republics recognized the impossibility of restoring the Aral Sea in its former volume and biologically activity in 1994. Since then, the governments of Kazakhstan and Uzbekistan have taken measures to stabilize the situation. This paper summarizes the work of the Interstate Commission for Water Coordination in Central Asia on possible ways forward.

Keywords: Aral Sea, desertification, system of polders, afforestation

1. Introduction

The Aral Sea case is world renowned as an example of human-induced catastrophe, as a result of which one of the largest inland seas on the globe has been transformed into a number of unproductive water bodies with highly concentrated waters. From 1960 to 2000, the Aral Sea lowering reached about 20 m in the Large Sea and about 14 m in the Small Sea, the coastline receded 130 km, the water volume decreased more than 70%, and the water surface area diminished from 68,000 km^2 to 23,000 km^2. With the exposure of the seabed, approximately 4 million hectares of environmentally hazardous zone has been created and the desertification process has intensified. A group of extremely unstable landscapes, comprising types of sand and solonchak occupies a huge 35% of the area. Salt transfer towards settlements and the movement of unfixed sand towards functioning roads and economic enterprises cause a direct risk. The zone of active salt and dust transfer from the deflation sources that threatens those enterprises extends to 50–70 km. Degradation of the resource potential in the Aral Sea has become irreversible, and its preservation and restoration is not possible in the nearest future.

The impossibility of restoring the Aral Sea in its biologically active form and original volume was recognized by the Central Asian states through "The Concept for Socio-Economic and Environmental Development in the Aral Sea and its Coastal Zone (Priaralie)", which was adopted by the Heads of State in January 11, 1994. The Governments of Kazakhstan and Uzbekistan took respective measures and decisions that largely contributed to stabilization of the situation in Priaralie and to the development of its

---

* To whom correspondence should be addressed. e-mail: dukh@icwc-aral.uz

J.A.A. Jones et al. (eds.), *Threats to Global Water Security*,
© Springer Science + Business Media B.V. 2009

living and environmental capacities. Meanwhile, the future of the Aral Sea itself remains problematic and is still on the regional agenda, but both countries use their own scenarios for problem solution.

2. Researching options for rehabilitation

The Interstate Commission for Water Coordination in Central Asia (ICWC) concentrated attention on the two directions of work on the Aral Sea and Aral Sea coast:

1. To forecast the future of Aral Sea water bodies
2. To survey and evaluate the situation of the drying bottom of the Aral Sea in order to clarify the landscape changes, development of wetlands and afforestation process

The first direction is to develop forecasts of salt and water balances in the Aral Sea for various scenarios of water inflow to the coastal zone (the Priaralie zone) from the Amudarya and Syrdarya.

It is clear that the volume and timetable of water that can be expected at the mouths of the deltas of the two inflowing rivers, the Amudarya and Syrdarya, depend on two principal factors: the natural flow which can be formed on the upper watershed and scenarios of water use in the zones of flow dissemination (the irrigated area). These volumes should be reduced to the quantity of water that is needed for the satisfaction of the delta's requirements. For the Syrdarya delta the content of infrastructure is clear – the North Sea was completed by construction works 2 years ago, together with six lake systems in the delta (INTAS, 2006). The delta of the Amudarya has three different options (SIC ICWC, 2003).

As a result, three options of water inflow were considered, and for each of the options three water use scenarios (national, business as usual, and optimistic) were analyzed for two hydrological series – minimal and maximal. In total, there were 18 particular cases grouped into three inflow infra-structure options.

2.1. MAINTENANCE OF THE EXISTING INFRASTRUCTURE IN THE AMUDARYA DELTA WITHOUT RADICAL RECONSTRUCTION

In this option, water level in the Eastern bowl of the Sea will keep lowering to a maximum of 28 m in the national water use scenario and stabilizes between 21 and 31 in the optimistic water use scenario. Water salinity will increase to 150 g/l in all the scenarios by 2008, and then its trend will depend on water use scenarios and water availability. In the worst case, salinity keeps stable at 200 g/l. As to the optimistic scenario, salinity would go down to 25 g/l in 2020–2025.

In the Western bowl, water level decreases to 17.5–18.9 m d.s.l at various rates until 2020.

2.2. INFRASTRUCTURE DEVELOPMENT IN THE AMUDARYA DELTA

Construction of infrastructure under this project will improve delta productivity and reduce socio-economic and environmental damage but will aggravate the state of affairs in the Large Sea: the Eastern Sea level lowers to 27.5 m, the Western Sea level to 27.6 m d.s.l in the worst cases, while salinity reaches 370 g/l in the Eastern Sea, though it decreases to between 90 and 106 g/l in the optimistic scenarios. A better situation will be observed in the Western bowl, where water salinity averages 65 to 75 g/l, and decreases to 53 g/l in the optimistic scenario by 2025.

2.3. HYPOTHETICAL OPTION

Water supply to the Western bowl is improved through a newly constructed system of waterways, the Amudarya–Sudochie–Adjibai Bay. This system is completely focused on a deeper reservoir. In the Western bowl, the water level is set between 29 and 31 m, with a short-term minimum of 28 and a maximum of 32.3. The Eastern bowl is also stabilized at 26 to 27 m d.s.l., but salinity in the Eastern bowl increases to 380 g/l.

Taking into account the problematic situation with saving the present level and volume of two Aral Sea bodies beside the North Sea, the GTZ and ICWC organized field and remote sensing surveys of the dry bed of the Aral Sea (Dukhovny et al., 2007). The goals of this investigation, which took place in 2005–2007, were to create a map of this new desert and to delimit the area of unstable landscape. In this work, the Scientific Information Center of Interstate Commission for Water Coordination in Central Asia (SIC ICWC) has characterized vegetation cover as natural or artificial against the background of on-going changes, given a geo-botanical description of the terrain, based on the characteristics of vegetation cover and soil conditions. In addition, the research allows a more precise estimation of the status of measures undertaken to reduce the negative consequences of the sea drying up. Greater attention is paid to phyto-reclamation of the newly exposed ground.

Over the last 15–20 years, about 225,000 ha of the exposed seabed have been covered by artificial afforestation activities. A similar area was defined as self-protecting by natural process. The area of more than half a million hectares to be protected includes 57,600 ha with the highest priority for protection and another 60,000 ha that could be a zone of very high risk. Moreover, within a 50 km band northward of this zone, 466,000 ha pose heavy environmental risk, of which 368,000 ha are located in an area to be stabilized.

3. Conclusions

Thus, the future of the Aral Sea and its coast requires the concerted actions of all the countries in Central Asia and a single plan to undertake those actions. Priority should be given to water conservation at all levels of the water hierarchy and end-users that will promote increases in the volumes of environmental flows. This will allow more water for use in the lowlands and permit complex protection measures to be organized by a combination of developing wetlands, concentration on water delivery to the Western Bowl and the development of afforestation of the ecologically unstable zones.

References

Dukhovny, V.A., Wilps, H., Navratil, P., Ruziev, I. and Stulina, G., 2007: *Comprehensive remote sensing and ground based studies of the dried Aral Sea bed.* Tashkent, GTZ, ICWC, 158pp.

INTAS, 2006: *The rehabilitation of the ecosystem and bioproductivity of the Aral Sea under conditions of water scarcity.* Vienna-Tashkent, INTAS Project – 0511 REBASOWS Report, 275pp.

SIC ICWC, 2003: South Priaralie – new perspectives. Tashkent, ICWC, 150pp.

NATURAL DISASTER: PREVENTION OF DRINKING WATER SCARCITY

R.A. Khaydarov[1], R.R. Khaydarov[1*] and S.Y. Cho[2]
[1] *Institute of Nuclear Physics, Ulugbek, 100214, Tashkent, Uzbekistan*
[2] *Yonsei University, Seoul, South Korea*

Abstract. The paper deals with a novel technical approach to alternative drinking water supplies for inhabitants of the Aral Sea Region. The water treatment technique based on using special ion-exchange fibrous sorbents with immobilized silver nanoparticles is described. The newly developed sorbents are currently used in water treatment filters for removing heavy metal ions and organic contaminants from drinking water in the Aral Sea Region.

Keywords: Water supply, Aral Sea, nanoparticles, ion-exchange sorbents, water treatment

1. Introduction

The Aral Sea disaster which threatens the life, health, and habitat of the population is well known all over the world due to the efforts of the region's countries as well as public and international organizations. Due to a shortage of drinking water, most of the population in the Aral Sea Region drinks and otherwise uses water that does not meet the international quality standards recommended by the World Health Organization (WHO) (UNDP, 2007).

2. Water treatment technique

The local problem of removing water hardness and organic contaminants can be solved by using special home and industrial water treatment systems based on developed fibrous sorbents with immobilized silver nanoparticles. Polyacrylonitrile (PAN) cloth with a surface density of 1.0 kg/m^2 and thickness of 10 mm was utilized as the raw material for making ion-exchange sorbents. Experiments by Khaydarov and Khaydarov (2006) have shown that the optimum condition for obtaining cation-exchange sorbents is achieved by treatment of PAN cloth with a 20% $NH_2NH_2 \cdot H_2O$ solution at 70°C for 30 min and with a 5% solution of NaOH at 25–30°C for 1 h. The exchange capacity (Cu^{2+}) of the obtained sorbents is 3.5–4.0 meq/g. The treatment of

---

* To whom correspondence should be addressed. e-mail: renat2@gmail.com

the cation-exchange sorbents by a 1% solution of polyethylenimine at 25°C for 1 h was selected as the optimal condition for the anion-exchange sorbents production. Capacity (Cr^{6+}) of the sorbents is 3.0 meq/g. The fibrous sorbents created have a high exchange capacity comparable with that of resin. An advantage of the fibrous ion-exchange sorbents over resin is the high rate of the sorption process (about 100 times greater) and small value of pressure drop of the sorbent layer for purified water. The cation- and anion-exchange filters based on the chemically-modified sorbents produced can be used for removing metal ions (Zn, Ni, Cu, Sb, Co, Cd, Cr, etc.) and organic compounds (M-^{32}P, M-^{131}I, M-^{99}Mo+^{99m}Tc, etc.) from water.

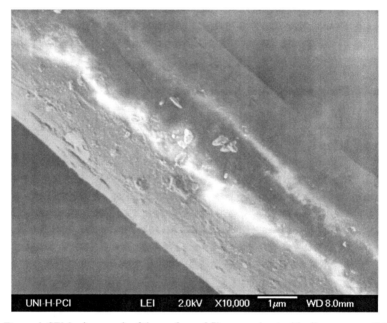

Figure 1. SEM micrograph of the surface of fibrous sorbent with silver nanoparticles

In order to prevent microbial growth and formation of a biofilm on the sorbents during their usage, we have impregnated silver nanoparticles into the surface of the fibrous ion-exchange sorbents. Silver nanoparticles have been obtained by an electrochemical technique based on electro-reduction of anodically solved silver ions in water (Khaydarov, R.R. et al., 2008). TEM and SEM measurements have shown that silver nanoparticles suspended in water solution were nearly spherical and their size distribution lies in the range of 2–20 nm, the average size being about 7 nm. We have impregnated fibrous sorbents with our nano-sized silver colloids by a method analogous to the technique used by H. J. Lee and co-workers (Lee and Jeongm, 2004) for treating textile fabrics. The immobilized silver particles had relatively good dispersibility, having sizes in the range of 50–300 nm

(Fig. 1). Most of the initial silver nanoparticles had agglomerated into clusters because of attractive interaction forces between them. The concentration of silver nanoparticles was optimized with respect to the best antibacterial and antifungal action on *Aspergillus niger, Penicillium phoeniceum, Staphylococcus aureus* and *Escherichia coli*. Microbiological studies with Petri dishes containing solidified MPA nutritional medium have clearly demonstrated that 20 mg/kg silver concentration provides reliable prevention of the growth of these microorganisms on the surface of the ion-exchange fibrous sorbents.

At the present time, newly developed sorbents are used in household drinking water filters and in treatment systems for removing heavy metal ions, radionuclides and organic contaminants. The productivity of the household filters is 20 l/h and that of the treatment systems up to 5 m^3/h, and the total pressure drop is very small (approximately 0.02 MPa). In addition, the purification and regeneration processes do not consume any additional energy. Thanks to the financial support of CNCP Grant (Project #265) (CNCP, 2007) 3,000 household water treatment filters and nine minipurification systems were produced in 2007 for distribution within the Aral Sea Region.

References

CNCP, 2007: NewsLetter. Klykov S. et al. (eds), **5**: 71–74.

Lee, H.J. and Jeongm, S.H., 2004: *Textile Research Journal*, **74**: 442–447.

Khaydarov, R.R., Khaydarov, R.A., Gapurova, O., Estrin, Y., and Scheper, T., 2008: J. Nanoparticle Res, doi: 10.1007/s11051-008-9513-x.

Khaydarov, R.A. and Khaydarov, R.R., 2006: Purification of drinking water from ^{134,137}Cs, ^{89,90}Sr, ^{60}Co and ^{129}I. In: *Medical Treatment of Intoxication and Decontamination of Chemical Agents in the Area of Terrorist Attack*, edited by C. Dishovsky Springer, The Netherlands, pp. 171–181.

UNDP Energy and Environment Unit, 2007: *Water—Critical Resource for Uzbekistan's Future.* New York and Geneva, United Nations.

ESTIMATION OF ECOLOGICAL RISK OF TRANSBOUNDARY POLLUTION OF THE AMU RIVER

L. Kondratjeva* and N. Fhisher
Institute of Water and Ecology Problems FEB RAS, Kim Yu Chen St. 65, 680000 Khabarovsk, Russian Federation

Abstract. Much of the population's health in Priamurje depends on the "health" of the Amu River. It is difficult to manage impacts on the Amu ecosystems, due to shortage of information on water pollution and lack of longterm studies of aquatic biology. However, the ecological situation demands urgent scientific attention.

Keywords: Amu River, ecological risk, transboundary pollution, bioindication, polytoxicosis of fish

1. Introduction

Ecological risk (ER) from water pollution may be considered from several viewpoints: scientific, engineering, economic and social. The social aspect is expressed in a form of possible consequences for the health of the population (Tal and Linkov, 2004; Kondratjeva, 2005).

For a long time, ecological risk in the Amu River was associated with phenol pollution (1995–2002). For many indigenous people, fish is the main food and the fish smell was called "phenol smell". Studies of natural substrates (water, bottom sediments and fish) and human samples (blood, urine, human milk, placenta) revealed various toxic substances. Benzene and its derivatives were found in fish and human milk and placenta (Ryabkova, 2002).

2. Polytoxicosis of fish

In order to select priority indicators of ecological status, seasonal surveys of the Amu water and fish quality were undertaken in 2002. Bio-indication was combined with physical and chemical methods (IR and UV spectroscopy, gas and liquid chromatography, atomic-absorption spectrometry). Transboundary pollution of the Amu River from adjacent areas of China, wastes from power sources, river transportation and forest fires were the most important factors.

---

* To whom correspondence should be addressed. e-mail: kondrlm@rambler.ru

In summer and winter, all fish quality indices meet health and hygiene standards, except for mercury. In winter, mercury concentrations in fish from the main river were twice as high as in tributaries. Table 1 lists major impacts on the fish. Maximum diversity of toxic components is usually found in fish caught in the Amu mainstream. Changes in the organoleptic properties of fish occurs in the freeze-up period under complex contamination of aquatic medium by chlorine-containing pesticides and heavy metals.

TABLE 1. Fish quality assessment in the Amu River during freeze-up

Parameter	Parameter values (max)	Fish
Smell intensive	Acrid	*Leuciscus waleckii, Lota lota, Parasilurus asotus, Hemibarbus maculates*
3-methyl-amine	2.4–6.2 mg/kg	*Leuciscus waleckii, Lota lota, Parasilurus asotus, Hemibarbus maculates*
Volatile azoth substances	290–410 mg/kg	*Abramis brama, Hemibarbus labeo*
Histamine	19.7–22.0 mg/kg	*Abramis brama, Hemibarbus labeo*
Heavy metal composition in fish muscles (mg/kg) Pb Hg Zn Cu	0.02–0.13 0.56–0.72 10.2–10.8 1.18–1.41	*Lota lota, Hemibarbus labeo, Hemibarbus maculates, Parasilurus asotus, Cyprinus carpio, Abramis brama*
Chlor-organic pesticides	0.0623–0.0751 mkg/kg	*Parasilurus asotus, Hemibarbus labeo*

Our data show that concentrations of hexachlorocyclohexane and DDT pesticides in fish caught in the Amu in winter are comparable to similar data from British rivers, before industrial effluents were limited in 1972. Agricultural complexes in China along the right bank of the Amu supply most of the pesticides.

Research in 2002–2005 helped formulate a polytoxicosis hypothesis (Kondratiyeva et al., 2003), and showed that priority should be given to: PAHs, PCBs, nitrosamines, benzene derivatives, organochlorine pesticides, methylated amines, cadmium, mercury and lead.

3. Transboundary industrial pollution

Benzene derivatives are considered risk factors for Amu fish. The tech-nogenic accident in Jiling province, China, in November 2005 proved this viewpoint. Since the accident ecological risks for the Amu hydrobionts have been linked to benzene and chloroform derivatives. Their concentrations significantly exceeded Russian norms for fish industry waters. At the time of the spill, nitrobenzene concentrations exceeded the norm 20 times. Chloroform content in the Sungari after the accident was 600 times higher than the permissible level. Our studies revealed that nitrobenzene pollution was coupled with a significant pollution with toxic chlorine-containing substances of different volatility levels (Table 2).

Four months after the spill (March 2006), no nitrobenzene remained in water or bottom sediments. But it was in fish. Other volatile benzene derivatives like toluene, xylene and ethylbenzene were also in fish. These are ER factors with a long-lasting impact.

TABLE 2. Toxic substances in various components of the Amu River ecosystem after industrial accident in the Sungari River Basin

Components	Substances
	Freeze-up period (2005–2006)
Water	Nitrobenzene, benzene, toluene, xylene, ethylbenzene, chloroform, dichloromethane, dichlorobenzene, PAH, heavy metals
Bottom sediments	Chloroform, tetrachloromethane, chlorobenzenes, PAH, heavy metals
Fish	Nitrobenzene, benzene, toluene, xylene, ethylbenzene, heavy metals, chlorine containing pesticides
	May, June 2006
Water	Phtalat, dibutilphtalat, chlorphenols, pesticides, acetochlorine, atrazine, PAH, heavy metals
Bottom sediments	Diethyl phthalate, dioctyl phthalate, methylphenol, benzo(b)fluorantene, perylene, benzopyrene, benzene, xylene, toluene, chlorphenols, methylbenzene
Fish	Benzothyozole, anisole, naphthalene, phthalate, xylene, toluene, ethylbenzene, heavy metals, phtor-phosphor containing pesticides

The results of the research project to assess the impact on the Amu in July 2006 supervised by the Khabarovsk Krai Ministry of Natural Resources show that other pollutants besides benzene derivatives accumulated in the bottom sediments, fish and molluscs. Phthalate, naphthalene, highly toxic anisole and benzothyozole, acetochlorine and triazine pesticides as well as new-generation phtor-phosphor containing pesticides were identified for the first time. The lack of reliable information about pollution from developing areas in China makes Russian efforts less effective.

References

Kondratjeva, L., 2005: *Ecological Risk of Water Ecosystem Pollution.* Vladivostok: Far Eastern of Sciences.

Kondratiyeva, L., Chukhlebova, L. and Rapoport, V., 2003: Ecological aspects of fish organoleptic index changes in frozen up Amu River. In: *V. Ya. Levanidov's Biennial Memorial Meetings*, Vladivostok: Far Eastern of Sciences, **2**: 113–118.

Ryabkova, V., 2002: Persistent organic compounds, influence of the Amu River biota and health of people of Priamurje. In: *Regions of new development*, Vladivostok-Khabarovsk: FEB RAS, **2**: pp. 89–92.

Tal, A. and Linkov, I., 2004: Role of Comparative Risk Assessment in Addressing Environmental Security in the Middle East. *Risk analysis*, **24**: 1243–1252.

ON THE DEVELOPMENT OF A STRATEGY FOR THE OPTIMAL USE OF THE UPSTREAM WATER RESOURCES OF THE AMUDARY BASIN IN THE NATIONAL INTERESTS OF THE TAJIK REPUBLIC

S. Navruzov[*]

Institute of Mathematics of Tajik Academy of Sciences, Dushanbe, Tajikistan

Abstract. The problems of the use of water resources in Central Asia are investigated by considering the sovereignty of the states and the increasing demand for water for economic development. A complex program is proposed and presented, with appropriate mathematical software, intended to aid in the calculation of the possible alternative strategies for using the Amudarya's upstream water resources to satisfy the apparently incompatible requirements of the national economies in irrigation and energy generation.

Keywords: Water resources, national economics, irrigation, transboundary, energy generation, optimal solution, program complex, up-stream, down-stream, compensating services

1. Essence of the conflict

The problems of joint management of transboundary water resources and energy generation for Central Asian republics are a major issue, because they affect practically all aspects of national and regional economics. Until recently, the Central Asian republics have exploited their water resource systems within the distributive scheme. This scheme considered the region as a space controlled and guided by the center. The former mechanism of centralized management of the water resources provided a reliable basis for work on all hydro-technical constructions, coordinating the solutions of problems of regional water use from year to year. This mechanism guaranteed stable functioning of the agricultural sector, based on irrigation, and safe flows during the winter–spring period.

However, presently the political and economic situation in the region has been radically changed. Since the proclamation of independence, each of the sovereign state tends first of all to maximize the use of all natural resources, taking into account their own economic and political interests. Thus, the main water potential of Central Asia is concentrated in the basin of two big rivers, the Syrdarya and the Amudarya, which provide the needs

---

[*] To whom correspondence should be addressed. e-mail: snavruzov@rambler.ru

of the population of five countries: Kazakhstan, Kyrgyzstan, Tajikistan, Turkmenistan and Uzbekistan.

There are certain disagreements between the states of the Amudarya and the Syrdarya transboundary basins relating to energy potentials in the upstream sections of the rivers. As an example, Kyrgyzstan completely determines the policy of use in the Syrdarya basin, because the water resources are formed within its territory. Other Central-Asian republics, using the water resources of the Syrdarya river are forced to accept this situation. This situation has initiated the introduction of compensating services between the republics of the region.

The situation in the Amudarya river basin can become more aggravated, if we take into account Afghanistan's intentions to act in the near future, with requirements to participation in water distribution in the Pyandj river basin. As a result, the volume of water will be reduced for Turkmenistan and Uzbekistan, especially in the Amudarya basin downstream.

The main problems existing between Central Asian republics in the field of water and energy regulation are:

- Conflicting interests of the states which are upstream and downstream on the river
- Absence of interstate structures in the sphere of joint management of the water-energy resources, between the appropriate corresponding authorities
- The recommendatory character of decision-making and absence of any responsibility for their execution at the level of the created regional management structures
- Inconsistency of actions at regional and national levels between management structures of the water economy and power engineering
- Absence of arbitration practice between countries which have conflicts concerning the distribution of water resources with no opportunities for real influence on the solution of this problem

2. Problem solution

For decision relating to these problems, it is important to take into consideration the following main positions:

- Formation of a coordinated regional policy of fair and reasonable development and use of water-energy resources, based on international law with the participation of all countries in the region
- Complex assimilation of hydropower energy potential and regulation of water and energy resources and maintenance of synchronicity in their decisions
- Production of effective economical and legal mechanisms for co-operation, which guarantee that the accepted decisions are carried out

- Working out the complex of mathematical models of the accepted decisions, allowing the modeling of various scenarios of water use in the basins of transboundary rivers

Proceeding from this, in 1992 the Institute of Mathematics of the Tajik Academy of Sciences presented to the Tajik Government a proposal for developing a sound strategy for using the water resources of the Vakhsh River. That proposal was accepted and a series of governmental institutions were brought together to develop the project as suggested. The Institute of Mathematics prepared the computer programs for calculating the possible alternatives for using the Vakhsh River's water resources, with regard to satisfying the seemingly incompatible requirements of the national economy in terms of irrigation and power engineering (Institute of Mathematics, 1986).

The final aim was to create a state strategy for the rational use of water by the upstream Central Asian republics in the Amudarya basin. The principal aspects of the strategy should be brought to the attention of the other Central Asian republics, in order to conclude an interstate agreement on the use of the Amudarya water resources. In this connection, the problem of a complex use of Central Asia water resources will be completely decided by taking into consideration the sovereignty of the states and their market interrelations.

The project suggested by the author is directed to achieving the purpose mentioned. While realizing the project, the following results will be obtained:

- The computer software needed to calculate acceptable alternatives for utilizing water resources of the upper part of the Amudarya basin by managing the activity of the existing reservoir cascade
- The computer software for conducting an similar multi-variant calculation if the Rogun reservoir is constructed and added to the reservoir cascade for a combined regulation system

The software will allow the following:

- Carrying out an expert selection of reasonable alternative strategies for using the water resources
- Developing recommendations for the Tajik Government to develop a new water use policy

To get these results, the following actions are planned:

- Gathering, systematizing and integrating hydrologic, hydro-geological and morphometric data on the upper Amudarya basin and existing modes of the water units
- Creating a database of the information

- Developing the program system for the management of the Vakhsh cascade with regard to the priority of the Tajik Republic's requirements in electricity – in two variants (1) for the existing cascade of reservoirs and (2) for the cascade extended by the Rogun reservoir
- Realization of computing experiments for a qualitative estimation of the efficiency of different managing decisions for the functioning the Vakhsh cascade of reservoirs
- Creating information system of the Amudarya transboundary river using a Geographical Information System (GIS)

3. Mathematical software

Three levels of mathematical models are offered: *analytical, optimization and imitational*. Within the framework of analytical models (Navruzov, 1991), the theoretical game models for the distribution of water resources between the states of the Amudarya basin are considered. A number of characteristic examples of games with non-opposite interests are discussed.

Optimization models (Navruzov, 1990) of the reservoir management of the upstream water resources of the Amudarya are offered, which use as a base the "block-hierarchical" principle, which at the initial stage provides for the development of different national models and their further coordination within the framework of regional models. According to this, we can divide the territory of the Amudarya basin into two zones: the zone of water demand and the zone of water production.

The technique of finding a compromise solution among the needs of the states, in terms of the volumes of water consumed at the level of the coordination of management between zones of consumption and formation is offered.

Construction of imitational models is carried out on example the Vakhsh–Amudarya cascade of reservoirs (Navruzov, 1986, 1991). The cascade includes three large reservoirs, such as Rogun, Nurek and Tuymuyn. Mean while, these reservoirs are located in the territory of different Central-Asian republics, namely: Rogun and Nurek belong to Tajikistan and Tuymuyn to Uzbekistan.

4. Decision-making

Program realization of mathematical models, data base, ways of reception and transfer information, method of organization dialogue with users, visualization means based on GIS are decision-making support system (DMSS). Methodology of creation of the DMSS includes questions: *computer modeling of managing decision-making; solving problem of multi-purpose management of water resources; carrying out decision on results of imitation modeling.*

The specialized applied program of DMSS uses of program software MS Office operation system Windows XP & GIS Arc Info/Arc View. Information database is constructed based on MS Access and mathematical models are used for finding of optimal solution. Analysis and representation of spatial data is carried out based on GIS technology.

References

Institute of Mathematics of the Tajik Academy of Sciences, 1986: *Forecasting the working size of the Nurek reservoir.* Research publication of Academy of Sciences Computing Center.

Navruzov S.T., 1986: *Calculation of management rules for a reservoir cascade for the purpose of irrigation and power.* Moscow, Computer Center of the USSR Academy of Sciences, 20pp.

Navruzov S.T., 1990: On a method of constructing a zone guaranteed return for a linear cascade of reservoirs. *Reports of Tajik Academy of Sciences*, **33(3)**: 153–157.

Navruzov S.T., 1991: Qualitative research of a problem of optimum control to a cascade of reservoirs. *Izvestiya of Tajik Academy of Sciences*, **116(2)**: 10–15.

INTEGRATED MANAGEMENT STRATEGY FOR TRANSBOUNDARY WATER RESOURCES IN CENTRAL ASIA

I. Sh. Normatov[*] and G.N. Petrov
Institute of Water Problems, Hydropower and Ecology, Academy of Sciences of Republic of Tajikistan, 12 Parvin St., Dushanbe 734002, Tajikistan

Abstract. For the rational use of water resources in view of the reduction of the area of glaciation and degradation of glaciers in the Aral Sea basin, the paper offers the Integrated Water Resources Management of the region and describes the stages of its introduction via individual sub-basins.

Keywords: Transboundary water resources, Central Asia

1. Introduction

The Central Asia Region includes the republics of Kazakhstan, Kyrgyzstan, Tajikistan, Turkmenistan and Uzbekistan. Hydrographically the region is distinguished as the Aral Sea basin, which consists of two basins – the Syrdarya and the Amudarya Rivers. The macroeconomic parameters of the countries of the region and the water resources of the Aral Sea basin are presented in Tables 1 and 2.

2. Dynamics of irrigation and hydropower development in the Central Asia Region

The principal spheres of water resources use in Central Asia today are irrigated agriculture and hydropower. At the beginning of the 20th century, about 3.5 million hectares were irrigated in the region. By the 1990s the total irrigated area in the region increased to 8.8 million hectares. The total established capacity of all electric power stations in the region increased to 37.8 million kilowatts and at that time the capacity of hydropower stations in the region reached 11.31 million kilowatts (Petrov et al., 2006).

Unfortunately, all these impressive results led to the same great negative consequences. The intensity of ecological destruction in the region, which became especially apparent in the Aral Sea zone, has sharply increased, salinization and desertification have spread and the quality of water especially in the lower stretches of rivers has worsened (Normatov et al., 2006).

---

[*] To whom correspondence should be addressed. e-mail: inomnor@mail.ru

J.A.A. Jones et al. (eds.), *Threats to Global Water Security*,
© Springer Science + Business Media B.V. 2009

TABLE 1. Macroeconomic development of the Central Asia Region

Country	Area 1,000 km²	Population millions	Per capita gross output $1,000s/person	Per capita energy consumption tons fuel/person
Kazakhstan	2,636.20	14.95	3.56	3.67
Kyrgyzstan	198.50	4.90	0.68	0.66
Tajikistan	143.10	6.20	0.99	0.84
Turkmen-istan	488.00	4.70	1.52	3.30
Uzbekistan	447.36	24.60	2.26	2.70
Central Asia	3,913.16	55.35	2.22	2.64

In November 2006 a Regional Workshop on "Assessment of Snow-Glacier and Water Resources in Asia" was held in Kazakhstan. Participants, including experts and professionals from Central Asia together with international experts, noted that changes in glaciers in the world's largest and highest mountain system will have significant effects on nearly 1.5 billion people. They recognized that glaciers are key indicators in detecting climate change.

TABLE 2. Surface water resources of the Aral Sea basin

Country	Amudarya basin km³/year	Syrdarya basin km³/year	Aral Sea basin	
			km³/year	%
Kazakhstan	–	4.50	4.50	3.9
Kyrgyzstan	1.90	27.4	29.30	25.3
Tajikistan	62.9	1.1	64.00	55.4
Turkmenistan	2.78	–	2.78	2.4
Uzbekistan	4.70	4.14	8.84	7.6
Afghanistan	6.18	–	6.18	5.4
Central Asia	78.46	37.14	115.6	100.0

Tajikistan is a mountainous country with 93% of its territory occupied by mountains and in Tajikistan there are more 8,400 glaciers with a total area of 8,476.2 km², about 6% of the country. The main center of glaciation is the Fedchenko glacier – the largest mountain glacier in the world (Fig. 1).

TABLE 3. Characteristics of the Fedchenko glacier

Length	77 km
Average width	2.5 km
Maximal width	5 km
Area with all tributaries	~652 km²
Ice thickness	~1 km

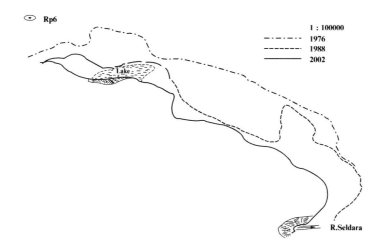

Figure 1. Decline of the Fedchenko glaciers

Zaallay and Kaindi of the Academy of Sciences have noted that just over the last 40 years 14 smaller glaciers have disappeared in the mountain range with a total former area of 7.6 km². The average speed of movement of glaciers has decreased from 72 to 69 cm daily related to loss of mass. Over the 20th century the glacier system has lost about 12–15 km³ of ice.

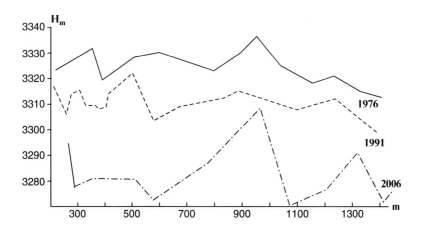

Figure 2. Cross section change of the profile of No. 4 of the Fedchenko glacier

The expedition on the Fedchenko ice field in September 2006 demonstrated that the glacier continues to be reduced non-uniformly at a speed of 8–10 m in a year (Fig. 2).

3. Integrated management of water resources

The goal of Integrated Management of Water Resources (IMWR) is to unite all aspects of water resources control under a single leadership at the basin level. Normatov and Petrov (2007) described the process of testing IMWR principles on sub-basins of the Aral Sea drainage system before expanding these to the whole basin. The existing water organizations have to a considerable degree inherited their structures and functions from the former Soviet Union. Their vertical structures reflect Soviet command economy. These vertical structures work against IMWR and real integration is possible only on the general national level. At the some time, Ministries and departments are not giving the necessary attention to problems of integrated control of resources. Integration is even more limited at the inter-state level. The opportunity of integration is discussed only at a restricted number of chance meetings between ministries.

During the recent years of independence in the Central Asia Region, a few legislative acts have been developed supporting integration in the water sector. However, the effectiveness of these integration reforms is too restricted; as in former years, no two sectors discuss their problems on a regular basis. The first aim of IMWR should be to establish formal mechanisms for integration at a minimum level of accessibility according to two main models: (a) to organize Working Groups, and (b) to found Basin Commissions to establish the basinwide method of water resources control.

3.1. STRATEGY OF IMWR FOR THE VAKHSH BASIN

The main goal of the Strategy of IMWR for the Basin of the Vakhsh is development of a Tajik national strategy for integrated water resources control of Vakhsh part of Amudarya River transboundary basin to be affirmed by the Government for the years 2015–2020. The strategy of IMWR is to achieve three goals of ecologically sound development: economic development, social development, and protection of the environment. This strategy of IMWR for the Vakhsh River sub-basin is based on the international right of Tajikistan as an independent sovereign state to use all the water resources within its territory according to national legislative standards, but also taking account of the interests of other states within the transboundary basin of the Amudarya.

In this case priority will be given to: (1) a multi-branched approach to control of water resources, and protection of all freshwater sources, involving introducing more water-efficient irrigation and widening provision of sewage farms; (2) planning the stable and rational use, protection, economy and control of water resources based on the demands and priorities of society within the political framework of national development taking account of economical effectiveness, social usefulness and necessity of completing projects.

3.2. PROPOSED MEASURES TO IMPROVEMENT WATER RESOURCES
      CONTROL

It is proposed to establish a Commission of the Republic which comprises other Working Groups within its structure to accelerate the introduction of IMWR. The main tasks of the Commission are the harmonization inter-departmental interests, the planning of water resources utilization, the solution of practical problems of introducing integrated control of water resources, and cooperation between the various bodies administering the basin.

The main tasks of Commission are:

- Long-term planning of water resources use and planning of development
- Drawing up recommendations on definition and agreement political aspects on sphere of using and protecting water resources
- Consideration of suggestion on improvement of regional and interstate water dividing, improvement of regimes of water resources using
- Informing society about development plans and water resources using

All the activities of the Commission and Working Group must be directed towards improvement of water resources control, and the gradual achieve-ment of independence of basin Commissions and Committees by the active participation of all interested parties and sectors.

Short-term measures: creating at the outset an active Working Group to prepare the necessary documents and proposals for the Republican Commission; implementing the decisions adopted by Government for the operational channels of administration; reforming the Administration of the Irrigation System according to the hydrographical principles of water resources control and providing representations of all interested parties of water users and society. There will be elaborate and co-ordinated basinwide plans for the control, use and protection of water resources.

Medium-term measures: by the end of the period, the structure of the main ministries with interests in the water sector will be redesigned. This requires centralization of water resources at the basin level and creation of a system of complex monitoring of natural resources and the necessary data base of statistics.

Long-term measures: completion of mastering water-energetic resources of the Vakhsh River basin and creation of a regional market for the water resources.

References

Normatov, I.Sh. and Petrov, G.N., 2007: *Country Report: Vakhsh River Basin (VRB), Tajikistan.* Asian G-WADI Meeting Network, China, 35pp.

Normatov, I.Sh., Petrov, G.N., Froebrich, J., Bauer, M and Olsson, O., 2006: Improved dam operation in the Amu Darya basin including transboundary aspects. *Proceedings of International Symposium "Dams in the Societies of 21st Century"*, Barcelona, Spain, 216–221.
Petrov, G.N., Shermatov, N. and Normatov, I.Sh., 2006: Modern water – energetic resources use problems and perspectives of Tajikistan on conditions of global non-stability climate. *Proceedings of 22nd National Convention of Environmental Engineering and National Seminar on Rainwater Harvesting and Water Management*, Nagpur, India, 154–160.

LaVergne, TN USA
18 August 2009
155111LV00004B/91/P

9 789048 123438